DIY Science

Illustrated Guide to Home Chemistry Experiments

All Lab, No Lecture

First Edition

Robert Bruce Thompson

O'REILLY®

BEIJING · CAMBRIDGE · FARNHAM · KÖLN · PARIS · SEBASTOPOL · TAIPEI · TOKYO

Illustrated Guide to Home Chemistry Experiments
All Lab, No Lecture

by Robert Bruce Thompson

Copyright © 2008 Robert Bruce Thompson. All rights reserved. Printed in U.S.A.

Published by Make:Books, an imprint of Maker Media, a division of O'Reilly Media, Inc. 1005 Gravenstein Highway North, Sebastopol, CA 95472.

O'Reilly books may be purchased for educational, business, or sales promotional use. For more information, contact our corporate/institutional sales department: 800-998-9938 or corporate@oreilly.com.

Print History
April 2008
First Edition

Publisher: Dale Dougherty
Associate Publisher: Dan Woods
Executive Editor: Brian Jepson
Editor: Tom Sgouros
Creative Director: Daniel Carter
Designer: Alison Kendall
Production Manager: Terry Bronson
Copy Editor: Nancy Kotary
Indexer: Patti Schiendelman
Cover Photograph: Jason Forman

The O'Reilly logo is a registered trademark of O'Reilly Media, Inc. The *DIY Science* series designations, *Illustrated Guide to Home Chemistry Experiments: All Lab, No Lecture*, and related trade dress are trademarks of O'Reilly Media, Inc. The trademarks of third parties used in this work are the property of their respective owners.

Important Message to Our Readers: Your safety is your own responsibility, including proper use of equipment and safety gear, and determining whether you have adequate skill and experience. Chemicals, electricity, and other resources used for these projects are dangerous unless used properly and with adequate precautions, including safety gear. Some illustrative photos do not depict safety precautions or equipment, in order to show the project steps more clearly. These projects are not intended for use by children.

Use of the instructions and suggestions in *Illustrated Guide to Home Chemistry Experiments: All Lab, No Lecture* is at your own risk. O'Reilly Media, Inc. and the author disclaim all responsibility for any resulting damage, injury, or expense. It is your responsibility to make sure that your activities comply with applicable laws, including copyright.

ISBN-10: 0-596-51492-1
ISBN-13: 978-0-596-51492-1

Dedication

Carl Wilhelm Scheele
(1742 – 1786)

To Carl Wilhelm Scheele, one of the first true chemists, who did so much with so little. As a practicing pharmacist without access to the advanced laboratory equipment available to many of his contemporaries, Scheele discovered numerous chemical elements and compounds—including oxygen, nitrogen, chlorine, barium, manganese, molybdenum, tungsten, citric acid, glycerol, the pigment Scheele's Green (cupric hydrogen arsenite), and many others—debunked the phlogiston theory, and was among the first to establish the rigorous, standardized, consistent quantitative procedures that are the hallmark of modern chemistry. Scheele died at age 43, apparently from mercury poisoning contracted as a result of his unfortunate habit of tasting the new compounds he prepared.

Contents

Preface

Christmas morning, 1964. I was 11 years old. My younger brother and I arose at the crack of dawn and noisily rushed downstairs to find out what was under the tree. Our parents followed us, bleary-eyed.

Santa had been good to us that year. Colorfully wrapped presents were scattered—not just under the tree, but across most of the living room floor. Being boys, we started tearing open the presents with no thought at all for the care that had gone into wrapping them. We were after the loot.

There were the inevitable disappointments: sweaters from Grandma, school clothes from Aunt Betty, and hand-knitted stocking caps for both of us from Pete and Sarah, our elderly next-door neighbors. But there was plenty of good stuff, too. Sports equipment and a cap pistol for my younger brother. A battery-powered Polaris nuclear submarine that actually fired small plastic missiles. A bicycle for my brother and a BB gun for me! Lots of books, the kind we both liked to read. A casting set, with a lead furnace and molds to make toy soldiers.

As we opened the packages, my brother and I mentally checked off items against our wish lists. We'd both gotten everything we asked for. Almost. One item had been at the top of every iteration of my wish list since the Sears Christmas Wish Book had arrived, and that item was nowhere to be found. I searched frantically through the piles of discarded wrapping paper, hoping I'd overlooked a box. It wasn't there.

My parents had been watching my brother and me ripping through gifts like Tasmanian Devils. Just as I'd decided that I hadn't gotten the one gift that I really, really wanted, my mom and dad called me into the kitchen. There it sat, on the kitchen table: exactly what I'd been hoping for. It was already unboxed and spread wide open to show the contents. My father said, "This is from your mother and me. It is not a toy."

It was a Lionel/Porter/Chemcraft chemistry set, and the exact model I'd asked for. The biggest one, with dozens of chemicals and hundreds of experiments. Glassware, an alcohol lamp, a balance, even a centrifuge. Everything I needed to do real chemistry. I instantly forgot about the rest of my presents, even the BB gun. I started reading the manual, jumping from one experiment to another. I carefully examined each of the chemical bottles. The names of the chemicals were magical. Copper sulfate, sodium carbonate, sulfur, cobalt chloride, logwood, potassium ferricyanide, ferrous ammonium sulfate, and dozens more.

I used the balance to weigh something for the first time. I put an object in one of the balance pans and carefully added weights to the other pan until the needle was centered. As I was about to jump on to something else, my dad brought me to a screeching halt. "Write it down," he said. "A scientist records what he observes. If you don't work methodically and write down what you observe, you're not a scientist. You're just playing around." I've been recording my observations ever since.

I soon lost interest in the other gifts, but getting that chemistry set was a life-changing experience. My mother told me years later that she and my dad had hoped that the chemistry set would hold my interest for at least a few weeks. As it turned out, it held my interest a bit longer. With my dad's help, I built a chemistry workbench in the basement, and later a photographic darkroom. I scrounged equipment and chemicals from every source I could think of, and saved up for things that required cash. I spent every spare moment in that lab, and went on to major in chemistry in college and graduate school. Even now, more than 40 years later, I have a chemistry lab in the basement. It's a much better lab than the one I had back in the 1960s, but the work habits I learned then stand me in good stead now.

What I experienced that Christmas morning was repeated in millions of other homes through the years as boys (and, alas, only a few girls) opened their first chemistry sets. From the 1930s through the 1960s, chemistry sets were among the most popular Christmas gifts, selling in the millions. It's said that in the 1940s and 1950s there was a chemistry set in nearly every household where there was a child. Even as late as the 1970s, chemistry sets remained popular and were on display in every toy store and department store. And then something bad happened. By the 1980s, chemistry sets had become a dying breed. Few stores carried them, and most of those sets that remained available were pale shadows of what chemistry sets had been back in the glory days.

The decline of chemistry sets had nothing to do with lack of interest. Kids were and are as interested as ever. It was society that had changed. Manufacturers and retailers became concerned about liability and lawsuits, and "chemical" became a dirty word. Most chemistry sets were "defanged" to the point of uselessness, becoming little more than toys. Some so-called "chemistry sets" nowadays are actually promoted as using "no

heat, no glass, and no chemicals," as if that were something to be proud of. They might just as well promote them as "no chemistry."

Even the best chemistry set that is still sold, the $200 Thames & Kosmos Chem C3000, is an unfortunate compromise among cost, liability, and marketability. The Chem C3000 kit lacks such essential equipment as a balance and a thermometer, provides little glassware, and includes only the tiny amounts of chemicals needed to do unsatisfying micro-scale chemistry experiments. Despite these criticisms, the C3000 kit is a good choice for giving late elementary school or early middle school students their first exposure to hands-on chemistry lab work. It allows kids to produce bright colors and stinky smells, which after all are the usual hooks that draw kids into chemistry. The problem is that that's not enough.

Laboratory work is the essence of chemistry, and measurement is the essence of laboratory work. A hands-on introduction to real chemistry requires real equipment and real chemicals, and real, quantitative experiments. No existing chemistry set provides anything more than a bare start on those essentials, so the obvious answer is to build your own chemistry set and use it to do real chemistry.

Everything you need is readily available, and surprisingly inexpensive. For not much more than the cost of a toy chemistry set, you can buy the equipment and chemicals you need to get started doing real chemistry. Of course, the main reason for that is the absence of the hidden liability surcharge. If you buy a chemistry set and burn yourself with the alcohol lamp, you might sue the maker of the chemistry set. If you buy an alcohol lamp by itself and burn yourself, you have no one else to blame.

So what about the very real dangers involved in serious chemistry lab work? After all, some of the experiments in this book use concentrated acids, flammable liquids, corrosives, and poisons. In one experiment we manufacture napalm, for heaven's sake. Will readers of this book be dropping like flies, blowing themselves up, burning the house down, or growing extra arms? Of course not. Dangers can be dealt with. One of the recurring lessons throughout this book is the importance of assuming personal responsibility for useful but dangerous actions—understanding the specific risks and taking the necessary steps to minimize or eliminate them.

I set out to write this book after a conversation with our friend and neighbor Jasmine Littlejohn. At age 14, Jasmine is a bright kid who's interested in science as a career. I asked her one day how much science she was learning in school. "Hardly any," she replied. "On a typical day, we spend hours on math, social studies, English, and other stuff, and about 15 minutes on science." Although Jasmine attends a good public school, like most schools it devotes little time and few resources to science and has only limited lab facilities. No doubt the school would list money and safety concerns as reasons, but such excuses do nothing to help Jasmine.

With her mom's approval, I could give Jasmine access to my basement chemistry lab, but that would solve only part of the problem. If Jasmine was to do more than make pretty colors and stinky smells, if Jasmine was to do real chemistry, she'd need more than just access to a lab. She'd need detailed instructions and some sort of structured plan to guide her through the learning process. She'd need to learn how to use the equipment and how to handle chemicals safely. She'd need well-designed experiments that focused on specific aspects of laboratory work. In other words, she'd need a home chemistry lab handbook, one devoted to serious chemistry rather than just playing around.

My first thought was to get Jasmine one of the classic home chemistry books published back in the '30s, '40s, or '50s. Some of those were excellent, but all of them required chemicals—such as benzene, carbon tetrachloride, salts of mercury, lead, and barium, concentrated nitric acid, and so on—that were once readily available but are now very expensive or difficult to obtain.

In one sense, that wasn't really a problem. I already had most of that stuff in my lab. But even the best of those old books would have required some serious red-lining before I'd have turned Jasmine loose with it. One, for example, suggested tasting highly toxic lead acetate (also known as "sugar of lead") to detect its sweetness. Others were a bit casual about handling soluble mercury compounds or contained experiments that were potentially extremely dangerous.

I concluded that the only good solution was to write a new book, one devoted to learning real chemistry at home, and one that would also be useful for the many thousands of other people out there—young people and adults—who wanted to experience the magic of chemistry just as I'd done on that long-ago Christmas morning, and to do so on a reasonably small budget with readily available equipment and chemicals. And so the *Illustrated Guide to Home Chemistry Experiments* was born.

WHO THIS BOOK IS FOR

This book is for anyone, from responsible teenagers to adults, who wants to learn about chemistry by doing real, hands-on laboratory experiments.

DIY hobbyists and science enthusiasts can use this book to master all of the essential practical skills and fundamental knowledge needed to pursue chemistry as a lifelong hobby. Home school students and public school students whose schools offer only lecture-based chemistry courses can use this book to gain practical experience in real laboratory chemistry. A student who completes all of the laboratories in this book has done the equivalent of two full years of high school chemistry lab work or a first-year college general chemistry laboratory course.

And, finally, a word about who this book is *not* for. If you want to make fireworks and explosives—or perhaps we should say if *all* you want to make is fireworks and explosives—this book is not for you. If your goal is to produce black powder or nitroglycerine or TATP, you'll have to look elsewhere. Neither will you find instructions in this book for producing methamphetamine in your home lab or synthesizing other illegal substances. In short, if you plan to break the law, this book is not for you.

HOW THIS BOOK IS ORGANIZED

The first part of this book is made up of narrative chapters that cover the essential "book learning" you need to equip your home chemistry lab, master laboratory skills, and work safely in your lab:

1. Introduction
2. Laboratory Safety
3. Equipping a Home Chemistry Lab
4. Chemicals for the Home Chemistry Lab
5. Mastering Laboratory Skills

The bulk of the book is made up of seventeen hands-on laboratory chapters, each devoted to a particular topic. Most of the laboratory chapters include multiple laboratory sessions, from introductory level sessions suitable for a middle school or first-year high school chemistry laboratory course to more advanced sessions suitable for students who intend to take the College Board Advanced Placement (AP) Chemistry exam:

6. Laboratory: Separating Mixtures
7. Laboratory: Solubility and Solutions
8. Laboratory: Colligative Properties of Solutions
9. Laboratory: Introduction to Chemical Reactions & Stoichiometry
10. Laboratory: Reduction-Oxidation (Redox) Reactions
11. Laboratory: Acid-Base Chemistry
12. Laboratory: Chemical Kinetics
13. Laboratory: Chemical Equilibrium and Le Chatelier's Principle

ACKNOWLEDGMENTS

Although only my name appears on the cover, this book is very much a collaborative effort. It could not have been written without the help and advice of my wife, Barbara Fritchman Thompson. My editors, Tom Sgouros and Brian Jepson, were with me every step of the way, contributing numerous helpful suggestions. As always, the O'Reilly design and production staff, who are listed individually in the front matter of this book, worked miracles in converting my draft manuscript into an attractive finished book.

Finally, special thanks are due to my friends and technical reviewers, Dr. Mary Chervenak and Dr. Paul Jones. Mary, who holds a Ph.D. in organic chemistry from Duke University, is a research chemist for the Dow Chemical Company. Paul, who also holds a Ph.D. from Duke University in organic chemistry, is a professor of organic chemistry at Wake Forest University. Mary and Paul outdid themselves as technical reviewers, flagging my mistakes and contributing innumerable useful suggestions and comments. With the help of this pair-o'-docs (see Figure 0-1), this is a much better book than it might otherwise have been. Thanks, guys.

THANK YOU

Thank you for buying the *Illustrated Guide to Home Chemistry Experiments: All Lab, No Lecture*. I hope you enjoy reading and using it as much as I enjoyed writing it.

FIGURE 0-1: *Paradoxybenzene, also known as 1,4-diphdbenzene*

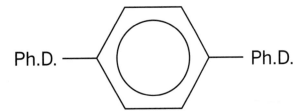

HOW TO CONTACT US

We have verified the information in this book to the best of our ability, but you may find things that have changed (or even that we made mistakes!). As a reader of this book, you can help us to improve future editions by sending us your feedback. Please let us know about any errors, inaccuracies, misleading or confusing statements, and typos that you find anywhere in this book.

Please also let us know what we can do to make this book more useful to you. We take your comments seriously and will try to incorporate reasonable suggestions into future editions. You can write to us at:

Maker Media
1005 Gravenstein Hwy N.
Sebastopol, CA 95472
(800) 998-9938 (in the U.S. or Canada)
(707) 829-0515 (international/local)
(707) 829-0104 (fax)

Maker Media is a division of O'Reilly Media devoted entirely to the growing community of resourceful people who believe that if you can imagine it, you can make it. Consisting of *Make* magazine, *Craft* magazine, Maker Faire, as well as the Hacks, Make:Projects, and DIY Science book series, Maker Media encourages the Do-It-Yourself mentality by providing creative inspiration and instruction.

For more information about Maker Media, visit us online:

MAKE: *www.makezine.com*
CRAFT: *www.craftzine.com*
Maker Faire: *www.makerfaire.com*
Hacks: *www.hackszine.com*

To comment on the book, send email to:
bookquestions@oreilly.com

The O'Reilly web site for the *Illustrated Guide to Home Chemistry Experiments: All Lab, No Lecture* lists examples, errata, and plans for future editions. You can find this page at:
http://www.makezine.com/go/homechemlab

For more information about this book and others, see the O'Reilly web site: *www.oreilly.com*

To contact the author directly, send mail to:
robert@homechemlab.com

I read all mail I receive from readers, but I cannot respond individually. If I did, I'd have no time to do anything else. But I do like to hear from readers.

I also maintain a dedicated web site to support the *Illustrated Guide to Home Chemistry Experiments: All Lab, No Lecture*. The home page contains corrections and errata, supplemental material that didn't make it into the book, updated lists of sources for equipment and chemicals, additional experiments, links to electronic books and other resources, and so on. Visit this site before you buy any equipment or chemicals and before you do any of the experiments. Revisit it periodically as you use the book.

Book's home page: *www.homechemlab.com*

The Home Chem Lab discussion forums are open to anyone who is interested in home chemistry. You can read the forums anonymously, but you must register before you can post or reply to articles to the forums. Registration is free, and all registration information is kept private.

Discussion forums: *forums.homechemlab.com*

You can sign up to have the Home Chem Lab newsletter delivered directly to your inbox by sending a blank email to the following address. If you sign up, please add the address *newsletter@homechemlab.com* to your spam filter whitelist. If newsletter email to your address bounces, that address is removed automatically from the mailing list. If possible, subscribe using a "real" (ISP) email address rather than an email address from Yahoo, Gmail, or a similar free email service. Such addresses often have delivery problems with email newsletters. The newsletter is published sporadically, whenever there's something worth writing about. Registration for the newsletter is free, and all registration information is kept private.

Newsletter: *hcl-subscribe@homechemlab.com*

Introduction

1

What you take away from an experience depends on how you approach it and what you hope to get out of it.

If you're a hobbyist, no problem. Simply read the book from start to finish and do the labs in the order they're presented. You'll have a lot of fun and learn a lot about chemistry along the way.

If you're a home school student or a public school student who wants a laboratory chemistry course to supplement and enhance a lecture-based high school chemistry course, you'll need to do a bit more planning. The lab sessions in this book cover at least two years' worth of high school chemistry lab work, including a full set of labs appropriate for a first-year chemistry lab course and a second group of labs appropriate for a second-year or AP (Advanced Placement) chemistry lab course.

The laboratory sessions in this book are organized topically by chapter, but their order may not correspond to the order in which topics are covered in your chemistry text book. That's not really a problem, because you can do these labs in whatever order matches that of your chemistry text. Choosing which labs to do is another matter.

LOOK IT UP

Because this book focuses on chemistry lab work rather than chemistry theory, if you aren't using this book in conjunction with a chemistry textbook there may be times when you come across a term that's unfamiliar to you. Usually, the term will be clear from context. If it isn't clear, don't guess. Look it up, either on-line or in a good general chemistry textbook.

One good general chemistry textbook is **Chemistry, The Central Science** (Brown, et al., Prentice Hall, 2002), but there are many others available. Chemistry textbooks are often available very inexpensively in used bookstores, and even a copy that's several editions out of date is fine. Chemistry textbooks are frequently updated with very minor changes, but even a copy that's 10 or more years old covers all the fundamentals. General chemistry just doesn't change much over the years.

Students who will go on to major in college in nonscience disciplines need only a first-year chemistry course with some exposure to basic chemistry lab procedures. For these students, one 60- to 90-minute chemistry lab period per week suffices. The following labs, some of which require two sessions, are a good starting point:

6.1 – Differential Solubility: Separate Sugar and Sand
6.5 – Chromatography: Two-Phase Separation of Mixtures (first part only)
7.1 – Make Up a Molar Solution of a Solid Chemical
8.3 – Observe the Effects of Osmotic Pressure
9.1 – Observe a Composition Reaction
9.3 – Observe a Single-Displacement Reaction
11.1 – Determine the Effect of Concentration on pH
13.1 – Observe Le Chatelier's Principle in Action
14.1 – Observe the Volume-Pressure Relationship of Gases (Boyle's Law)
15.1 – Determine Heat of Solution
15.2 – Determine the Specific Heat of Ice
16.1 – Produce Hydrogen and Oxygen by Electrolysis of Water
18.1 – Observe Some Properties of Colloids and Suspensions
19.1 – Using Flame Tests to Discriminate Metal Ions
19.2 – Using Borax Bead Tests to Discriminate Metal Ions

Students who will go on to major in college in science disciplines need a first-year chemistry course with much more exposure to chemistry lab procedures. For these students, allocate two 90-minute to 2-hour chemistry lab periods per week or one 3- to 4-hour lab period. (Regularly scheduled weekend lab sessions, when other classes do not interfere, are often the most suitable time for home school students and the only practical time for public school students.) The following lab sessions are a good starting point:

6.1 – Differential Solubility: Separate Sugar and Sand
6.2 – Distillation: Purify Ethanol
6.3 – Recrystallization: Purify Copper Sulfate
6.4 – Solvent Extraction
6.5 – Chromatography: Two-Phase Separation of Mixtures
7.1 – Make Up a Molar Solution of a Solid Chemical
7.2 – Make Up a Molal Solution of a Solid Chemical
7.3 – Make Up a Molar Solution of a Liquid Chemical
7.4 – Make Up a Mass-to-Volume Percentage Solution
8.1 – Determine Molar Mass by Boiling Point Elevation
8.2 – Determine Molar Mass by Freezing Point Depression
8.3 – Observe the Effects of Osmotic Pressure
9.1 – Observe a Composition Reaction
9.2 – Observe a Decomposition Reaction
9.3 – Observe a Single-Displacement Reaction
9.4 – Stoichiometry of a Double Displacement Reaction

10.1 – Reduction of Copper Ore to Copper Metal
10.2 – Observe the Oxidation States of Manganese
11.1 – Determine the Effect of Concentration on pH
12.1 – Determine the Effect of Temperature on Reaction Rate
12.2 – Determine the Effect of Surface Area on Reaction Rate
12.3 – Determine the Effect of Concentration on Reaction Rate
13.1 – Observe Le Chatelier's Principle in Action
13.2 – Quantify the Common Ion Effect
14.1 – Observe the Volume-Pressure Relationship of Gases (Boyle's Law)
14.2 – Observe the Volume-Temperature Relationship of Gases (Charles's Law)
14.3 – Observe the Pressure-Temperature Relationship of Gases (Gay-Lussac's Law)
15.1 – Determine Heat of Solution
15.2 – Determine the Specific Heat of Ice
15.3 – Determine the Specific Heat of a Metal
16.1 – Produce Hydrogen and Oxygen by Electrolysis of Water
18.1 – Observe Some Properties of Colloids and Suspensions
18.2 – Produce Firefighting Foam
18.3 – Prepare a Gelled Sol
19.1 – Using Flame Tests to Discriminate Metal Ions
19.2 – Using Borax Bead Tests to Discriminate Metal Ions

A NOTE ON THE AP CHEMISTRY EXAM

While we were discussing the AP Chemistry exam, Dr. Paul Jones made a profound comment that is worth serious consideration. Many of Paul's first-year organic chemistry students pass the AP Chemistry exam with a 4 or 5 score and skip first-year college general chemistry. Paul thinks that's a mistake, because almost none of those students actually got the full equivalent of first-year college general chemistry in their AP courses.

As Paul said, it's fine for a history major to take the AP Chemistry exam and test out of taking a first-year college chemistry course, and it's fine for a chemistry major to take the AP History exam and test out of taking a first-year college history course. But the history major shouldn't test out of taking the first-year history course, and the chemistry (or other science) major shouldn't test out of taking the first-year chemistry course.

That's not to say that future chemistry (or physics, biology, or other hard science) majors shouldn't take the AP Chemistry course itself. Worst case, by taking the intro course in your major even though you already have the AP course under your belt, you end up with an easy A, impress your professors, and hit the ground running.

For second-year (AP) chemistry students, allocate at least two 2-hour chemistry lab periods per week or one 4-hour lab period. The following lab sessions (which assume that the preceding group of labs has been completed in a first-year course) are a good starting point:

6.6 – Determine the Formula of a Hydrate
7.5 – Determine Concentration of a Solution by Visual Colorimetry
11.2 – Determine the pH of Aqueous Salt Solutions
11.3 – Observe the Characteristics of a Buffer Solution
11.4 – Standardize a Hydrochloric Acid Solution by Titration
12.4 – Determine the Effect of a Catalyst on Reaction Rate
13.3 – Determine a Solubility Product Constant
14.4 – Use the Ideal Gas Law to Determine the Percentage of Acetic Acid in Vinegar
14.5 – Determine Molar Mass from Vapor Density
15.4 – Determine the Enthalpy Change of a Reaction
16.2 – Observe the Electrochemical Oxidation of Iron
16.3 – Measure Electrode Potentials
16.4 – Observe Energy Transformation
16.5 – Build a Voltaic Cell

16.6 – Build a Battery
17.1 – Photochemical Reaction of Iodine and Oxalate
19.3 – Qualitative Analysis of Inorganic Anions
19.4 – Qualitative Analysis of Inorganic Cations
19.5 – Qualitative Analysis of Bone
20.1 – Quantitative Analysis of Vitamin C by Acid-Base Titration
20.2 – Quantitative Analysis of Chlorine Bleach by Redox Titration
20.3 – Quantitative Analysis of Seawater
21.1 – Synthesize Methyl Salicylate from Aspirin
21.2 – Synthesize Rayon Fiber
22.1 – Use the Sherlock Holmes Test to Detect Blood
22.2 – Perform a Presumptive Test for Illicit Drugs
22.3 – Reveal Latent Fingerprints
22.4 – Use the Marsh Test to Detect Arsenic or Antimony

Table 1-1 summarizes how the lab sessions in this book map to the experiments recommended by the College Board for the AP Chemistry exam. Note that some of the recommended experiments are completed in the first year and need not be repeated in the second.

Table 1-1:

AP Recommended Experiments mapped to laboratory sessions in this book

#	AP Recommended Experiment	#	Corresponding Laboratory Session(s)
1	Determination of the formula of a compound	9.2	Observe a Decomposition Reaction
2	Determination of the percentage of water in a hydrate	6.6	Determine the Formula of a Hydrate
3	Determination of molar mass by vapor density	14.5	Determine Molar Mass from Vapor Density
4	Determination of molar mass by freezing point depression	8.2	Determine Molar Mass by Freezing Point Depression
5	Determination of the molar volume of a gas	14.4	Use the Ideal Gas Law to Determine the Percentage of Acetic Acid in Vinegar
6	Standardization of a solution using a primary standard	11.4	Standardize a Hydrochloric Acid Solution by Titration
7	Determination of concentration by acid-base titration, including a weak acid or weak base	20.1	Quantitative Analysis of Vitamin C by Acid-Base Titration
8	Determination of concentration by oxidation-reduction titration	20.2	Quantitative Analysis of Chlorine Bleach by Redox Titration
9	Determination of mass and mole relationship in a chemical reaction	9.4	Stoichiometry of a Double Displacement Reaction
10	Determination of the equilibrium constant for a chemical reaction	13.3	Determine a Solubility Product Constant

#	AP Recommended Experiment	#	Corresponding Laboratory Session(s)
11	Determination of appropriate indicators for various acid-base titrations; pH determination	11.1	Determine the Effect of Concentration on pH
		11.2	Determine the pH of Aqueous Salt Solutions
12	Determination of the rate of a reaction and its order	12.1	Determine the Effect of Temperature on Reaction Rate
		12.2	Determine the Effect of Surface Area on Reaction Rate
		12.3	Determine the Effect of Concentration on Reaction Rate
		12.4	Determine the Effect of a Catalyst on Reaction Rate
13	Determination of enthalpy change associated with a reaction	15.4	Determine the Enthalpy Change of a Reaction
14	Separation and qualitative analysis of cations and anions	19.3	Qualitative Analysis of Inorganic Anions
		19.4	Qualitative Analysis of Inorganic Cations
		19.5	Qualitative Analysis of Bone
15	Synthesis of a coordination compound and its chemical analysis	21.2	Synthesize Rayon Fiber
16	Analytical gravimetric determination	20.3	Quantitative Analysis of Seawater
17	Colorimetric or spectrophotometric analysis	7.5	Determine Concentration of a Solution by Visual Colorimetry
		17.1	Photochemical Reaction of Iodine and Oxalate
18	Separation by chromatography	6.5	Chromatography: Two-Phase Separation of Mixtures
19	Preparation and properties of buffer solutions	11.3	Observe the Characteristics of a Buffer Solution
20	Determination of electrochemical series	16.1	Produce Hydrogen and Oxygen by Electrolysis of Water
		16.2	Observe the Electrochemical Oxidation of Iron
		16.3	Measure Electrode Potentials
21	Measurements using electrochemical cells and electroplating	16.4	Observe Energy Transformation
		16.5	Build a Voltaic Cell
		16.6	Build a Battery
22	Synthesis, purification, and analysis of an organic compound	21.1	Synthesize Methyl Salicylate from Aspirin

MAINTAINING A LABORATORY NOTEBOOK

A *laboratory notebook* is a contemporaneous, permanent *primary record* of the owner's laboratory work. In real-world corporate and industrial chemistry labs, the lab notebook is often a critically important document, for both scientific and legal reasons. The outcome of zillion-dollar patent lawsuits often hinges on the quality, completeness, and credibility of a lab notebook. Many corporations have detailed procedures that must be followed in maintaining and archiving lab notebooks, and some go so far as to have the individual pages of researchers' lab notebooks notarized and imaged on a daily or weekly basis.

If you're just starting to learn about chemistry lab work, keeping a detailed lab notebook may seem to be overkill, but it's not. Although this book provides tables for recording data and spaces for answering the questions it poses, that's really for the convenience of hobbyist readers. If you're using this book to prepare for college chemistry, and particularly if you plan to take the Advanced Placement (AP) Chemistry exam, you should keep a lab notebook. Even if you score a 5 on the AP Chemistry exam, many college and university chemistry departments will not offer you advanced placement unless you can show them a lab notebook that meets their standards.

LABORATORY NOTEBOOK GUIDELINES
USE THE FOLLOWING GUIDELINES TO MAINTAIN YOUR LABORATORY NOTEBOOK:

- The notebook must be permanently bound. Looseleaf pages are unacceptable. Never tear a page out of the notebook.

- Use permanent ink. Pencil or erasable ink is unacceptable. Erasures are anathema.

- Before you use it, print your name and other contact information on the front of the notebook, as well as the volume number (if applicable) and the date you started using the notebook.

- Number every page, odd and even, at the top outer corner, *before* you begin using the notebook.

- Reserve the first few pages for a table of contents.

- Begin a new page for each experiment.

- Use only the righthand pages for recording information. The lefthand pages can be used for scratch paper. (If you are lefthanded, you may use the lefthand pages for recording information, but maintain consistency throughout.)

- Record all observations as you make them. Do not trust your memory, even for a minute.

- Print all information legibly, preferably in block letters. Do not write longhand.

- If you make a mistake, draw one line through the erroneous information, leaving it readable. If it is not otherwise obvious, include a short note explaining the reason for the strikethrough. Date and initial the strikethrough.

- Do not leave gaps or whitespace in the notebook. Cross out whitespace if leaving an open place in the notebook is unavoidable. That way, no one can go back in and fill in something that didn't happen. When you complete an experiment, cross out the whitespace that remains at the bottom of the final page.

- Incorporate computer-generated graphs, charts, printouts, photographs, and similar items by taping or pasting them into the notebook. Date and initial all add-ins.

- Include only procedures that you personally perform and data that you personally observe. If you are working with a lab partner and taking shared responsibility for performing procedures and observing data, note that fact as well as describing who did what and when.

- Remember that the ultimate goal of a laboratory notebook is to provide a permanent record of all the information necessary for someone else to reproduce your experiment and replicate your results. Leave nothing out. Even the smallest, apparently trivial, detail may make the difference.

LABORATORY NOTEBOOK FORMAT
USE THE FOLLOWING GENERAL FORMAT FOR RECORDING AN EXPERIMENT IN YOUR LAB NOTEBOOK:

Introduction
The following information should be entered before you begin the laboratory session:

Date
Enter the date at the top of the page. Use an unambiguous date format, such as 2 September 2008 or September 2, 2008 rather than 2/9/8 or 9/2/8. If the experiment runs more than one day, enter the starting date here and the new date in the procedure/data section at the time you actually begin work on that date.

Experiment title
If the experiment is from this or another laboratory manual, use the name from that manual and credit the manual appropriately. For example, "Quantitative Analysis of Chlorine Bleach by Redox Titration (Illustrated Guide to Home Chemistry Experiments, #20.2)". If the experiment is your own, give it a descriptive title.

Purpose
Write one or two sentences that describe the goal of the experiment. For example, "To determine the concentration of chlorine laundry bleach by redox titration using a starch-iodine indicator."

Introduction (optional)
Any preliminary notes, comments, or other information may be entered in a paragraph or two here. For example, if you decided to do this experiment to learn more about something you discovered in another experiment, note that fact here.

Balanced equations
Write down balanced equations for all of the reactions involved in the experiment, including, if applicable, changes in oxidation state.

Chemical information
Important information about all chemicals used in the experiment, including, if appropriate, physical properties (melting/boiling points, density, etc.), a list of relevant hazards and safety measures from the MSDS (the Material Safety Data Sheet for the chemical), and any special disposal methods required. Include approximate quantities, both in grams and in moles, to give an idea of the scale of the experiment.

Planned procedure
A paragraph or two to describe the procedures you expect to follow.

Main body
The following information should be entered as you actually do the experiment:

Procedure
Record the procedure you use, step by step, as you actually perform the procedures. Note any departures from your planned procedure and the reasons for them.

Data
Record all data and observations as you gather them, inline with your running procedural narrative. Pay attention to significant figures, and include information that speaks to accuracy and precision of the equipment and chemicals you use. For example, if one step involves adding hydrochloric acid to a reaction vessel, it makes a difference if you added 5 mL of 0.1 M hydrochloric acid from a 10 mL graduated cylinder or 5.00 mL of 0.1000 M hydrochloric acid from a 10 mL pipette.

Sketches
If your setup is at all unusual, make a sketch of it here. It needn't be fine art, nor does it need to illustrate common equipment or setups such as a beaker or a filtering setup. The goal is not to make an accurate representation of how the apparatus actually appears on your lab bench, but rather to make it clear how the various components relate to each other. Be sure to clearly label any relevant parts of the set up.

Calculations
Include any calculations you make. If you run the same calculation repeatedly on different data sets, one example calculation suffices.

Table(s)
If appropriate, construct a table or tables to organize your data. Copy data from your original inline record to the table or tables.

Graph(s)
If appropriate, construct a graph or graphs to present your data and show relationships between variables. Label the axes appropriately, include error bars if you know the error limits, and make sure that all of the data plotted in the graph are also available to the reader in tabular form. Hand-drawn graphs are preferable. If you use computer-generated graphs, make sure that they are labeled properly and tape or paste them into this section.

Conclusion
The following information should be entered after you complete the experiment:

Results
Write a one- or two-paragraph summary of the results of the experiment.

Discussion
Discuss, if possible quantitatively, the results you observed. Do your results confirm or refute the hypothesis? Record any thoughts you have that bear upon this experiment or possible related experiments you might perform to learn more. Suggest possible improvement to the experimental procedures or design.

Answer questions
If you've just completed a lab exercise from this or another book, answer all of the post-lab questions posed in the exercise. You can incorporate the questions by reference rather than writing them out again yourself.

2

Laboratory Safety

This is a short chapter, but a very important one. Many of the experiments described in this book use chemicals, such as strong acids and bases, that are dangerous if handled improperly. Some experiments use open flame or other heat sources, and nearly all of the experiments use glassware.

Working in a home chemistry lab has its dangers, but then so does driving a car. And, just as you must remain constantly alert while driving, you must remain constantly alert while working in a home chemistry lab.

It's important to keep things in perspective. More serious injuries occur every year among a few hundred thousand high-school football players than have ever occurred in total among millions upon millions of home chemists in the 200-year history of home chemistry labs. Statistically, students are much, much safer working in a home chemistry lab than they are out skateboarding or riding bicycles.

Most injuries that occur in home chemistry labs are minor and easily avoidable. Among the most common are nicks from broken or chipped glassware and minor burns. Serious injuries are very rare. When they do occur, it's nearly always because someone did something incredibly stupid, such as using a flammable solvent near an open flame or absentmindedly taking a swig from a beaker full of a toxic liquid. (That's why one of the rules of laboratory safety is never to smoke, drink, or eat in the lab.)

The primary goal of laboratory safety rules is to prevent injuries. Knowing and following the rules minimizes the likelihood of accidents, and helps ensure that any accidents that do occur will be minor ones.

DR. PAUL JONES COMMENTS:

Everyone rightly treats strong acids with great respect, but many students handle strong bases casually. That's a very dangerous practice. Strong bases, such as solutions of sodium hydroxide, can blind you in literally seconds. Treat every chemical as potentially hazardous, and always wear splash goggles.

LABORATORY SAFETY RULES WE RECOMMEND

PREPARE PROPERLY

ALL LABORATORY ACTIVITIES MUST BE SUPERVISED BY A RESPONSIBLE ADULT

- Direct adult supervision is mandatory for all of the activities in this book. This adult must review each activity before it is started, understand the potential dangers of that activity and the steps required to minimize or eliminate those dangers, and be present during the activity from start to finish. Although the adult is ultimately responsible for safety, students must also understand the potential dangers and the procedures that should be used to minimize risk.

FAMILIARIZE YOURSELF WITH SAFETY PROCEDURES AND EQUIPMENT

- Think about how to respond to accidents before they happen. Have a fire extinguisher and first-aid kit readily available and a telephone nearby in case you need to summon assistance. Know and practice first-aid procedures, particularly those required to deal with burns and cuts. Paul Jones notes, "Since getting my cell phone, I've started to always have it on me in the lab. Seems easy enough to do and then I wouldn't have to find a phone if something bad happened."

- One of the most important safety items in a home lab is the cold water faucet. If you burn yourself, immediately (seconds count) flood the burned area with cold tap water for several minutes to minimize the damage done by the burn. If you spill a chemical on yourself, immediately rinse the chemical off with cold tap water, and keep rinsing for several minutes. Ideally, every lab should have an eyewash station, but most home chemistry labs do not. If you get any chemical in your eyes, immediately turn the cold tap on full and flood your eyes until help arrives.

- Keep a large container of baking soda on hand to deal with acid spills, and a large container of vinegar to deal with base spills.

ALWAYS READ THE MSDS FOR EVERY CHEMICAL THAT YOU WILL USE IN A LABORATORY SESSION

- The MSDS (Material Safety Data Sheet) is a concise document that lists the specific characteristics and hazards of a chemical. Always read the MSDS for every chemical that is to be used in a lab session. If an MSDS was not supplied with the chemical, locate one on the Internet. For example, before you use potassium chromate in an experiment, do a Google search using the search terms "potassium chromate" and "MSDS."

ORGANIZE YOUR WORK AREA

- Keep your lab bench and other work areas clean and uncluttered—before, during, and after laboratory sessions. Every laboratory session should begin and end with your glassware, chemicals, and laboratory equipment clean and stored properly.

DRESS PROPERLY

WEAR APPROVED EYE PROTECTION AT ALL TIMES

- Everyone present in the lab must at all times wear splash goggles that comply with the ANSI Z87.1 standard. Standard eyeglasses or shop goggles do not provide adequate protection, because they are not designed to prevent splashed liquids from getting into your eyes. Eyeglasses may be worn under the goggles, but contact lenses are not permitted in the lab. (Corrosive chemicals can be trapped between a contact lens and your eye, making it difficult to flush the corrosive chemical away.)

WEAR PROTECTIVE GLOVES AND CLOTHING

- Never allow laboratory chemicals to contact your bare skin. When you handle chemicals, particularly corrosive or toxic chemicals or those that can be absorbed through the skin, wear gloves of latex, nitrile, vinyl, or another chemical-resistant material. (Ansell gloves has a pretty good table that ranks the chemical resistance of various gloving materials: *http://www.ansellpro.com/download/Ansell_ 7thEditionChemicalResistanceGuide.pdf*). Wear long pants, a long-sleeved shirt, and leather shoes or boots that fully cover your feet (*not* sandals). Avoid loose sleeves. To protect yourself and your clothing, wear a lab coat or a lab apron made of vinyl or another resistant material. Wear a disposable respirator mask when you handle chemicals that are toxic by inhalation.

AVOID LABORATORY HAZARDS

AVOID CHEMICAL HAZARDS

- Never taste any laboratory chemical or sniff it directly. (Use your hand to waft the odor toward your nose.) Never use your mouth to fill a pipette. When you heat a test tube or flask, make sure the mouth points in a safe direction. Always use a boiling chip or stirring rod to prevent liquids from boiling over and

DR. PAUL JONES COMMENTS:

Although none of the lab sessions in this book require putting glassware under a vacuum, if you do so for one of your own experiments, always use a polycarbonate explosion shield between yourself and the glassware. Even high-quality glassware that appears undamaged can implode under vacuum, spraying glass shards at high velocity.

being ejected from the container. Never carry open containers of chemicals around the lab. Always dilute strong acids and bases by adding the concentrated solution or solid chemical to water slowly and with stirring. Doing the converse can cause the liquid to boil violently and be ejected from the container. Use the smallest quantities of chemicals that will accomplish your goal. In particular, the first time you run a reaction, do so on a small scale. If a reaction is unexpectedly vigorous, it's better if it happens with 5 mL of chemicals in a small test tube than 500 mL in a large beaker.

AVOID FIRE HAZARDS
- Never handle flammable liquids or gases in an area where an open flame or sparks might ignite them. Extinguish burners as soon as you finish using them. Do not refuel a burner until it has cooled completely. If you have long hair, tie it back or tuck it up under a cap, particularly if you are working near an open flame.

AVOID GLASSWARE HAZARDS
- Assume all glassware is hot until you are certain otherwise. Examine all glassware before you use it, and particularly before you heat it. Discard any glassware that is cracked, chipped, or otherwise damaged. Learn the proper technique for cutting and shaping glass tubing, and make sure to fire-polish all sharp ends.

DON'T DO STUPID THINGS

NEVER EAT, DRINK, OR SMOKE IN THE LABORATORY
- All laboratory chemicals should be considered toxic by ingestion, and the best way to avoid ingesting chemicals is to keep your mouth closed. Eating or drinking (even water) in the lab is very risky behavior. A moment's inattention can have tragic results. Smoking violates two major lab safety rules: putting anything in your mouth is a major no-no, as is carrying an open flame around the lab.

NEVER WORK ALONE IN THE LABORATORY
- No one, adult or student, should ever work alone in the laboratory. Even if the experimenter is adult, there must at least be another adult within earshot who is able to respond quickly in an emergency.

NO HORSING AROUND
- A lab isn't the place for practical jokes or acting out, or for that matter for catching up on gossip or talking about last night's football game. When you're in the lab, you should have your mind on lab work, period.

NEVER COMBINE CHEMICALS ARBITRARILY
- Combining chemicals arbitrarily is among the most frequent causes of serious accidents in home chemistry labs. Some people seem compelled to mix chemicals more or less randomly, just to see what happens. Sometimes they get more than they bargained for.

DON'T MAKE EXPLOSIVES
- Yes, I know. One thing that nearly all home chemists have in common is the gene that compels us to make stuff that goes boom, and the louder the better. Resist the temptation. In addition to the obvious danger of losing some fingers—or your head—you risk having DHS agents kick down your door and cart you off to prison. Years ago, it was a rite of passage for home chemists to manufacture explosives, from black powder to nitroglycerin to acetone peroxide. Most (not all) of us survived unscathed, and thought no more about it. The authorities weren't thrilled about kids blowing stuff up, but they generally resigned themselves to the fact that "boys will be boys." No more. If you're caught making explosives nowadays—and you probably will be caught if you try it—the best you can hope for is a big fine, and that's only if you can afford a good lawyer. Just don't do it.

Laboratory safety is mainly a matter of common sense. Think about what you're going to do before you do it. Work carefully. Deal with minor problems before they become major problems. Keep safety constantly in mind, and chances are good that any problems you have will be very minor ones.

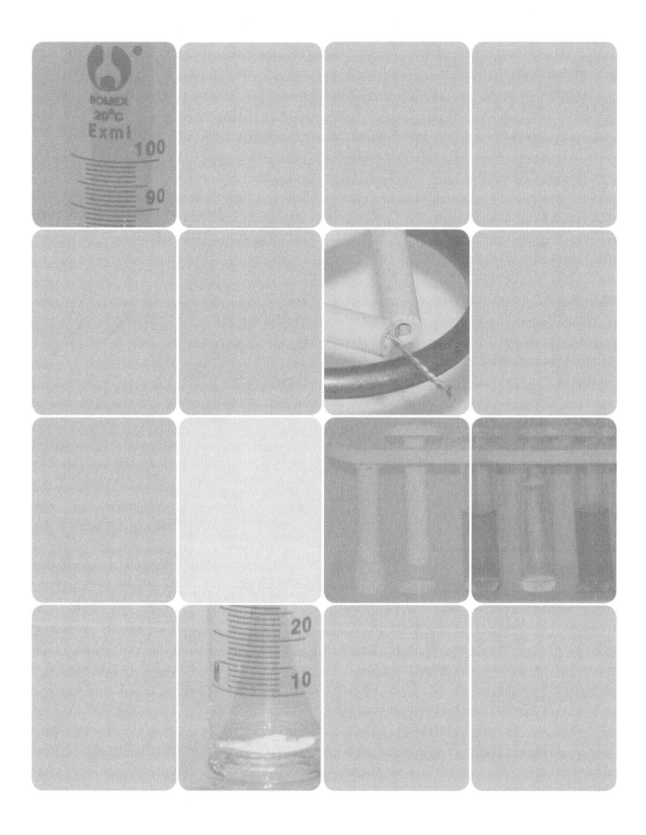

3

Equipping a Home Chemistry Laboratory

Although it doesn't take much special equipment to play around with chemistry, if you want to do serious chemistry, you'll need some real equipment. The essence of chemistry is measuring things—masses, volumes, temperatures, voltages, and so on—and doing that with reasonable accuracy demands equipment designed for the task.

Fortunately, although equipping a professional chemistry lab is very costly indeed, equipping a home chemistry lab is relatively inexpensive. Professional labs may have $10,000 balances and $1,500 pH meters, but we can learn just as much using a $50 or $100 student-grade balance and a $25 student-grade pH meter. The same is true for glassware and other laboratory equipment.

By following the guidelines in this chapter, you can have a well-equipped home chemistry lab for less than the cost of a decent set of golf clubs or a good bookshelf audio system. If that is still more than you care to spend, there are numerous compromises and substitutions you can make to keep the cost of your laboratory down

This chapter describes the purpose and function of basic laboratory equipment, lists the types and amounts of equipment you need, recommends laboratory equipment suppliers that focus on the needs of home chemists, and concludes with some thoughts about where to locate your lab and how to keep things organized.

CHEAPER BY THE CASE

Do not overlook the advantages of banding together with other home schoolers or like-minded hobbyists to purchase laboratory equipment—particularly glassware. Many vendors offer significant discounts on box or case quantities of various items. For example, although you probably don't need an entire case of test tubes, buying a full case may cost only half as much as buying the same number of test tubes individually or in small quantities.

GENERAL PURPOSE GLASSWARE AND PLASTICWARE

Any chemistry lab requires a good assortment of glassware to mix, store, measure, and dispense solutions. All of it needn't be actual labware. To some extent, you can substitute ordinary household items such as drinking glasses, measuring cups, empty soft drink bottles, and so on. But you'll also need at least some formal labware for tasks that require it.

When Robert built his first home chemistry lab in the early 1960s, some laboratory glassware was still made from ordinary flint glass. Nowadays, with the exception of glass tubing, some volumetric (measuring) glassware, and similar items, nearly all laboratory glassware is made from Pyrex or other heat-resistant borosilicate glasses like Kimax, Endural, or Bomex. Still, it's worth checking. Flint-glass labware is ill-suited for any task that involves heat, whether that heat is applied externally or produced by a reaction.

Nowadays, a lot of "glassware" is actually made of plastic. Plastic has some advantages over glass; notably, that it usually doesn't break if you drop it. Plastic labware is often (but not always) less expensive than glass equivalents. Because plasticware is produced by injection molding, it's easy and inexpensive to produce complex shapes that are expensive to produce in glass.

Against these advantages, plasticware has several disadvantages. Obviously, plasticware is less heat-resistant than glassware. Although good plasticware stands up to autoclaving, it cannot be heated over a flame or used for reactions that produce high temperatures. Plasticware may be damaged by organic solvents, oxidants, and some other chemicals. Because plasticware does not possess the vitreous surface of glassware, it is more difficult to clean thoroughly and may be permanently stained or otherwise contaminated by some chemicals, as Figure 3-1 shows.

On balance, we prefer to use glassware for most purposes. We suggest that you devote most of your glassware budget to actual glassware, and buy plasticware only for tasks where its advantages outweigh its disadvantages.

Labware, whether glass or plastic, is commonly available in at least two quality grades. **Laboratory-grade** glassware—also called **professional-grade**—is more expensive, better finished, often more precisely graduated, of heavier construction, better annealed, and is usually made in the United States, Europe, or Japan from Pyrex or other name-brand borosilicate glass. **Student-grade** glassware—also called **economy-grade** or **educational-grade**—is less expensive (sometimes much less), cruder in appearance, often less precisely graduated, may use thinner glass, often has annealing of questionable quality, and is usually made in China from Bomex or other inexpensive borosilicate glass. There are similar differences in grades of plasticware.

FIGURE 3-1:

A plastic graduated cylinder damaged by chemicals

DROP THE FLASK AND TURN AROUND SLOWLY

Before you order any laboratory glassware, make sure that it's legal for you to possess it. We know that sounds strange, but legislators in some jurisdictions—in an attempt to eliminate underground meth labs—have passed laws that make it illegal to possess some pretty innocuous items. For example, in Texas it's illegal to possess an Erlenmeyer flask, unless you have a permit for it. We are not making this up.

BE CAREFUL WITH ALL GLASSWARE

Even the best glassware can shatter unexpectedly, particularly if it is heated very strongly or if when hot it contacts a cold liquid. Inexpensive, student-grade glassware is more likely to shatter, because it may use thinner glass and may not be annealed as well as more costly name-brand glassware. Such accidents are very unlikely, even with inexpensive glassware, but you should always keep the possibility in mind. When you work with glassware, and particularly if you heat it, *always* wear splash goggles and protective clothing.

We use student-grade glassware in this book, solely for cost reasons. For example, we paid $1.80 for one Chinese Bomex student-grade flask. Similar name-brand laboratory grade flasks made of Pyrex or Kimax cost two or three times as much. If budget is not an issue, buy laboratory-grade glassware. But if the budget is tight, don't hesitate to buy student-grade labware.

TEST TUBES

For most people, the test tube (shown in Figure 3-2) is the one piece of equipment that defines a chemistry lab, and rightly so. Test tubes are used so often and for so many purposes that it's hard to imagine a chem lab without them. Most of the time, you'll use test tubes to mix solutions, heat samples, observe reactions, and perform other similar tasks. But test tubes have many other uses. With a one- or two-hole rubber stopper, a glass tube or two,

HOME CHEMISTRY ON A BUDGET

If you're on a very tight budget, consider ordering the $36 Basic Chemistry Equipment Kit (CE-KIT01) from Home Science Tools. This kit contains an assortment of essential glassware and hardware for a home chem lab, including an alcohol lamp and tripod stand, 250 mL and 600 mL beakers, a 250 mL Erlenmeyer flask, 10 mL and 100 mL graduated cylinders, a funnel, test tubes, a test tube rack and holder, an eye dropper, a stirring rod, a thermometer, rubber stoppers, glass and plastic tubing and a pinch clamp, and a vial of pH indicator paper (*http://www.homesciencetools.com*).

This kit contains many of the items you need to complete the laboratory sessions in this book, although you'll need to supplement it with other individual items, either purchased or homemade. (See the lists of glassware and equipment for some ideas.)

and some rubber or plastic tubing, you can convert a test tube into a miniature gas-generating apparatus or distilling setup or reaction vessel. Inverted in a pan of water, a test tube can be used to capture gases. With a solid cork or rubber stopper, a test tube serves to store samples. You can probably imagine many other uses without much effort.

Test tubes are readily available in sizes from 10×75 mm to 25×200 mm. For general use, the best sizes are 15×125 mm, which holds 14 mL, 16×150 mm (20 mL), or 18×150 mm (27 mL). One nice thing about test tubes is that they're cheap. High-quality test tubes in standard sizes sell for $4 to $6 a dozen, and there are often significant discounts for buying larger quantities. That means that you needn't worry about damaging test tubes, as you might with more expensive glassware. You can abuse your test tubes when necessary, and treat badly soiled test tubes as disposable. A dozen test tubes is a good starting point for a home lab.

BEAKERS

Beakers are among the most commonly used items of laboratory glassware; they're used when test tubes aren't large enough. Beakers are flat-bottomed, cylindrical containers, usually equipped with a pouring spout, and are used for routine mixing, measuring, heating, and boiling of liquids. Beakers are available in glass (usually Pyrex or a similar borosilicate glass), polypropylene, and other materials, and are available in capacities from 10 mL to 5,000 mL (5 L) or more.

Polypropylene beakers are popular because they are unbreakable, but they cannot be used for heating or boiling liquids and they may discolor or become cloudy when used with some organic solvents or strongly colored compounds. Polypropylene

FIGURE 3-2: *Test tubes in a test tube rack*

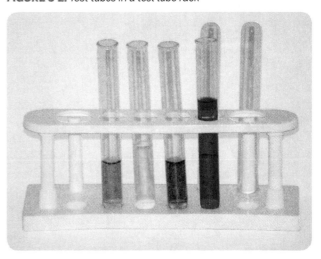

beakers are also more difficult to clean thoroughly and may have to be discarded in situations where a glass beaker could simply be washed.

Most beakers are graduated with painted, etched, or raised lines to indicate the approximate volume the beaker contains at various fill levels. For example, the 250 mL beaker shown in Figure 3-3 has markings every 25 mL. Graduated beakers are typically accurate to within ±5%, and can be used to measure and dispense liquids when high accuracy is not required.

Laboratory beakers are available in two styles. **Berzelius beakers,** also called **high-form beakers** are tall and slender. **Griffin beakers**, also called **low-form beakers**, are shorter and squatter. For a home laboratory, either style is fine.

If you use a beaker for short-term storage of a solution, cover it to prevent evaporation or contamination. Use a **watch glass**, an ordinary piece of glass, or plastic film wrap. If you use a beaker to boil a solution, always add a boiling chip or leave a stirring rod in the beaker to prevent the solution from boiling over, as shown in Figure 3-4.

A well-equipped home chemistry lab should have a good selection of beakers, most mid-sized, but with a few larger and smaller ones for special purposes. The 150 mL or 250 mL size is the most useful for most home labs. Many laboratory equipment suppliers sell assortments of various-size beakers at a discounted price. Many also sell boxes of six to twelve beakers of the same size at a significant discount relative to the single-unit price.

BEAKERS VERSUS FLASKS

We use beakers and flasks pretty much interchangeably. In effect, a beaker is just a wide-mouth flask, and a flask a narrow-mouth beaker. Flasks are better for swirling or heating solutions, when the container must be sealed or is part of an apparatus, or when the contents are volatile. Beakers make it easier to add large amounts of solids to a solution, and their lips make them more convenient for pouring, filling burettes, and so on.

We wondered if there were other good reasons to use one or the other in particular situations, so we asked our advisor Dr. Paul Jones, who is an organic chemistry professor. He confirmed our thoughts on the matter, and added that he seldom uses beakers in his labs, but instead uses Erlenmeyer flasks almost exclusively. Paul added, only partially in jest, that he uses flasks when he cares about what they contain and beakers when he doesn't. Which is probably pretty good advice.

FLASKS

Like beakers, flasks are used routinely for storing, mixing, measuring, heating, and boiling liquids. The primary difference is that beakers have wide mouths and flasks have narrow mouths, which means it's easy to seal a flask using a rubber or cork stopper. Because they are easily sealed, flasks are commonly used as reaction vessels and for constructing various apparatuses such as gas-washing bottles, gas generators,

FIGURE 3-3: *A standard laboratory beaker*

FIGURE 3-4: *Use a stirring rod to prevent boiling over*

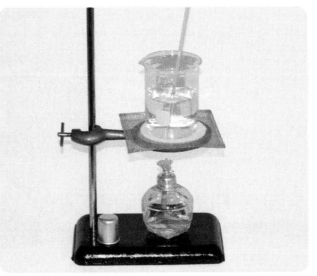

receivers for a distillation apparatus, and so on. In addition to the volumetric flasks detailed later in this chapter, there are three main types of general-purpose laboratory flasks.

ERLENMEYER FLASKS

An **Erlenmeyer flask**, shown in Figure 3-5, has a wide, flat base and a conical cross section, which allows it to sit on the lab bench without risk of tipping. We frequently use an Erlenmeyer flask, also called a **conical flask**, for a task that requires a vessel larger than a test tube.

A well-equipped home chemistry lab should have Erlenmeyer flasks in various sizes. You'll need at least one 500 mL and two 250 mL Erlenmeyer flasks, but it's useful to have several 125 mL or 250 mL flasks. Like beakers, Erlenmeyer flasks can often be purchased in boxes of six or twelve at a significant discount, and are available in assortments of various sizes at a discount.

FLORENCE (BOILING) FLASKS

A **Florence flask**, also called a **boiling flask**, is specifically designed for vigorous boiling. Florence flasks are more robust than Erlenmeyer flasks, and are less likely to break from physical or thermal shock. A Florence flask is a good reaction vessel for distillations, refluxing, and similar operations. There are two varieties of Florence flask. A flat-bottom Florence flask, shown in Figure 3-6, has a small flat area on the bottom of the flask, which means that it can sit stable on a flat surface and can be heated on a wire mesh. A round-bottom Florence flask must be supported by a clamp at all times. In most home chem labs, an Erlenmeyer flask can be used instead of a Florence flask for occasional distillations and similar operations. If you plan to do frequent distillations, particularly on a larger scale or with high-boiling

DANGEROUS ASSUMPTIONS

In order to make Figure 3-5 more visually appealing, we filled the flasks with attractively colored solutions. The 50 mL flask contains tap water with a few drops of green food coloring. The 125 mL flask contains tap water with a few drops of red food coloring. The 250 mL flask contains tap water with a few drops of yellow food coloring. The 500 mL flask contains tap water with . . . several grams of a toxic copper compound.

Or does it? We may have used blue food coloring in the 500 mL flask and a toxic chromium compound in the 50 mL flask. The point is that we can't tell just by looking, and neither can you.

Never judge the contents of a flask or other container by visual appearance alone.

liquids, have at least one Florence flask on hand. The 250 mL and 500 mL sizes are the most useful.

FILTERING FLASKS

A **filtering flask**, shown in Figure 3-7, looks like an Erlenmeyer flask with a side arm. In use, tubing connects a vacuum source to the side arm, establishing a vacuum in the flask. A formal laboratory provides connections to a central vacuum pump. In a home laboratory, the vacuum source is usually a hand vacuum pump or an aspirator connected to a water faucet. The suction

FIGURE 3-5: *Erlenmeyer flasks in various sizes*

FIGURE 3-6: *A 250 mL flat-bottom Florence flask*

draws liquid through a Büchner funnel atop the flask, greatly increasing the speed of filtering. If you have an aspirator or other source of constant suction, you can also use the filtering flask to dry the material trapped by the filter paper by continuing to draw air through the Büchner funnel after all of the liquid has been filtered off.

Filtering flasks are readily available in sizes from 125 mL to 1,000 mL or more. A 250 mL or 500 mL filtering flask is generally the most useful size for a home chem lab. For most filtering operations you can get by without a filtering flask, using gravity filtration with a standard funnel and filter paper, as long as you're willing to wait as long as it takes for gravity to do its work. Of course, depending on what you're filtering, gravity filtration may take several minutes or even several hours. In some cases, a very finely divided precipitate may clog the filter paper, bringing gravity filtration to a dead stop. If that happens, suction is the only cure.

FUNNELS

Standard funnels are used to transfer liquids into narrow-mouth containers. A **powder funnel** resembles a standard funnel, but has a shorter, wider stem to allow solid chemicals to flow freely through the funnel.

Filtering funnels are available in two common varieties. One resembles a standard funnel, but has vertical ridges or channels inside the body of the funnel that are designed to keep the filter paper from adhering to the funnel. A **Büchner funnel**, shown in Figure 3-7 atop a filtering flask, is a special type of filtering funnel, designed to be used with suction to speed up filtration.

A **safety funnel**, also called a **thistle tube**, is shown in Figure 3-8. A safety funnel comprises a long tube with an open bulb on the top end, which may or may not be stoppered. A safety funnel is usually inserted in a rubber stopper and used to introduce additional reagents to a reaction vessel during the reaction.

A **separatory funnel** (or **sep funnel**), shown in Figure 3-9, is used to separate multiple layers of immiscible liquids by draining off the lower layer or layers through a stopcock. Sep funnels are commonly used, particularly in organic chemistry experiments, for extraction and washing.

BOTTLES AND VIALS

Any chemistry lab needs a variety of bottles and vials to store and dispense chemicals. Although bottles seem mundane, some home chemists become fascinated with the great variety available. More than one home chemist has developed a serious secondary hobby of collecting bottles, some obsessively. I haven't been bitten that badly, but will confess that I do have a collection of old bottles in various fascinating shapes and unusual

FIGURE 3-7: *A filtering flask with a Büchner funnel*

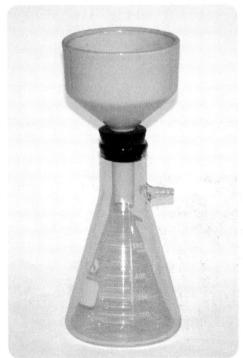

FIGURE 3-8: *A safety funnel or thistle tube*

FIGURE 3-9: *A separatory funnel*

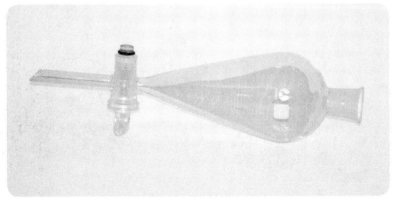

colors. But even if your interest in bottles is merely utilitarian, you'll probably want to acquire a lot of them in many different sizes and types.

STORAGE BOTTLES

Storage bottles, shown in Figure 3-10, are used for stock solutions and other general storage. If you work mostly with chemicals from their original containers, you'll need fewer storage bottles; if you make up stock solutions of commonly used chemicals, you'll need more. I prefer the latter method, because it's faster, more convenient, and more precise to measure small quantities of stock solutions rather than weighing out individual chemicals each time they're needed.

You don't have to buy storage bottles, of course. You can recycle empty household containers. For larger bottles, one good source is empty 500 mL soda bottles, which are made of resistant polyethylene terephthalate plastic and have tight-fitting caps. (Don't use them to store organic solvents, concentrated acids or bases, or other chemicals that might react with the plastic, but dilute stock solutions of most inorganic chemicals are fine.) Before you reuse them, wash recycled bottles thoroughly with soapy water, soak them in a bleach solution (one part chlorine bleach to five parts water), and then rinse them thoroughly and allow them to dry until no odor of bleach (or anything else) remains.

If you decide to buy storage bottles, you'll find them readily available in glass and various plastics in sizes ranging from 15 mL (0.5 ounce) or smaller to 1 quart/liter and larger. For general use, I prefer standard flint-glass narrow-mouth bottles in the styles called French Square and Boston Round, with plastic caps. Such bottles are commonly available in colorless or brown (amber) glass, and, less commonly, in blue glass. (Blue glass bottles were historically used to store poisons.) Brown glass bottles are usually less expensive than similar bottles in colorless glass, presumably because making colorless glass requires using purer and more expensive materials. That's just as well, because some stock solutions are light sensitive, and should be stored in brown or amber glass. It doesn't hurt to store any solution in a brown glass bottle, so that's what I generally use.

Before you buy storage bottles, think about the quantities of stock solutions you're likely to make up for your lab. For my lab, I decided to make up 100 mL of most stock solutions, so I ordered most of my bottles with that in mind. As it turned out, 4-ounce (118 mL) bottles were considerably less expensive than 100 mL bottles, so I ordered 100 4-ounce bottles. I also wanted to make up 500 mL of commonly used stock solutions such as dilute acids and bases, so I decided to order two dozen 500 mL bottles as well. Once again, pint (473 mL) bottles turned out to cost less than the 500 mL metric equivalent, so I ordered pint bottles instead of the 500 mL ones. Pint bottles are too small for 500 mL

IMPLOSION DANGER

Filtering flasks are considerably more expensive than standard Erlenmeyer flasks. Don't attempt to save money by building your own filtering flask with a standard Erlenmeyer flask and a two-hole rubber stopper. Filtering flasks are built of thick glass to resist air pressure when they're evacuated. A standard Erlenmeyer flask is much more fragile, and may implode when evacuated, scattering broken glass and chemicals all over the lab (and you).

Some experimenters construct filtering flasks with standard Florence flasks, which are more strongly built than Erlenmeyer flasks and less likely to implode, but we strongly recommend against this practice. If you need a filtering flask, buy a filtering flask. Don't attempt to make do.

Always clamp a filtering flask securely while it's in use. Otherwise, it may tip over and shatter, scattering glass and chemicals.

CHEAPER LOCALLY AND BY THE CASE

General laboratory supply vendors sell bottles, but they are often quite expensive, particularly if you buy only a few at a time. Also, shipping charges may be very high. Check your local Yellow Pages under "Bottles" for local vendors, and you may be surprised. For example, we bought a case of 4-ounce amber glass storage bottles with plastic caps for less than $0.40 each.

FIGURE 3-10: *A selection of storage bottles*

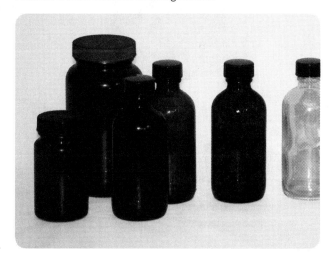

of solution, but that's no problem. For example, if I make up 500 mL of 1 M hydrochloric acid, I simply store part of it in a 4-ounce bottle and the remainder in a pint bottle, from which I refill the smaller bottle as needed.

Every bottle should be labeled indelibly with its contents. If you're patient and want to do things the traditional way, you can use an oilstone or whetstone to frost an area on a glass bottle that can then be labeled with a marking pen or wax pencil. I prefer to use my laser printer to print sheets of sticky labels, which adhere quite well to glass and plastic bottles. After applying the label, I cover it with clear tape to protect it. It's also fine to hand-label bottles, as shown in Figure 3-11, as long as you print neatly and use an indelible marking pen. Include at least the name of the chemical and, if it's a solution, its concentration. Many reagents have limited shelf lives, so date the bottle when you make up the contents.

BARNES BOTTLES

A **Barnes bottle** is a dropping bottle that is used to store an **indicator** or another solution that is typically used drop-wise in small quantities. The rubber bulb of the dropper in a Barnes bottle is large enough to plug the mouth of the bottle. When you need a few drops of the solution contained in the Barnes bottle, you remove the dropper, dispense the solution, and then reinsert

the dropper in the bottle, which seals it. It's handy to have half a dozen or so Barnes bottles on hand, filled with phenolphthalein and other indicators, dilute acids and bases, and so on. You can, of course, substitute any dropping bottle that you recycle from the bathroom or kitchen.

WASH BOTTLES

A **wash bottle** is a soft plastic bottle that allows you to dispense a liquid through a tube in a fine stream by squeezing the bottle. Washing bottles are used for rinsing glassware, adding small amounts of liquid to a reaction vessel in a controlled manner, and so on. You'll want at least one washing bottle, filled with distilled or deionized water for general use around the lab. You may want additional washing bottles to contain alcohols and other commonly used liquids. Note that washing bottles are designed to dispense liquids, but are not intended to be used for long-term storage. An empty, thoroughly washed shampoo bottle with a squeeze top is an adequate substitute, as is a boutique spring water bottle with a squeeze top.

MISCELLANEOUS GENERAL GLASSWARE

In addition to the test tubes, beakers, flasks, bottles, and other general glassware, a well-equipped lab requires an assortment of miscellaneous glassware.

FIGURE 3-11: *Make it a habit to label and date all storage bottles*

FIGURE 3-12: *Barnes bottles*

FIGURE 3-13: *A wash bottle*

GLASS AND STRONG BASES DON'T MIX

Although glass is impervious to most solutions used in a home chem lab, an exception is concentrated solutions of strong bases, such as sodium hydroxide and potassium hydroxide. These solutions literally dissolve glass, given enough time. Store concentrated base solutions in plastic bottles that are rated for use with strong bases.

Glass tubing

An assortment of *glass tubing* is useful for constructing various apparatuses, including gas-generating bottles, distillation setups, and so on. Look for tubing with an outside diameter of 5 mm, which is the size accepted by all but the smallest-holed rubber stoppers. Glass tubing is available in borosilicate (Pyrex) and standard flint glass. The former is useful for its resistance to heat, but the latter is more generally useful, because it can be bent, stretched, and otherwise manipulated after being heated in a standard alcohol lamp flame.

Stirring rods

Stirring rods are solid glass rods that are available in various lengths and diameters. In addition to their nominal purpose, stirring rods can be used to decant liquids from one container to another without spillage, and to prevent boiling liquids from "bumping." Have several stirring rods available, and consider buying rods that include a "rubber policeman" on one end. (A rubber policeman is a rubber scraper that fits snugly on the end of the stirring rod.) These are useful for scraping crystals from the sides of flasks and beakers and for other similar tasks. Choose the length and diameter of your stirring rods according to the sizes of glassware you use. Rods much smaller than 5 mm are relatively fragile, particularly longer ones. I generally use stirring rods that are 5 mm in diameter and 8" (200 mm) long.

Evaporating dishes

An *evaporating dish* is a shallow bowl, usually made of porcelain or borosilicate glass. They are used for evaporating the solvent from a sample, either in the open air at room temperature or in a desiccator or an oven. The 75 mm porcelain models are a good size for a home chem lab. You can substitute a Pyrex saucer or similar heat-resistant container.

Watch glasses

Watch glasses are round convex/concave glass plates that are used as beaker covers and for evaporating solvent from small samples. I keep on hand a couple of watch glasses in sizes that fit my beakers. For covering a beaker, plastic film wrap works just as well.

Crucibles

Crucibles are small ceramic or metal pots with lids, used when you need to heat a sample to a very high temperature. Although it's often possible to reuse crucibles, you should discard it if the reaction causes any staining or other noticeable change. Fortunately, crucibles are cheap—at least the Chinese-made ones are. I confess that I prefer those, because if I'm going to ruin a crucible, I'd sooner ruin a $0.75 Chinese one than a $5.00 U.S.-made Coors crucible.

AVOID TOXIC DUST

Always wear splash goggles, gloves, and a respirator when you use a mortar and pestle to grind chemicals. To minimize the amount of dust that escapes into the air, make a temporary cover with a sheet of thin cardboard or heavy paper large enough to cover the mortar. Punch a hole in the cardboard large enough to fit the pestle, and keep the mortar covered as you grind the chemical. Wear a respirator mask any time you grind a chemical.

Mortar and pestle

A *mortar and pestle* is used to grind dry chemicals to a fine powder. A laboratory mortar and pestle may be made of ceramic or glass or such metals as stainless steel or nickel. Many kitchen supply stores carry mortars and pestles intended for kitchen use that are also fine for laboratory use (but use them in one place or the other, *NOT* both). As a teenager, my mortar and pestle looked very much like a 4" steel pipe cap and a large steel bolt.

VOLUMETRIC GLASSWARE

Volumetric glassware is used to measure liquids with accuracy ranging from moderate to very high. Many standard beakers and flasks have graduations accurate to ±5%. Although this level of accuracy is sufficient for many routine tasks, you'll often need to measure liquids more precisely. Volumetric glassware allows you to measure liquids with accuracy ranging from ±1% to ±0.1% or less, depending on its type, design, and quality.

Class A volumetric glassware provides the highest accuracy, but is expensive and is used primarily in professional laboratories. Class A volumetric glassware complies with the Class A tolerances defined in ASTM E694, must be permanently labeled as Class A, and is supplied with a serialized certificate of precision. (Some manufacturers have begun supplying "generic" Class A vessels, which meet Class A tolerances, but are not technically Class A because they are not serialized or certified.) All Class A volumetric glassware is actually glass; volumetric plasticware is not eligible for Class A status.

Class B volumetric glassware has tolerances twice those of Class A (except graduated cylinders, which have rules of their own), but is considerably less expensive and is adequate for most home laboratories. Class B volumetric glassware must comply with the Class B tolerances defined in ASTM E694 and must be permanently labeled as Class B. Many inexpensive volumetric glassware items, particularly those made in China, are unlabeled and so cannot be guaranteed to meet Class B standards. In practice, most of them seem to meet Class B standards, or at least come close.

Table 3-1 lists the Class A and Class B tolerances for the four common types of volumetric glassware used in home chem labs.

Depending on its type and design, volumetric glassware may be rated To Contain (TC) or To Deliver (TD). A vessel marked TC contains the amount specified when it is filled to the graduation line. A vessel marked TD delivers the amount specified when it is filled to the graduation line and then emptied using the proper procedure. The difference arises because of drainage holdback error. For example, volumetric flasks are rated TC. If you fill a 500 mL volumetric flask to the graduation line, it contains exactly 500 mL of solution (within its tolerance, and assuming that the ambient temperature corresponds to that at which the flask was rated, usually 20°C). If you empty that flask into another container, a bit less than 500 mL will transfer. That's because some of the solution remains in the flask, wetting its inner surface.

GRADUATED CYLINDERS

A *graduated cylinder* is a tall, slender cylinder with numerous graduation lines from near the bottom to near the top. You

TABLE 3-1: *Comparison of Class A and Class B tolerances for volumetric glassware*

	Graduated cylinder		Pipette	
Nominal Volume	Class A	Class B	Class A	Class B
1 mL	n/a	n/a	±0.006 mL	±0.012 mL
5 mL	n/a	n/a	±0.01 mL	±0.02 mL
10 mL	±0.08 mL	±0.1 mL	±0.02 mL	±0.04 mL
25 mL	±0.14 mL	±0.3 mL	±0.03 mL	±0.06 mL
50 mL	±0.20 mL	±0.4 mL	±0.05 mL	±0.10 mL
100 mL	±0.35 mL	±0.6 mL	±0.08 mL	±0.16 mL
250 mL	±0.65 mL	±1.4 mL	n/a	n/a
500 mL	±1.10 mL	±2.6 mL	n/a	n/a
1000 mL	±2.00 mL	±5.0 mL	n/a	n/a
2000 mL	n/a	±10.0 mL	n/a	n/a

EVERYTHING YOU ALWAYS WANTED TO KNOW ABOUT VOLUMETRIC GLASSWARE

One of our readers, Michael Dugan, has written an excellent introduction to volumetric glassware.
To learn more about this subject, visit Mike's website at *http://tinyurl.com/egkj8.*

ISO VERSUS U.S. LABELING

Volumetric glassware made in the United States often uses the traditional TC and TD markings. Other than products intended for export to the United States, most volumetric glassware made in Europe and Asia uses the ISO-standard labeling of "IN" for To Contain and "EX" for To Deliver.

FIGURE 3-14: *10 mL and 100 mL graduated cylinders*

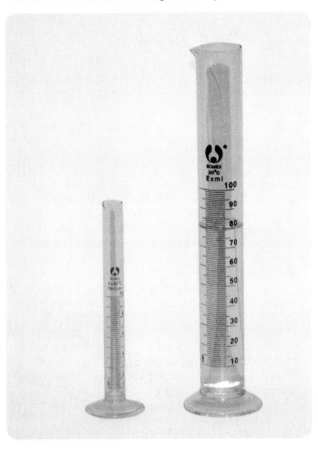

use a graduated cylinder to measure liquids with moderate to moderately high accuracy—from ±1% to ±0.2% of full capacity, depending on the capacity and quality of the cylinder. Graduated cylinders are made in glass and various plastics. Although they are breakable and usually more expensive than similar plastic models, I prefer glass cylinders because they are much more resistant to organic solvents, oxidizers, and other chemicals, and because they are much easier to keep clean. Figure 3-14 shows typical 10 mL and 100 mL glass graduated cylinders.

Single-scale graduated cylinders are graduated from zero at the bottom of the cylinder to the nominal capacity near the top. *Double-scale graduated cylinders* have a second set

TABLE 3-1 *(continued)*: *Comparison of Class A and Class B tolerances for volumetric glassware*

	Burette		Volumetric flask	
Nominal Volume	Class A	Class B	Class A	Class B
1 mL	n/a	n/a	n/a	n/a
5 mL	n/a	n/a	n/a	n/a
10 mL	±0.02 mL	±0.04 mL	±0.02 mL	±0.04 mL
25 mL	±0.03 mL	±0.06 mL	±0.03 mL	±0.06 mL
50 mL	±0.05 mL	±0.10 mL	±0.05 mL	±0.10 mL
100 mL	±0.10 mL	±0.20 mL	±0.08 mL	±0.16 mL
250 mL	n/a	n/a	±0.12 mL	±0.24 mL
500 mL	n/a	n/a	±0.20 mL	±0.40 mL
1000 mL	n/a	n/a	±0.30 mL	±0.60 mL
2000 mL	n/a	n/a	±0.50 mL	±1.00 mL

of graduations, with zero at the top, which makes it easier to calculate the amount dispensed from the cylinder. Graduated cylinders are available in capacities from 10 mL to 2000 mL. The most useful sizes in a home chem lab are the 10 mL and 100 mL versions.

PIPETTES

A graduated *pipette* (also spelled *pipet*) is a slender glass tube that is used to measure and dispense liquids with a very high degree of accuracy and precision. Three types of graduated pipettes are commonly used in chemistry labs.

Volumetric pipette

A *volumetric pipette*, also called a *transfer pipette*, is easily recognized by the bulge in its middle. Standard volumetric pipettes have only one graduation line that corresponds to the nominal capacity of the instrument, and so can be used only to measure that specific quantity. For example, a 50 mL volumetric pipette can measure exactly 50 mL (within its tolerance), but cannot measure smaller quantities because it has no graduation lines for them. Volumetric pipettes are available in Class A and Class B, and may be calibrated To Contain or To Deliver. A few volumetric pipettes are dual-calibrated TC *and* TD, with two corresponding graduation lines. Volumetric pipettes are specialized instruments that aren't needed in most home chem labs. Instead, use a Mohr or serological pipette to measure and transfer small quantities of liquids, and a volumetric flask for larger quantities.

Mohr pipette

A *Mohr pipette*, also called a *measuring pipette*, is graduated from a zero line near the top of the pipette to a baseline near the bottom. Those graduations allow a Mohr pipette to be used to measure any amount of liquid from its smallest increment up to its maximum capacity. The fineness of the graduations depends on the capacity of the pipette. For example, a 10.0 mL Mohr pipette is typically graduated by 0.1 mL markings, while a similar 1.0 mL model is graduated by 0.01 mL markings. You can interpolate between markings. For example, if the fluid level in a 10.0 mL Mohr pipette is about halfway between the 5.1 mL and 5.2 mL marks, you can record that as 5.15 mL, give or take.

NEVER PIPETTE BY MOUTH

In the Bad Olde Days, many chemists used their mouths to suck liquids into pipettes and to blow out the last droplet. Many of them dropped dead, too. *Never put your mouth in contact with a pipette*. Not even if you're pipetting pure water. If you get into the habit of filling pipettes by mouth, the day will inevitably arrive when you suck something nasty into your mouth. Just don't do it. Use a pipette bulb or pipette pump instead.

A Mohr pipette is a TD device, in the sense that when a Mohr pipette has delivered its maximum calibrated volume, some liquid remains in the tip. Do not blow out this remaining liquid, which is accounted for in the calibration.

Serological pipette

A *serological pipette*, also called a *blow-out pipette*, can be thought of as a TC version of a Mohr pipette. The only real difference is that the serological pipette is calibrated TC rather than TD, which means you have to blow out the liquid that remains in the tip after the serological pipette drains. Serological pipettes are made in glass and plastic versions, and have at best Class B accuracy. They are also inexpensive and sufficiently accurate for nearly any task in a home chem lab. Serological pipettes are readily available in capacities of 0.2 mL, 0.5 mL, 1.0 mL, 2.0 mL, 5.0 mL, and 10.0 mL. Most serological pipettes are color-coded for easy identification. Glass models sell for $0.50 to $3.00 each, depending on capacity and quantity, and disposable plastic models for less. Keep at least one 1.0 mL and one 10.0 mL glass serological pipettes on hand.

Various types of ungraduated or roughly graduated pipettes are also useful in a chemistry lab, including medicine droppers, the *Pasteur pipette* (essentially a large medicine dropper with a long, fine tip), and the *Beral pipette*, which is described later in this chapter.

FIGURE 3-15: *10.0 mL (top) and 1.0 mL serological pipettes*

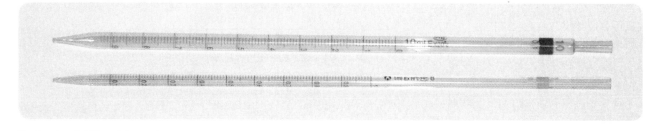

BURETTES

A *burette* (also spelled *buret*) is used to dispense controlled small amounts of a liquid with great precision. Burettes are used to perform titrations for quantitative analyses, determining accurate concentrations of stock solutions, and so on.

Burettes are available in glass and plastic models. Although glass burettes are breakable, we prefer them, because they are easier to clean and are unaffected by organic solvents. The least expensive burettes use a rubber tube and pinchcock to control the flow of the reagent. More expensive models use a ground glass stopcock, which must be kept scrupulously clean and stored disassembled to prevent the ground glass joint from binding. The most expensive models use a Teflon stopcock to control the flow, and are much easier to clean and maintain. You'll want a burette if you plan to do quantitative analyses. A 50 mL model is the best compromise of capacity, flexibility, accuracy, and general usefulness.

VOLUMETRIC FLASKS

A *volumetric flask*, shown in Figure 3-16, is used to make up a precise volume of solution. It has one graduation line that indicates the nominal volume. To use a volumetric flask to make up a solution, you fill it partially with distilled water, add the chemical or chemicals to be dissolved or diluted, swirl the flask until the solute is dissolved, and then fill the flask to the graduation line.

Unless a volumetric flask is otherwise labeled, it is designed to be accurate at 20°C. Small variations in room temperature are no cause for concern, but dissolving or diluting some chemicals, such as strong mineral acids, is highly exothermic, and raises the temperature of the solution enough to affect accuracy. (Although it is less common, dissolving or diluting some other chemicals can be endothermic, and can lower the temperature of the solution enough to affect accuracy.) If you use a volumetric flask to make up solutions of these exothermic or endothermic solutes, allow the solution to come to room temperature before you top up the flask to the graduation line.

Although some glass volumetric flasks are made of Pyrex or similar borosilicate glass, many are of flint glass. For those flasks in particular (and even for Pyrex flasks), we prefer to do the initial dissolving of exothermic solutes in a beaker or Erlenmeyer flask. For example, if we're making up 500 mL of a stock solution of sodium hydroxide, which releases a great deal of heat as it dissolves, we start with about 400 mL of distilled water in a 600 mL beaker, add the sodium hydroxide gradually, and stir until it dissolves. Only after that solution has cooled to room temperature do we transfer it to the volumetric flask (rinsing the beaker several times with distilled water to make sure we've done a quantitative transfer) and then top off the volumetric flask.

EXOTHERMIC DANGERS

When you make up a solution using a concentrated or solid strong acid or base, always gradually add the acid or base to the water, not vice versa. If you add water to the strong acid or base, the solution may boil almost instantly, ejecting the chemical forcefully from the container.

THE CASE OF THE MISSING STOPPER

Some volumetric flasks are sold with ground-glass stoppers or plastic snap caps included, but for others you must order the stopper or cap separately. Using your thumb as a stopper is considered poor practice. It can also be hard on your thumb. I confess, though, that at times I've found myself stopperless and have used a clean-gloved thumb as a makeshift stopper.

FIGURE 3-16: *A 100 mL volumetric flask*

Volumetric flasks are available in glass (Class A or Class B) and plastic (Class B only), in sizes from 10 mL to 2000 mL or larger. A home lab should ideally have a set that includes 25 mL, 50 mL, 100 mL, 250 mL, 500 mL, and 1000 mL flasks. But volumetric flasks, particularly good ones, aren't cheap. If you can afford only one volumetric flask, choose a 100 mL flask. If you can afford two volumetric flasks, add a 500 mL flask for making up stock solutions that you use in larger amounts, such as dilute acids and bases. If you can afford three, add a 25 mL flask for mixing up small quantities of stock solutions of expensive chemicals, such as silver nitrate, expensive indicators with short shelf lives that you buy by the gram, and so on. If you're on a very tight budget, you can do without a volumetric flask and use a graduated cylinder instead. You'll give up some accuracy, but your results will be accurate enough for all of the labs in this book.

ACCURATE, ACCURATER, ACCURATEST?

With the exception of graduated cylinders, many people think of volumetric flasks as the least accurate volumetric instruments, with burettes being more accurate, and pipettes most accurate of all. In reality, although graduated cylinders are in fact less accurate, the Class A and Class B standards are remarkably similar for the other three types of volumetric glassware, if you compare identical capacities. Of course, pipettes are made in much smaller capacities than burettes or volumetric flasks. Those small pipettes have extremely high accuracy, but only for measuring correspondingly small quantities of liquids.

MICROSCALE EQUIPMENT

The recent trend in chemistry labs, particularly school and university labs, is to substitute *microscale chemistry* equipment and procedures for traditional semi-micro or macroscale equivalents. Microscale chemistry, often called *microchemistry*, is just what it sounds like. Instead of using standard test tubes, beakers, and flasks to work with a few mL to a few hundred mL of solutions, you use miniaturized equipment to work with solution quantities ranging from 20 µL (microliters, where one µL equals 0.001 mL) to a couple mL.

Using microscale equipment and procedures has many advantages. Microscale equipment and procedures are less expensive than standard equipment and procedures, which is a major reason for the popularity of microscale chemistry. Using microscale equipment and procedures means that chemicals are needed in very small quantities, which are safer to work with and easier to dispose of properly. Microscale also makes it economically feasible to do experiments with very expensive chemicals, such as gold, platinum, and palladium salts. Setup and teardown is faster, allowing more time for actual experiments, and cleanup usually requires only rinsing the equipment and setting it aside to dry.

Against these advantages, there are several disadvantages to microscale chemistry. First and foremost, everything is on such a small scale that it can be difficult to see what's going on. For example, you may need a magnifier to examine a precipitate (or even to determine whether there *is* a precipitate). Because of the small scale, measuring or procedural errors so small that they would have no effect on a traditional scale experiment can greatly affect the outcome of a microscale experiment. With traditional equipment and procedures, a milligram balance (or even a centigram model) is sufficiently accurate to do reasonably precise quantitative work, but equivalent precision in microscale requires a very expensive tenth-milligram (0.0001 g) or hundredth-milligram (0.00001 g) analytical balance. Finally, although it sounds odd, like most people we find microscale experiments less satisfying emotionally than larger-scale experiments. Watching two drops of reagent combine in a reaction plate to yield a precipitate smaller than a grain of sand just isn't as thrilling as combining 10 mL of each reagent in a test tube and watching the precipitate settle to the bottom of the tube, where it can be isolated, purified, and used for subsequent experiments and tests.

Still, the ubiquity of microscale chemistry means that it's important for you to become familiar with basic microscale equipment and procedures. Most of the experiments in this book use semi-microscale, such as test tubes and small beakers and flasks. If you want to gain experience with microscale procedures, you can substitute microscale equipment and techniques. The only microscale equipment you need to do the experiments in this book is described in the following sections.

REACTION PLATES

A *reaction plate*, shown in Figure 3-17, is a flat sheet of glass, plastic, or porcelain with an array of small wells that typically contain 0.25 mL to 4 mL each. Reaction plates are available in many sizes with from half a dozen to 96 or more wells. Reaction plates are widely used in microchemistry to substitute for test tubes and other containers, but they are equally useful in a general chem lab when you need to test one reagent against many unknowns or many reagents against one unknown. For example, in one of the experiments in the Forensic Chemistry chapter, we use Marquis reagent (a presumptive test used by drug enforcement agents) to test many substances found around the home. If we performed these tests in standard test tubes, we'd need a fairly large amount of Marquis reagent (and a lot of test tubes, all of which we'd have to wash after we completed the tests). Using a reaction plate instead means that we need only a few drops of Marquis reagent for each test, and we have only one item to wash when we complete the tests.

It's useful to have both white and black reaction plates. White plates make it easier to see color changes. Black plates make it easier to see precipitates. Alternatively, a clear plastic or glass reaction plate serves both purposes, according to the color of the surface that you rest it on.

Reaction plates are available in deep-well models that have cylindrical, flat-bottom wells and shallow-well models with wells with rounded bottoms. Models with a large number of wells generally have wells of lower capacity, and vice versa. One of the most useful and common sizes is a 24-well deep-well model that holds 3 mL per well. At a buck or two each, these are cheap enough that every home chem lab should

have at least a couple on hand. (Having at least two available makes it easier to do comparative procedures such as colorimetric comparisons.)

DROPPERS AND DISPOSABLE PIPETTES

Droppers and *disposable pipettes* (also called *Beral pipettes*) are used to transfer or dispense small amounts of liquids in a controlled fashion. The traditional eyedropper, with its glass tube and rubber bulb, remains useful for these purposes, particularly when the liquid in question is an organic solvent or other liquid that would damage or react with a plastic pipette. (Of course, that same solvent may damage a plastic reaction plate, which is an argument for having at least one glass or porcelain reaction plate available.)

Beral pipettes, shown in Figure 3-18, are single-piece injection-molded plastic dropper pipettes. Beral pipettes are cheap, disposable, and have many different uses around the lab. In addition to their intended use, I have used Beral pipettes as miniature reaction vessels, miniature filtering funnels, thistle tubes, and, by cutting their bulb ends at a diagonal, as disposable miniature chemical scoops to avoid contaminating our stocks of reagent-grade dry chemicals.

In bags of 100, Beral pipettes sell for five or six cents each, versus two or three times that much if you buy in small quantities. Beral pipettes come in various sizes, and are rated by the amount they contain and the number of drops they dispense per mL. I prefer the standard 150 mm by 5 mm models, which contain about 3 mL, dispense 25 drops per mL, and are graduated in mL.

FIGURE 3-17: *A reaction plate*

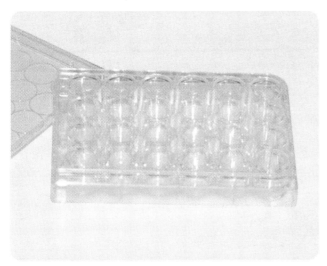

FIGURE 3-18: *Beral pipettes*

RECOMMENDED LABORATORY GLASSWARE

Table 3-2 lists my recommendations for a basic set of laboratory glassware. With only these items, you can complete all of the laboratories in this book.

If you're on a tight budget, you can get along with less than the recommended basic glassware by washing and reusing glassware during a lab or by modifying the quantities we used or the equipment setups I describe. For example, in one lab you will distill 100 mL of ethanol from one flask to another. By reducing the quantity of ethanol to 10 mL, you can substitute a setup using test tubes rather than flasks. Similarly, many of the labs can be done microscale using reaction plates and Beral pipettes rather than the test tubes or beakers that I used. Even if you can't afford to buy all the basic glassware I recommend, don't let that stop you. You can manage, one way or another, by substituting and making do.

I maintain a current list of vendors of laboratory glassware, equipment, and chemicals at *http://www.homechemlab.com/ sources.html.* Check that page before you order any glassware, equipment, or chemicals.

At the expense of some inconvenience and accuracy, you can economize on glassware by using nonstandard procedures. For example, rather than buying a volumetric flask or flasks, you can use your 10 mL and 100 mL graduated cylinders as ad hoc volumetric flasks. Your results won't be quite as accurate, but they won't be terribly far off, either. Similarly, if there's no room in the budget for a burette to do titrations, you can substitute a $1 Mohr or serological pipette. The pipette is less convenient to use for a titration than the burette, but if you work carefully, your results can be as accurate with the pipette as they would have been with the burette. You can even substitute a calibrated Beral pipette, and do your titration by counting drops.

The items listed in Table 3-3 are useful additions to the basic glassware kit. None of them are absolutely necessary, but all of them make things easier, both for the laboratory sessions in this book and for other experiments you may do on your own. You will also find it helpful to increase the numbers of the frequently used items in Table 3-2, especially beakers, Erlenmeyer flasks, test tubes, pipettes, and graduated cylinders. For example, my home lab has at least half a dozen beakers and Erlenmeyer flasks in each of the sizes I use most often, a dozen serological pipettes, several dozen test tubes, and half a dozen graduated cylinders.

Having more than the required minimum on hand allows for some breakage, and also speeds up lab work. For example, if you have only one 10 mL graduated cylinder, you might have to wash and dry it in the midst of a lab session. If you have extras, you can simply rinse and set them aside as you use them and wash all of them thoroughly at the end of the session.

TABLE 3-2: *Recommended basic laboratory glassware*

Description	Qty	Notes
Beaker, 50 mL	1	Useful as a weighing boat, for making up small amounts of solutions, and so on.
Beaker, 150 mL	6	Most generally useful size for typical home lab work; can substitute 250 mL.
Beaker, 250 mL	2	Use when a 150 mL beaker isn't quite large enough.
Beaker, 600 mL	1	Useful for making up stock solutions, for general mixing, as an ice bath, and so on.
Bottle, storage, amber glass	*	As required; use to store stock solutions; choose appropriate sizes from 25 mL to 500 mL.
Bottle, wash, 500 mL	1	Use to store and dispense distilled water for making up solutions, rinsing glassware, and so on.
Burette, 50 mL	1	Use for titrations, standardizing solutions, and so on.
Crucible with cover	1+	To fit clay triangle described in equipment section.
Cylinder, graduated, 10 mL	1	Use for fast, moderately accurate measurement of liquids.
Cylinder, graduated, 100 mL	1	Use for fast, moderately accurate measurement of liquids and as a substitute volumetric flask.
Dish, evaporating	1	Porcelain or Pyrex glass.
Flask, Erlenmeyer, 250 mL	2	Most generally useful size for typical home lab work.
Flask, Erlenmeyer, 500 mL	1	Use as reaction vessel for larger-scale experiments, as a distillation vessel, gas-washing bottle, and so on.
Flask, volumetric, 100 mL	1	Use for making up solutions accurately.
Funnel, general-purpose	1+	Assorted sizes useful; may be glass or plastic (resistant to boiling temperatures).
Pipette, eyedropper, glass	3+	Use with solvents that would damage polyethylene Beral pipettes.
Pipette, Beral, disposable	10+	Use for microscale procedures, as a substitute burette, and so on.
Pipette, Mohr or serological, 1.0 mL	1+	Use for microscale titrations and for exact measurement of small quantities of liquids.
Pipette, Mohr or serological, 10.0 mL	1+	Use for microscale titrations and for exact measurement of small quantities of liquids.
Plate, reaction	1+	Use for microscale procedures; choose 24-well transparent plastic or ceramic model.
Rod, stirring, glass	3+	Use for stirring and to prevent bumping when boiling solutions.
Tubes, capillary	12+	Use for chromatography spotting.
Tube, test	12+	16 x 150 mm or 18 x 150 mm.
Tubing, flint glass, 5 mm	1	12" or 300 mm; use for building apparatuses.
Watchglass	1+	Large enough to cover 250 mL and 600 mL beakers.

TABLE 3-3: *Recommended supplemental laboratory glassware*

Description	Qty	Notes
Bottle, Barnes dropping, 25 mL	12+	Use to store indicator solutions and other solutions that are often used dropwise.
Bottle, gas-generating	1	Buy or make your own with an Erlenmeyer flask, 1- or 2-hole stopper, and tubing.
Bottle, gas-washing	1	Buy or make your own with an Erlenmeyer flask, 2-hole stopper, and tubing.
Flask, filtering, 250 mL or 500 mL	1	Use for faster filtration; requires hand vacuum pump or other vacuum source.
Flask, Florence (flat-bottom), 250 mL	1	Better than Erlenmeyer flask for high-temperature distillations and similar procedures.
Flask, volumetric, 500 mL	1	Use for making up larger quantities of bench and stock solutions accurately.
Flask, volumetric, 25 mL	1	Use for making up small quantities of indicators and solutions of expensive chemicals accurately.
Funnel, Büchner, with stopper	1	Use with filtering flask.
Funnel, separatory, 100 mL to 250 mL	1	Use for separating aqueous and organic phases in extractions or washings.
Funnel, thistle tube	1	Useful for adding reactants to a stoppered flask.

LABORATORY EQUIPMENT AND SUPPLIES

In addition to glassware, a home chem lab requires a variety of laboratory equipment and supplies. Some of it you can make yourself, but much of it must be purchased.

BALANCE

A good balance is an essential tool for any chemistry lab. When I built my first home chem lab in the mid-1960s, the only options were mechanical balances, which were relatively inaccurate and very expensive. (I ended up building a knife-edge balance that was usable but never entirely satisfactory.) Fortunately, there are many accurate, inexpensive electronic balances available nowadays.

When you choose a balance for your home chemistry lab, pay close attention to the following important characteristics:

Resolution/Readability

Resolution or *readability* is the minimum increment of weight that can be determined by the balance. For example, a balance may be able to weigh samples to within 0.1 g, 0.01 g (one *centigram* or *cg*), 0.001 g (one *milligram* or *mg*), or 0.0001 g (100 *microgram* or *µg*). All other things being equal, a balance with higher resolution costs more than one with lower resolution. The least expensive balances suitable for use in a home lab cost $60 or so and can weigh samples to within

0.01 g. Balances with resolution of 0.001 g start at about $175, and those with 0.0001 g resolution cost $750 and up.

Reasonably high resolution is important, because it allows you to work to a particular level of accuracy while using smaller amounts of materials. For example, if 0.01 g resolution is sufficient to allow you to make up 100 mL of a stock solution to the desired degree of accuracy, if you instead used a 0.1 g balance, you would have to make up 1,000 mL of that stock solution to achieve the same degree of accuracy. Similarly, if you used a 0.001 g balance, you could make up only 10 mL of that stock solution to the same degree of accuracy.

Capacity

Capacity specifies the maximum weight that a balance can handle, and is related to the resolution and price of the balance. For example, a $25 centigram balance may have a maximum capacity of 25 g, and a $100 centigram model may have a maximum capacity of 200 g. Note that some balances also have a minimum capacity, below which their readings may be inaccurate.

Reasonably high capacity is important, because you will often *tare* and *weigh by difference*. For example, before you begin a procedure, you may weigh the beaker in which the reaction will take place. Rather than weighing reagents directly, you place the beaker on the balance and use the *tare function* to "zero out" the balance, so that the balance shows a weight of zero with the beaker on it. You then add the reagents to the beaker. You can read the weight of the first reagent you add directly on the display. You can determine the weight of the second reagent you add by subtracting the original weight from the new weight, and so on. When the reaction is complete, you can filter the precipitate and determine the weight of the product.

Accuracy and Precision

Accuracy and *precision* (also called *repeatability*) are separate but related characteristics. Accuracy specifies how closely a balance indicates the true weight of a sample. For example, if you weigh a calibration weight with a known weight of 100.0000 g, an accurate balance indicates that weight to within its resolution (e.g., an accurate centigram balance reads 100.00 g, and an accurate milligram balance reads 100.000 g).

Precision or repeatability specifies how closely repeated weighings of the same sample agree. For example, if you weigh the calibration weight nine times, a precise but inaccurate balance may report weights of 100.83 g seven times, 100.82 g once, and 100.84 g once. Although the indicated weights are far from the true weight, which means the balance is inaccurate, the separate weighings agree closely, which means the balance is precise. Conversely, an accurate but imprecise balance may report weights of 99.96 g, 99.97 g, 99.98 g, 99.99 g, 100.00 g, 100.01 g, 100.02 g, 100.03 g, and 100.04 g. No two weighings gave the same result, which means the balance is imprecise, but none of the results differed much from the actual weight, which means the balance is accurate.

Repeatability is more important than accuracy. An inaccurate balance can be calibrated by using a known weight and adjusting the display accordingly. An imprecise balance cannot be adjusted to be more precise.

Linearity

Linearity specifies the accuracy and precision of a balance across its range of capacities from minimum to maximum. For example, a centigram balance with a 200 g capacity should have similar accuracy and precision whether you are weighing a 1 g sample or a 195 g sample. The best way to test the linearity of a balance is to weigh multiple items individually and then together. For example, you might weigh three labeled bolts

and get results of 34.18 g for bolt A, 44.77 g for bolt B, and 61.29 g for bolt C. When you weigh bolts A and B together, a balance with good linearity will read 78.95 g (34.18 g + 44.77 g = 78.95 g). When you weigh bolts A, B, and C together, a balance with good linearity will read 140.24 g. Other than the least-expensive models, most decent electronic balances have good linearity through most of their range of capacities.

Of course, price is often a major consideration for a home lab, but I suggest you don't skimp too much. If you're on a tight budget, choose a $60 model like the My-Weigh Dura 50 Digital Scale (0.01 g/50 g). If you can afford more, choose the $100 My-Weigh iBalance 201 (0.01 g/200 g) shown in Figure 3-19, which is what I used for the lab sessions in this book. If you need a milligram scale, the least expensive model I can recommend is the $170 My-Weigh GemPro 250 (0.001 g/50 g), although I think the $220 Acculab VIC-123 Milligram Scale (0.001 g/120 g) or the $270 Acculab VIC-303 (0.001 g/300 g) is a much better choice.

Most laboratory supply places carry balances suitable for home labs. The best source we know of in terms of selection and price is Precision Weighing Balances (*www.balances.com*).

FIGURE 3-19: *My-Weigh iBalance 201 digital balance*

SERIOUS SCIENCE

The major factor that relegates even the best chemistry sets to the realm of scientific toys rather than serious tools for doing real science is their lack of a decent balance. There's only so far you can go in learning chemistry if your measurement methods consist of using a scoop of this or three drops of that. Chemistry in general and lab work in particular is founded on accurate measurements, which makes a good balance indispensable.

HEAT SOURCES

Although modern chemistry sets make a point of using "no flame," it's impossible to imagine a real wet-chemistry laboratory without heat sources. Heat sources are essential for warming and boiling solutions, bending and drawing glass tubing, drying samples, and so on. A well-equipped lab should have several heat sources, including the following:

Alcohol lamp

An *alcohol lamp*, shown in Figure 3-20, has been a staple of home laboratories since there have been home laboratories. Alcohol lamps are inexpensive, widely available, reliable, relatively safe to use, and provide a gentle heat that is useful in a wide variety of laboratory procedures. Even if you have other heat sources, you'll want an alcohol lamp for general heating tasks. I got my alcohol lamp as a part of a glassware assortment offered by United Nuclear, but you can buy one from any laboratory supply house.

Gas burner

Formal laboratories invariably have burners fueled by natural gas, which is impractical for most home chem labs. Fortunately, there are some good, inexpensive alternatives. Most laboratory supply houses sell gas microburners (Figure 3-21) that do not require a connection to a natural gas supply. Instead, they have a self-contained fuel tank that can be refilled from standard butane cylinders sold by drugstores and tobacconists.

One advantage of a gas burner is that it provides a much hotter, more focused flame than an alcohol lamp. Sometimes, such as when you are heating a solution in a test tube, this hotter flame is actually undesirable. Other times, for example when you are trying to reduce a metal ore with carbon, the higher temperature of the gas flame may be necessary to initiate and sustain the reaction. (If you have no gas burner, you can increase the temperature of an alcohol lamp flame significantly by using a blowpipe or a piece of glass tubing to blow air into the flame.)

Another advantage of a gas burner is that most models produce much more heat than an alcohol lamp. For example, an alcohol lamp takes a long time to bring 250 mL of a solution to a boil. A gas burner accomplishes the task much faster.

You may already have a suitable gas burner in your garage or workshop. Bernz-o-Matic and other companies produce inexpensive burners that use disposable propane cylinders.

KEEP YOUR BALANCE CLEAN

Never weigh chemicals directly on the balance pan. Instead, tare (zero out) the balance with a weighing paper, beaker, or similar container in place, and add the chemical to the weighing paper or container. Although you can buy purpose-made weighing papers or weighing boats from laboratory supply vendors, that's a needless expense. Instead, just use a piece of waxed paper about the size of the balance pan. A roll of waxed paper, a pair of scissors, and a few minutes is all you need to make a large supply of weighing papers.

OLDIES AREN'T ALWAYS GOODIES

You may be tempted to buy a traditional triple-beam (0.1 g) or quadruple-beam (0.01 g or centigram) balance. Although these mechanical balances have some advantages over the top-loading electronic balance that we recommend, they also have a serious drawback. The damping time of a mechanical balance—the time it takes to settle down and provide an accurate reading—is much longer than the damping time of a decent electronic balance, particularly with the quad-beam models that provide 0.01 g resolution. Although you can weigh samples just as accurately with a mechanical balance as with an electronic balance that has the same resolution, the process is much more awkward and time-consuming with the mechanical balance.

On the other hand, a mechanical centigram balance does have one major advantage. By carefully noting the exact position of the centigram slider, you can interpolate the weight of a sample to 0.002 g on most mechanical centigram balances, and to 0.001 g (1 mg) on some.

FIGURE 3-20: *An alcohol lamp*

Various accessories are available, including stands and heat spreaders, that make these burners very useful in a home lab.

Hotplate

As useful as alcohol lamps and gas burners are, there are times when you need a heat source that doesn't use flame. An open flame can be very dangerous, particularly if you are working with a flammable liquid or gas. For those situations, the best solution is an electric hotplate. The best models have electronic controls and are specifically designed for laboratory use. Many laboratory hotplates also include a built-in magnetic stirrer, which can be very useful. Unfortunately, laboratory hotplate/stirrers aren't cheap. Fortunately, an ordinary kitchen hotplate is inexpensive and serves the purpose nearly as well.

SUPPORT STANDS, RINGS, AND CLAMPS

A *ring stand*, also called a *support stand*, is a heavy metal base, usually of cast iron, that supports a vertical rod to which various rings and clamps can be attached. Ring stands vary in price, construction quality, base size, rod diameter and length, and other factors, but essentially any ring stand is suitable for use in a home chem lab.

Support rings are generally made from cast iron or steel, and are available in various diameters. Some require a separate clamp to attach them to the stand, but most include a built-in clamp. A ring is not normally used alone when heating a container. For heating flasks and beakers, a wire gauze with a ceramic center is used between the ring and the base of the container to spread the heat evenly. Figure 3-22 shows a ring stand with ring and wire gauze.

AVOID OPEN FLAMES

If you are working with any flammable liquid or gas, *never* use an alcohol lamp, gas burner, or any other heat source that produces a flame. No matter how careful you are, a catastrophic fire or explosion could occur without warning.

A ring stand and ring are also useful to support a filter funnel or sep funnel.

Crucibles are designed to be used at very high temperatures, and are normally heated by a gas flame directly on them. Despite their temperature resistance, crucibles are relatively fragile physically, and might fracture if supported directly by a metal ring. Instead, as shown in Figure 3-23, a ring is used to support a clay triangle, which in turn supports the crucible. The crucible contacts only the clay pipes, which insulate it from rapid temperature changes that could cause it to crack. In addition to rings, various types of clamps can be connected to the support rod. These clamps are used to secure glassware and any other apparatus to prevent them from tipping over or coming apart during a procedure. Among the most useful of these is the general-purpose *utility clamp*, shown in Figure 3-24, which is used to secure flasks, test tubes, and other glassware. Other general-purpose clamps include *right-angle clamps* and *adjustable-length clamps*. A *thermometer clamp* can be used for its nominal purpose, but is also useful for securing glass tubing. A *three-finger clamp* is usually used to support a separatory funnel, but is also useful for other purposes. A

FIGURE 3-21: *A butane microburner with fuel*

FIGURE 3-22: *A ring stand with a ring and wire gauze*

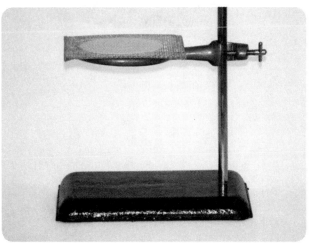

FIGURE 3-23: *A ring with clay triangle*

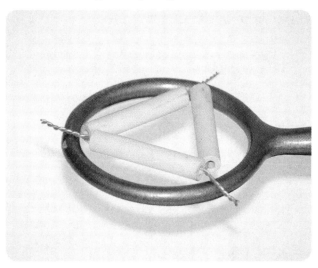

FIGURE 3-24: *A utility clamp supporting an Erlenmeyer flask*

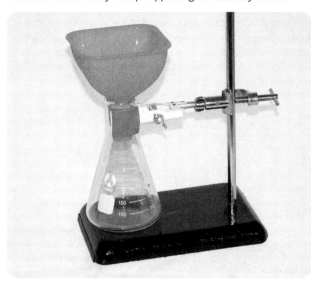

DON'T MELT YOUR CLAMPS

Many clamps have plastic or cork insulation on their jaws. If you use a clamp to secure a vessel that's being heated, make sure that the insulation can take the heat. In general, plastic or cork insulation is usable if the temperature won't exceed that of boiling water. For procedures that use higher temperatures, such as destructive distillations, use either solid metal clamps or those with insulation designed to stand up to those temperatures.

burette clamp is most often used to support a burette, but is also useful for holding any slender glassware.

A *tripod stand* provides a stable elevated surface upon which to heat flasks and beakers, typically over an alcohol lamp. A tripod stand is more accident-prone than a ring stand because the vessel isn't clamped, but tripod stands are inexpensive and suitable for heating small containers if you're careful not to knock them over.

PIPETTERS

A *pipetter* is a device used to fill pipettes and then deliver the liquid in a controlled fashion. A *bulb pipetter* is the simplest and cheapest type. It's a plain rubber bulb, similar to the ones used on eyedroppers, but sized to the capacity of the pipette. Ordinarily, you use the suction of the bulb to fill the pipette above the zero mark, and then quickly remove the bulb and use your finger to control the flow. Better bulb pipetters have multiple valves. You squeeze one place on the bulb to fill the pipette, another place to deliver the solution in a controlled fashion, and still a third place to dump the contents quickly. A *pipette pump*, shown in Figure 3-25, is a plastic device that friction-fits the top of the pipette. A thumb wheel allows you to draw in liquid, which can then be released in controlled fashion. You'll want at least one pipetter that's appropriate in size for each size of pipette you have. Some pipetters can be used with more than one size of pipette.

PH MEASURING TOOLS

An aqueous solution can range from pH 0 (strongly acid) through pH 7 (neutral) to pH 14 (strongly basic). It's often important to have some idea of the pH of a solution. Sometimes, all you need to know is whether the solution is acidic or basic. Other times, you need to know the pH with some precision.

For a quick and dirty determination of pH, indicator papers are the traditional choice. *Litmus paper* is so familiar even to nonchemists that the phrase "litmus test" has become part of the vernacular. Litmus paper turns red at pH 4.5 and below, and blue at pH 8.3 and above. When wet with a solution that has a pH between 4.5 and 8.3, litmus paper turns an intermediate purplish color. Even in laboratories that have expensive, precise electronic pH meters available, litmus paper is still used when all that's needed is a quick test for acidity or alkalinity. Every lab needs at least two vials of litmus paper, one red and one blue.

When you need a more precise indication of pH, you can use *pHydrion paper* or *universal indicator paper*, which use a combination of indicator dyes to provide a wide range of color changes at different pH levels. Such papers are available in broad- and narrow-range forms. Broad-range papers test all or most of the range of pH values from 0 to 14, but with accuracy limited to ±0.5 pH. Narrow-range papers are available in various ranges (e.g., pH 7.0 to pH 10.0), with typical accuracy of ±0.2 to ±0.3 pH.

An electronic **pH meter** can determine pH very accurately. Expensive laboratory-grade pH meters are incredibly accurate, but are overkill for a home lab. The Milwaukee Instruments pH600 meter shown in Figure 3-26 sells for $30, fits in your pocket, runs for 700 hours on one battery, and covers the entire pH 0 to pH 14 range with resolution of 0.1 pH and accuracy of 0.2 pH. Unless you're on a very tight budget, we strongly recommend that you acquire an inexpensive pH meter.

THERMOMETERS

It's often important to know the temperature of a solution, so any chem lab needs at least one thermometer. A traditional **tube thermometer** is a glass tube with a narrow bore and a bulb on one end, filled with a fluid that expands and contracts with changes in the temperature. In the past, most tube thermometers used mercury as the indicating fluid, but environmental concerns have caused a wholesale shift away from mercury to dyed alcohol or similar fluids.

Alcohol-filled thermometers, also called **spirit thermometers**, are as accurate as mercury thermometers, but are limited to measuring a smaller range of temperatures. A typical wide-range spirit thermometer measures temperatures from −20°C to +150°C, whereas a wide-range mercury thermometer might have an upper limit of +250°C. (Spirit thermometers with a wider range are available, but are longer, typically 400 mm or more rather than the standard 300 mm.)

Standard-length thermometers with very wide ranges have lower resolution. For example, a thermometer with a very wide range might have 2°C gradations, while a similar thermometer with a narrower range might have 1°C gradations. For general lab work, a thermometer with a range of about −20°C to +150°C and 1°C resolution is suitable. For several of the experiments in this book, a second thermometer with a narrower range (say, −20°C to +50°C) allows more accurate measurements.

There are two general classes of tube thermometers, called **partial-immersion thermometers** and **full-immersion thermometers**. A partial-immersion thermometer, which is the more common type, is designed to measure temperatures accurately when a specific length of the tube (usually 3" or 75 mm) is submerged in the liquid being measured. "Full-immersion" is a misnomer, because it doesn't mean that you literally submerge the entire thermometer. A full-immersion thermometer reads accurately when the tube is submerged in the liquid to the level of the indicator fluid. (For example, if the full-immersion thermometer reads 50°C, the thermometer should be submerged until the level of the liquid being tested is at the 50°C mark on the thermometer.)

A **digital thermometer** is also an excellent choice. A wide variety of models is available, starting at around $10. The least expensive models are direct substitutes for glass thermometers. They have about the same range and accuracy as glass thermometers, but are less fragile. Somewhat more expensive models boast wider ranges and better accuracy, often to within 0.1°C. The accuracy of digital thermometers often varies according to the temperature being measured. For example, one $50 model we looked at had a range of −50°C to 280°C. The accuracy of that thermometer was 0.1°C in the range −10°C to 150°C and 1°C outside that range.

SCOOPS, SPATULAS, AND SPOONS

Scoops, spatulas, and spoons are used to transfer dry chemicals, typically from a reagent bottle to a weighing boat or weighing paper. Good laboratory practice requires that to prevent contamination, you never return unused chemicals to the original bottle. In theory, that's correct, but chemicals are expensive and the economic realities of a home chem lab are such that I routinely violate this rule in my own lab.

FIGURE 3-25: *A pipette pump*

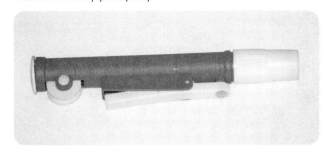

FIGURE 3-26: *An inexpensive pH meter*

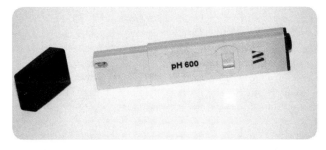

DON'T DIP

Beginners often dip the indicator paper into the solution to be tested, which may contaminate the solution. The proper method is to dip a clean stirring rod into the solution and then use it to transfer one drop of the solution to the indicator paper.

The risk of that practice is nil, because I always use a new scoop (or at least one that has been thoroughly washed) each time I transfer chemicals from the original bottle. For transferring small quantities of chemicals, use a modified Beral pipette. Trim the bulb end diagonally to make a disposable scoop. For transferring larger amounts of chemicals, use a disposable plastic spoon, which you can find in the picnic supplies section of the grocery store. (Or, if you're cheap like I am, save the ones you get at fast-food restaurants.)

Some chemicals tend to form solid masses in the bottle, making it difficult or impossible to scoop them out with one of these field-expedient plastic scoops. In that situation, use a standard stainless-steel spatula or an ordinary stainless-steel teaspoon or soup spoon. Wash the spatula or spoon afterward, rinse it in distilled water, and put it aside to dry thoroughly before you reuse it.

FILTER PAPER

Filter paper is used with a standard or Büchner funnel to separate solids (the filtrand) from the filtrate (the liquid you're filtering, also called the **supernatant** liquid). Filter paper is sold in precut circles of various sizes to fit different size funnels. The 9 cm or 11 cm size is most useful in home labs.

There are many varieties of filter paper with different characteristics. *Quantitative filter paper* is pure cellulose with almost no mineral content. It is used in quantitative tasks because when it is heated to combustion, it leaves almost zero ash to contaminate (and add weight to) the filtrand. *Qualitative filter paper* has a higher mineral content, but is less expensive and suitable for most uses.

Filter paper also differs in the tightness of its fibers, which determine both its flow rate (filtering speed) and its particle retention size. Coarse filter paper offers a very fast flow rate, typically captures particles 20 μm (micrometers, 0.001 mm) or larger, and is suitable for filtering coarse or gelatinous precipitates. Medium filter paper has a moderate flow rate, typically captures particles larger than 5 μm to 10 μm, and is suitable for general lab use. Fine filter paper has a low flow rate, typically captures particles larger than 1 μm to 3 μm, and is suitable for capturing the finest precipitates.

Ordinary unbleached coffee filters substitute adequately for formal filter paper for most home labs. Coffee filters have a high flow rate and correspond roughly to coarse laboratory filter paper.

CHROMATOGRAPHY PAPER

Chromatography paper has special characteristics that optimize it for chromatography. The type of fibers used and their length are carefully controlled, and the production process is designed to orient the fibers parallel to minimize spreading of the sample as the solvent carries it along the paper. Chromatography paper is sold in rolls, large sheets, and small strips, of which only the last is reasonably priced for home lab use.

If you don't have chromatography paper, you can substitute sheets or strips of ordinary 100% cotton bond writing paper, making sure to orient the grain of the paper parallel to the direction of the solvent flow. The results aren't quite as good as they would be if you used actual chromatography paper, but they're usable for most purposes in a home lab.

BATTERY AND ELECTRODES

To complete the electrochemistry labs in this book, you'll need a power supply, a selection of electrodes, and a set of leads with alligator clips. An ordinary 9V battery works fine for the power source. You can buy the electrodes or make your own, as follows:

Carbon
Use a bundle of several mechanical pencil "leads," which are actually graphite.

Aluminum
Cut strips from an aluminum beverage can. Use sandpaper or a similar abrasive to polish the surfaces to remove paint, oxidation, and plastic coatings.

Copper
Heavy-gauge copper wire salvaged from a scrap piece of Romex works well, as does a flattened section of copper tubing.

Iron
A large iron or steel nail or bolt works well. Polish the surface with sandpaper before you use it. Make sure that whatever you use is not galvanized, or you'll actually be using a zinc electrode instead of an iron or steel one.

You'll also need electrodes of lead, magnesium, nickel, tin, and zinc, which are best purchased from a laboratory supply vendor.

RUBBER STOPPERS AND CORKS

An assortment of rubber stoppers and corks is useful for sealing test tubes and flasks, constructing apparatuses, and so on. Rubber stoppers are popular because they are durable, reusable, provide a good seal, and are available in solid, 1-hole, and 2-hole versions.

Unfortunately, there is little standardization in the sizes of rubber stoppers or the stopper size required to fit vessels of particular sizes. One vendor's #2 stopper may be the same size as another vendor's #3 stopper, for example, and two 250 mL Erlenmeyer flasks from different vendors may require different stopper sizes.

Table 3-4 lists standard rubber stopper sizes used by most (but not all) U.S. vendors, the hole sizes for predrilled stoppers, the sizes of openings those stoppers fit, and the types of vessels they are typically used with. Some stoppers use different dimensions for the same stopper numbers, so it's important to verify the actual size of the stoppers before you order them. Either that, or order a large assortment that includes solid, 1-hole, and 2-hole stoppers in each size to make sure you always have a stopper that fits whatever vessel you need to use it with. Verify that the hole sizes of the stoppers you buy matches the outside diameter of your supply of glass tubing.

The one- and two-hole versions of #1 rubber stoppers usually have 4 mm holes, although we have seen two-hole #1 stoppers with 3 mm holes. The one- and two-hole versions of #2 and larger rubber stoppers all use the standard 5 mm hole size, except some #6.5 stoppers that use the 4 mm size. The oddball #6.5 size can be difficult to find, but is the best fit for a 3-liter soda bottle.

Corks are less durable and less convenient to use than rubber stoppers, but are also less expensive and more resistant to high temperatures. For historical reasons, corks use an entirely different numbering system from stoppers. In fact, cork measurements were originally specified in fractions of an inch rather than SI (metric), and only later were those inch denominations converted to SI.

Table 3-5 lists standard cork sizes used by most U.S. vendors. The actual size of a cork may vary slightly, depending on whether it is specified in U.S. customary units or SI. Corks are much more compressible than are rubber stoppers. Proper use of corks requires using a cork roller, which compresses the cork before it's inserted into the vessel. As the cork relaxes, it expands to seal the vessel. This results in a much better seal than is possible simply by pressing a cork into the opening using finger pressure.

Corks are usually sold without holes. If you want a 1-hole or 2-hole cork, use a cork borer to make it yourself. If you buy the cheapest corks, you'll probably discover why they were so cheap the first time you try to bore a hole in one. Cheap corks often have internal cavities and cracks, and are prone to crumble when you attempt to bore them. Better corks, usually sold as laboratory-grade, are solid throughout and less prone to crumbling.

TABLE 3-4: *U.S. standard rubber stopper sizes*

Stopper #	Hole size	Fits opening	Typically fits
00	3 mm	10 mm to 13 mm	Very small test tubes.
0	3 mm	13 mm to 15 mm	Small test tubes; many 25 mL flasks.
1	3 mm or 4 mm	15 mm to 17 mm	Some medium test tubes; many 50 mL flasks.
2	5 mm	16 mm to 18.5 mm	Some medium test tubes.
3	5 mm	18 mm to 21 mm	2-liter soda bottles; some large test tubes.
4	5 mm	20 mm to 23 mm	Some large test tubes.
5	5 mm	23 mm to 25 mm	Most 125 mL flasks; some very large test tubes.
6	5 mm	26 mm to 27 mm	Many 250 mL flasks; some 500 mL flasks.
6.5	4 mm or 5 mm	27 mm to 31.5 mm	3-liter soda bottles; some 500 mL flasks.
7	5 mm	30 mm to 34 mm	3-liter soda bottles; most 500 mL flasks.
8	5 mm	33 mm to 37 mm	Some 500 mL flasks.

TABLE 3-5: *Standard cork sizes*

Cork #	Top diameter		Bottom diameter		Length	
	inch	mm	inch	mm	inch	mm
000	1/4	6	5/32	4	1/2	13
00	5/16	8	7/32	5	1/2	13
0	3/8	10	9/32	7	1/2	13
1	7/16	11	21/64	8	5/8	16
2	1/2	13	3/8	10	11/16	17
3	9/16	14	27/64	11	3/4	19
4	5/8	16	15/32	12	13/16	20
5	11/16	17	17/32	13	7/8	22
6	3/4	19	37/64	15	15/16	24
7	13/16	21	5/8	17	1	25
8	7/8	22	43/64	17	1-1/16	27
9	15/16	24	47/64	18	1-1/8	29
10	1	25	49/64	20	49/64	31
11	1-1/16	27	53/64	21	53/64	31
12	1-1/8	29	57/64	22	57/64	31

RECOMMENDED LABORATORY EQUIPMENT AND SUPPLIES

Table 3-6 lists my recommendations for a basic set of laboratory equipment and supplies. With only these items (and perhaps a few things we've forgotten that you can scrounge around the house), you can complete all of the laboratories in this book. Most of these are described in the previous sections.

TABLE 3-6: *Recommended basic laboratory equipment and supplies*

Description	Qty	Notes
Aspirator, vacuum	1	Needed only if you use a filter flask and Büchner funnel; alternatively, use a manual vacuum pump.
Balance, electronic, 0.01 g/150 g or better	1	Higher resolution and capacity is always better, if you can afford it.
Barometer	1	Optional; you can use pressure readings from local TV or radio.
Burner, alcohol	1	Not absolutely necessary; the gas burner and hotplate can substitute for this item.
Burner, electric hotplate	1	For heating and distilling flammable liquids.

Description	Qty	Notes
Burner, gas	1	Purchase a butane microburner, or use a hardware store propane torch.
Brush, burette	1	For cleaning burettes.
Brush, test tube	1+	For cleaning test tubes.
Brush, utility cleaning	1+	For cleaning beakers, flasks, and other glassware.
Cables, patch with alligator clips, set of 2	3	50 cm length and 26-gauge or heavier wire are ideal.
Caliper	1	Optional, but useful.
Calorimeter	1+	Purchase, or build your own with nested foam cups.
Chips, boiling	*	Small bottle; used to prevent boiling over when heating liquids.
Chromatography jar and cover	3+	Large pickle jars or similar wide-mouthed containers are fine.
Clamp, burette	1	For supporting a burette on a ring stand.
Clamp, utility	1	For supporting funnels or flasks on a ring stand.
Cup, plastic, rigid	1	Should fit support ring loosely.
Digital multimeter (DMM)	1	Even a $10 or $15 model suffices.
Distillation apparatus	1	Purchase, or make with an Erlenmeyer flask, holed stopper, glass tubing, and rubber tubing.
Electrode set	1	Three each Cu and Mg electrodes; one each Al, C, Fe, Pb, Ni, Sn, Zn.
Gauze, wire with ceramic center	1	For protecting beakers and flasks from direct flame of gas burner.
Knife, utility	1	For general use around the lab.
Laser pointer, red	1	Use for Tyndall Effect tests of colloids and suspensions.
LED, red	1	For electrochemistry experiments.
Loop, inoculating	1	Platinum or Nichrome.
Magnet	1	For general use around the lab.
Magnifier or loupe	1	For general use around the lab.
Mortar and pestle	1	For grinding solid chemicals; can substitute pipe cap and large bolt.
Pad, protective	*	As needed to protect work surfaces; we use thick, nonslip pads sold in craft supply stores.
Pail, water	1	Use for drenching overenthusiastic reactions.
Paper, chromatography	*	Vial of 100 strips; several 15 cm sheets.
Paper, filter, 100 sheets	1	Choose medium flow, qualitative in a size appropriate for your Büchner funnel and gravity funnel.
Paper, litmus, red and blue, 100-strip vial	1	Use for quick determinations of acidity or alkalinity.

Description	Qty	Notes
Paper, pHydrion, 100-strip vial	1	Use for accurate determination of pH if you do not have a pH meter.
pH meter	1	Range pH 0–14 with 0.1 or 0.2 resolution and accuracy.
Pinchcock	1	Use with flexible tubing to control flow, to convert a pipette into a burette, and so on.
Pipetter, pump or bulb	1+	In sizes appropriate for your pipettes.
Reservoir, pneumatic trough	1	Use a plastic storage tub (Rubbermaid "S" 17-liter or similar).
Ruler, mm scale, 15 cm or 30 cm	1	For general length measurements.
Sand bed	1	Use when burning solids to contain heat and flame; large container with several kilograms of sand.
Ring stand	1+	5" x 8" (12.5 x 20 cm) or 6" x 9" (15 x 22.5 cm) with 20" (50 cm) support rod.
Ring, support, with clamp	2	4" (10 cm) and 5" (12.7 cm) sizes are most useful; make sure your funnels fit the ring.
Scoop, powder	1	Use for breaking up hardened masses of solid chemicals in bottles.
Spatula	1+	Use for removing small amounts of chemicals from bottles; substitute Beral pipettes or plastic spoons.
Stopper, rubber, assortment	*	As required, in solid, 1-, and 2-hole, to fit your test tubes and flasks.
Syringe, plastic, 10 to 50 mL, graduated, with cap	1	Used in gas chemistry experiments.
Test tube holder	1	Used to hold test tubes while heating.
Test tube rack	1+	To fit your test tubes.
Thermometer	1+	Digital thermometer preferred.
Timer	1	With second hand, for timing reactions.
Tongs, beaker	1	Use for handling hot beakers and other larger items.
Tongs, crucible	1	Use for handling hot crucibles and other small items.
Triangle, clay	1	To fit support rings, in size(s) appropriate for your crucibles.
Tubing, rubber or plastic, assortment	*	In various lengths and with an inside diameter to fit glass tubing.

You'll also need an assortment of items commonly found around the house, including an abrasive (such as sandpaper, emery board, or steel wool), aluminum foil, a 9V transistor battery, cotton balls and swabs, foam cups, empty tin cans and lids, empty soft drink bottles and caps, a stick of incense, strike-anywhere matches, paper clips and rubber bands, paper towels, pencils and marking pens of various colors and types, thread or string, toothpicks or tongue depressors, some stiff wire (such as a coat hanger), and so on.

Check out the current list of vendors at *http://www.homechemlab.com/sources.html* before you order any glassware, equipment, or chemicals.

MISCELLANEOUS EQUIPMENT
(JUNK COLLECTING)

Back in the mid-1960s, when I, a young teenager, was setting up my first home chemistry lab, my mother got used to things disappearing from her kitchen and laundry room. Now, 40 years later, my wife wonders about similar mysterious disappearances. There's no doubt that building a home chem lab can cause a Bermuda Triangle effect elsewhere in the house.

As you set up your own lab (and forever after) keep an eye out for anything that might be useful in the lab. For example, my chromatography jars are actually durable plastic one-quart containers that originally contained wonton soup from take-out Chinese food orders. My water bath began life as a Crock-Pot slow cooker, and two of my white plastic reaction plates look

> ### RECYCLE SODA BOTTLES
>
> Soda bottles in 500 mL, 1-liter, 2-liter, and 3-liter sizes are very useful around the lab, both as storage containers (label them clearly!) and as the basis for many kinds of ad hoc apparatus. With a 1- or 2-hole rubber stopper and some tubing, you can turn a soda bottle into a disposable, high-capacity gas-generating bottle or gas-washing bottle. With some thin-wall plastic tubing or metal tubing, you can even turn a soda bottle into a very capable condenser for distillations.

suspiciously like discarded ice cube trays. My first calorimeter —since replaced with a store-bought model—consisted of two large nested Styrofoam cups with a tight-fitting lid on the inner one. (The straw hole in the lid works just as well for a thermometer.) And so on.

If you're on a limited budget (or even if you're not), collect anything that looks like it might be useful, now or in the future. Chances are, it will be. I have drawers full of such stuff, and frequently end up using one thing or another as a field-expedient substitute for something I'd otherwise have had to order and await its arrival.

WORK AREA

Give serious consideration to the best location for your home lab. Of course, you may have to choose between a poor location and having no lab at all. In that case, do the best you can with what you have to work with. Here are some things to think about when you choose a location for your home lab.

Counter space

Even microscale experiments require a surprisingly large amount of workspace. In addition to space for the experiment setup itself, you'll need space for your balance, your lab notebook, temporary space for any chemicals that you are using, and so on. Consider 10 ft^2 (1 meter^2) of counter or table space as an absolute minimum—more is better. The ideal setup is arranged like a photographic darkroom or galley kitchen, with a wet side and a dry side. The wet side is used for doing the actual experiments, and the dry side for storing and weighing chemicals, recording observations in your lab notebook, and so on.

Storage

It's important to have secure storage space for your equipment and chemicals—particularly those chemicals that are toxic, flammable, or otherwise hazardous—and that your lab storage space not be shared with general household items, particularly food and beverages. Make absolutely certain that children and pets cannot get to dangerous equipment and chemicals. If you have only part-time possession of your work area, store your equipment and chemicals out of reach in Rubbermaid tubs or similar containers.

PROTECTING WORK SURFACES

Some of the chemicals you work with may stain or otherwise damage wooden or laminate work surfaces. I protect my work surfaces, which are standard kitchen laminate counters, by covering them with rubber nonslip mats that are available in various sizes and thicknesses at craft stores. I also put an old bath towel between the counter top and the rubber mat. The mat provides a smooth, level, chemical-resistant work surface, and the old towel absorbs any liquids that run off the mats.

My advisor, Dr. Mary Chervenak, is an expert on paints and coatings. I asked her and my other advisor, Dr. Paul Jones, if there was any kind of paint that could be used to protect surfaces from most laboratory chemicals. The short answer is "not really." Standard latex, polyurethane, and epoxy-based paints and coatings offer reasonably good protection against many reagents and solvents, including the dilute reagents used in most of the experiments in this book. However, they offer less (or no) protection against strong acids or bases or some organic solvents.

Still, as Dr. Jones commented, some protection is better than none, and in a sense you can think of these paints as ablative coatings. The coating itself may dissolve in or be eaten away by a strong chemical, but it may protect the underlying surface long enough for you to dilute, mop up, or neutralize the spill. If I used a wooden workbench or a similar surface, I'd put several thick coats of an epoxy-based deck or floor paint on it, and then protect it further with a rubber mat and towel.

Even if you take reasonable precautions and work carefully, it's almost inevitable that at some point you'll spill something nasty on your work surface. That's a good argument for choosing a work surface that's expendable. If you eat holes in a sheet of plywood or particle board, that's cheap and easy to replace. If you eat holes in your washer/dryer, you may have some explaining to do.

Ventilation

Some experiments produce smoke, strong odors, or even toxic fumes. It's important to have some means of ventilating the work area and exhausting these fumes to the outdoors. The ideal solution is a formal laboratory exhaust hood, but an open window with a portable exhaust fan can often serve the purpose. For experiments that produce a lot of smoke or fumes, work outdoors.

Lighting

It's dangerous to do chemistry in a poorly lit environment. My first home chem lab was located in a dark corner of the basement, and I had more than one near-accident before I installed some fluorescent fixtures. Make sure your work area is well lit. If the overhead lighting is inadequate or nonexistent, use table lamps, clamp lamps, or other portable light sources.

Electric receptacles

You can get along without electricity other than for lighting, but it's very convenient to have one or more electric receptacles within easy reach of your work surface. If those receptacles are not protected, install ground-fault circuit interrupters.

Water and sewer

Although it is not essential, it is extremely convenient for the home chem lab to have access to running water and a sewer connection. If that's not possible, acquire a large bottle to use as a water supply. The 5-gallon plastic carboys used in water coolers are ideal. With a stopper, some tubing, and a pinch clamp, you can build a siphon to deliver water as needed. Use a large plastic bucket, tub, or similar container for waste liquids, and empty it after each lab session.

Flooring

Even if you are extremely careful, sooner or later you are bound to spill something. Murphy's Law says that what you spill will be corrosive or strongly colored, and will fall exactly where you don't want it to fall. If the floor is carpet or another vulnerable material, your spouse, parents, roommate, or significant other will not be amused. Sheet vinyl or linoleum is the best flooring material for a home chem lab, although vitreous tile and similar resilient materials are also good. If the floor is concrete, consider painting it with epoxy-based floor paint to prevent it from being stained or absorbing spilled chemicals.

I have a seldom-used second kitchen in our basement guest suite that is an ideal home chemistry laboratory (and also serves well as a photographic darkroom). Not everyone is lucky enough to have a room or even a corner that can be dedicated to a laboratory. Fortunately, there are many alternatives.

Kitchen

Let's get this one out of the way first. The kitchen may seem to be an ideal location for a part-time chemistry laboratory. There's usually plenty of counter space and storage, hot and cold running water, good lighting, plenty of electrical outlets, resilient flooring, and the exhaust fan over the stove can serve as a field-expedient fume hood. But in fact, unless you limit yourself to using only nonhazardous chemicals, the kitchen is about the worst possible choice for a part-time chemistry lab. Food and laboratory chemicals are a dangerous combination.

Still, thousands of home chem labs over the years have been set up in kitchens, and the kitchen may be your only option. If so, take extreme care to prevent any contamination from occurring. Protect the counter or table surface with a resilient mat, and clean up thoroughly after each lab session. Make sure that all of your equipment, and particularly your chemicals, are returned to their storage containers and that none are missing.

Laundry room

Many laundry rooms are ideal candidates for a part-time chem lab. Laundry rooms are usually well equipped with electric power, water, drain, lighting, and have adequate ventilation. There is often sufficient storage and if not, it's usually easy to add more. There is probably a suitable work surface available, and if not it's usually easy to add a fold-down or other temporary work surface. The floor of most laundry rooms is concrete or vinyl, which minimizes the danger of damaging it with spills. In most homes, the laundry room is the best choice for a part-time lab. Of course, you have to clean up carefully after each lab session to avoid the risk of staining or damaging clothes.

Spare bathroom

A spare bathroom is often an excellent location for a part-time chemistry laboratory. It provides running water and a drain, plenty of electric receptacles (with ground-fault circuit interrupters), good lighting, and good ventilation. Many spare bathrooms are a bit short on storage and counter space, but that's often easy to remedy simply by constructing a collapsible framework that supports a plywood work surface over the tub. Particularly if you live in an apartment, a spare bathroom may be the best choice for a part-time chem lab.

Basement workshop

A basement workshop area is generally not an ideal location for a part-time chem lab, but may be the only option. On the plus side, storage and work surfaces are usually not a problem, the floor is often of concrete or another relatively impervious substance, and electric power is usually available. However, ventilation is often inadequate, and many basements have no water supply or drain.

TEMPERATURE MATTERS

If your work area is not heated and air-conditioned, give some serious thought to chemical storage. The shelf life specified for many chemicals assumes storage at normal room temperature. Allowing a chemical to bake in the summer heat of a storage shed or garage may reduce that shelf life from a year or more to a few weeks or even days. Less obviously, storing some chemicals at low temperatures may be a bad idea. Glacial acetic acid, for example, freezes at 16.7°C (62.1°F), and the solutes in solutions that are nearly saturated at room temperature will crystallize out at low temperatures. More importantly, some chemicals that are reasonably stable at room temperature may decompose or become unstable if they are stored at high temperature.

Garage or garden/storage shed

For some people, the garage or garden/storage shed will be the location of last resort. There are some good points to using such a location as a part-time lab (or even a full-time one). Storage and work surfaces are usually plentiful or can easily be made so, ventilation is often adequate, there is probably electric power available, and spills that damage the floor or work surfaces are of less concern than they would be in a finished area of the house. Weighed against this, it may be a long trek to the nearest sink and drain (although you may be able to use the garden hose as a water supply). Perhaps the worst aspect of using such a location is the lack of climate control. You'll swelter in the summer and freeze in the winter. Even if you can tolerate that, it's not the best environment for storing your equipment and supplies or for doing controlled experiments.

4

Chemicals for the Home Chemistry Lab

To state the obvious, a home chemistry lab requires chemicals. Years ago, laboratory chemicals were pretty easy to find. Dozens of small businesses specialized in supplying chemicals to home chemists, Fisher Scientific and other large specialty chemical suppliers were happy to sell to individuals, and even many corner drugstores did a side business in supplying home chemistry hobbyists.

Nowadays, laboratory chemicals are a bit harder to come by. Concerns about safety, liability, terrorism, and illicit drug labs have made it more difficult to obtain the chemicals that you need. Many vendors nowadays sell only to businesses, schools, and other institutions, refusing to sell chemicals to individuals. Fortunately, there are many exceptions. It's still possible to get every chemical you need, if you know where to look.

In this chapter, you'll learn everything you need to know about laboratory chemicals—from how they're named to how to handle them safely and where to buy them.

CHEAPER BY THE POUND

Do not overlook the advantages of banding together with other home schoolers or like-minded hobbyists to buy chemicals in bulk. For example, a vendor may charge $3 for 25g of a particular chemical, $5 for 100g, and $9 for 500g. If you need only small amounts of chemicals, you may be able to cut your chemical costs dramatically by arranging with other home-schooling families or hobbyists to order chemicals in larger quantities and divide them among you.

The cost advantage is particularly great for chemicals that incur hazardous shipping surcharges. For example, if you order 100 mL of concentrated nitric acid for $5, the vendor may add a $35 hazardous material shipping surcharge, for a total of $40. But if you order a 500 mL bottle of concentrated nitric acid for $15, the same surcharge applies, for a total of $50. If you divide that chemical with four friends, each of you gets 100 mL of concentrated nitric acid for only $10.

CHEMICAL NAMES

It's important for chemists to be able to identify a specific chemical unambiguously by its name or other designation. In the past, chemical naming was chaotic. Many chemicals had multiple names, and one name could refer to more than one chemical. Early names for specific chemicals, some of which originated with alchemists, have in recent years been largely superseded by modern, systematic names. Still, many of those older names are so ingrained that many chemists continue to use them.

For example, the following names and numbers all refer to the same inorganic compound, $CuSO_4 \cdot 5H_2O$:

Systematic (IUPAC) name: copper(II) sulfate pentahydrate
Common name: cupric sulfate, copper(II) sulphate, cupric sulphate
Vernacular name: copper sulfate, copper sulphate
Archaic name: sulphate of copper, blue vitriol, vitriol of copper, copper vitriol, bluestone
CAS number: 7758-99-8

Systematic name

The *systematic name*, also called the *IUPAC name*, is the "official" name of a chemical, as determined by the detailed naming rules published by IUPAC (International Union of Pure and Applied Chemistry). Every systematic name identifies exactly one chemical.

Common name

IUPAC uses the term *common name* differently than you might expect. To IUPAC, a common name is not merely one that is used casually or in nontechnical conversation. IUPAC defines a common name for a substance as one that identifies that substance unambiguously but does not comply with IUPAC naming conventions.

For example, the common name cupric sulfate is unambiguous because it uses the older (and deprecated) form "cupric" to identify the bivalent copper (Cu^{2+}) cation, which in the current IUPAC nomenclature is designed copper(II). Copper(II) sulphate is a common name rather than a systematic name because it uses the nonstandard "sulphate" spelling. Cupric sulphate is a common name rather than a systematic name for both reasons. Chemists, particularly those whose training predates the common use of IUPAC systematic names, frequently use common names for well-known compounds. The latest IUPAC standard approves the use of common names in addition to or instead of systematic names where the use of those common names would not lead to confusion.

Vernacular name

A *vernacular name* is one that is commonly used in trade or industry, but does not identify a substance unambiguously. For example, the vernacular name copper sulfate may refer either to copper(II) sulfate ($CuSO_4$) or to copper(I) sulfate (Cu_2SO_4), which is also known under the common name of cuprous sulfate.

Archaic name

An *archaic name* predates organized naming schemes. Chemists no longer use archaic names, although many are still in common use among nonchemists. For example, artists and potters may use "blue vitriol" or "bluestone" to mean copper(II) sulfate, and might have no idea that copper(II) sulfate, cupric sulfate, and copper sulfate refer to the same compound. Knowledge of archaic names can be useful to a home chemist, because many useful chemicals are sold under archaic names. For example, if you need hydrochloric acid it's useful to know that it's sold in hardware stores under the archaic name muriatic acid.

CAS number

A *CAS number* unambiguously identifies a particular chemical. CAS numbers are assigned to any chemical that has been described in the literature by the Chemical Abstracts Service, a part of the American Chemical Society, but are used worldwide. As of late 2007, more than 30 million chemicals had been

---

ALMOST UNIQUE

Actually, the name "cupric sulfate" *is* ambiguous, because that chemical exists in two distinct forms. The hydrated blue crystalline form contains five water molecules tightly bound to each cupric sulfate molecule as water of crystallization. The anhydrous white powder form contains no water of crystallization. In practice, "cupric sulfate" is unambiguous, because chemists who use this older style of nomenclature always mean the hydrated form unless they specify "anhydrous cupric sulfate."

assigned CAS numbers, with about 50,000 new CAS numbers added every week. A CAS number comprises three numeric strings separated by hyphens, the first of up to six numbers, the second of two numbers, and the third a single check-digit. CAS numbers are assigned sequentially, and therefore have no inherent meaning. CAS numbers are used primarily for searching the literature.

Most chemists use both systematic and common names, depending on the chemical in question. Many IUPAC names for well-known compounds are seldom or never used. For example, not even the most pedantic chemist uses the IUPAC name dihydrogen oxide for water or the IUPAC name 1,3,7-trimethyl-1H-purine-2,6(3H,7H)-dione for caffeine. But IUPAC

names for many chemicals are becoming more commonly used, particularly among younger chemists. For example, while many older chemists (and some younger ones) use the common names potassium ferrocyanide and potassium ferricyanide, many younger chemists (and some older ones) instead use the corresponding IUPAC names potassium hexacyanoferrate(II) and potassium hexacyanoferrate(III). Both naming styles are used interchangeably in this book, because it's good to be familiar with both styles.

CHEMICAL GRADES

Chemical grades define standards of purity and suitability for use for particular purposes. In order of decreasing purity (roughly), here are the chemical grades that you should be aware of.

Primary standard grade and specialized ultrapure grades
Primary standard grade chemicals are analytical reagents of exceptionally high purity that are specially manufactured for standardizing volumetric solutions, preparing reference standards, and running lot analyses to determine the purity of production runs of other chemicals. Specialized ultrapure grade chemicals are sold under various brand names by laboratory chemical producers and are used for specialized purposes such as spectroscopy and trace metal analyses. These chemicals are of the highest purity attainable, are extremely expensive, and are overkill for any home laboratory.

Reagent ACS Grade
Reagent ACS Grade chemicals meet or exceed the current American Chemical Society standards for purity, and are the purest chemicals used routinely in laboratories. These chemicals are sufficiently pure for any but the most demanding uses.

Reagent grade
Reagent grade chemicals are of very high purity, but have not been subjected to the exacting testing and certification standards required for labeling as ACS Grade. Reagent grade or better chemicals are required for reliable quantitative analyses, and highly desirable for qualitative analyses.

USP grade, NF grade, or FCC grade
USP grade (*United States Pharmocopeia*), *NF grade* (*National Formulary*), or *FCC grade* (*Food Chemical Codex*) chemicals are manufactured in facilities that comply with *current Good Manufacturing Practices* (*cGMP*) standards

and that meet the requirements of the USP, NF, or FCC, respectively. These chemicals are usually of a very high degree of purity, but may contain impurities that are insignificant for food or pharmaceutical use but may or may not be significant for qualitative or quantitative analyses and similar laboratory procedures. All of these grades of chemicals are suitable for general use in home laboratories.

Laboratory grade or purified grade
Laboratory grade chemicals are equivalent in purity to USP, NF, and FCC chemicals, but are not certified for pharmaceutical or food use. *Purified grade* chemicals are usually chemicals of

> **BETTER IS THE ENEMY OF GOOD ENOUGH**
>
> Reagent ACS Grade and reagent grade chemicals are ideal for use in a home lab, but are often much more expensive than lower-grade chemicals. Not always, though. They may even cost less. For example, when I was ordering chemicals for my home lab, I was about to order 100 g of a particular chemical in laboratory grade for $6.90 when I noticed that the same vendor was selling 100 g of the same chemical in reagent grade for $5.95. If the reagent grade chemical costs the same or little more than a lower grade, buy the reagent grade chemical. Otherwise, settle for the lower grade, unless you will use the chemical to perform critical qualitative analysis or quantitative analysis procedures.

high purity that were listed in earlier editions of the USP, NF, or FCC compendia, but are no longer listed in the current editions. Either of these grades of chemicals is suitable for general use in home labs.

Practical grade or chemically pure (CP) grade
Practical grade and *chemically pure grade* (*CP grade*) chemicals are of purity suitable for use in syntheses and other general applications, but are a step down in purity from laboratory or purified grade, and are not intended for analytical work. Either of these grades is acceptable for general use in home labs when purer grades are unavailable or unaffordable.

Technical grade
Technical grade chemicals are generally supplied in bulk quantities and are suitable for general industrial use, but should be used only as a last resort in a home chemistry lab.

Ungraded
Ungraded chemicals are produced for household, agricultural, and other uses where high purity is unimportant. The actual purity of ungraded chemicals can range from very poor to surprisingly good. To a large extent, the actual purity of these chemicals depends on the production process used to make them. For example, hydrochloric acid sold in hardware stores as "muriatic acid" for cleaning concrete is produced by reacting hydrogen and chlorine gases and dissolving the resulting hydrogen chloride gas in water, which may be ordinary tap water but is often distilled or deionized. Because there are few opportunities in the production process for the product to be contaminated, the muriatic acid you can buy in hardware stores is often of purity comparable to laboratory grade. Of course, by definition there are no guarantees of purity with ungraded chemicals.

Industry-specific grades
Many industries have industry-specific grading systems for chemicals, but the only one you're likely to encounter is *photo grade*. A photo grade chemical is one that contains only impurities that do not interfere with the intended use of the chemical in photographic processing. Photo grade chemicals may contain other impurities, which may or may not interfere with other uses. In general, photo grade chemicals are comparable in purity to laboratory grade or purified grade, and most are suitable for use in a home chemistry lab.

The best advice is to buy the purest chemicals you can afford, always keeping an eye on the intended use of those chemicals and the price differential between grades. If you have to control costs by buying only some chemicals in high-purity grades, we recommend that you buy purer grades of frequently used primary chemicals such as mineral acids, sodium hydroxide, and so on. Avoid anything less than laboratory/purified grade if you can possibly do so.

CHEMICAL RISK FACTORS AND SAFETY ADVICE

Any chemical can be dangerous, given the right (or wrong) conditions. People have died from acute water toxicity, which is to say from drinking too much water. Still, some chemicals are obviously more hazardous than others. Government and industry groups have created various labeling systems to warn users of the type and degree of danger associated with chemicals that present greater than normal hazards for storage, handling, and use. The following sections describe the safety information and hazard labeling systems you're most likely to encounter.

TOWER OF BABEL

It's up to the manufacturer or supplier to decide how to label a chemical. The same chemical from different suppliers may be labeled differently, using none, one, or multiple hazard-labeling methods. Furthermore, different suppliers may emphasize different hazards for the same chemical. For example, if a chemical is both toxic and flammable, one supplier may emphasize the former hazard and another supplier the latter hazard.

MATERIAL SAFETY DATA SHEETS (MSDS)
The *MSDS* (*Material Safety Data Sheet*) for a chemical presents pertinent safety information in a structured format that includes the following categories:

1. Product Identification
2. Composition/Information on Ingredients
3. Hazards Identification
4. First Aid Measures
5. Fire Fighting Measures
6. Accidental Release Measures
7. Handling and Storage

8. Exposure Controls/Personal Protection
9. Physical and Chemical Properties
10. Stability and Reactivity
11. Toxicological Information
12. Ecological Information
13. Disposal Considerations
14. Transport Information
15. Regulatory Information
16. Other Information

Most laboratory supply houses publish an MSDS for each chemical they carry, or at least for each chemical that presents any significant hazard. The MSDS may be produced by the vendor or by the company that actually made the chemical.

The MSDS should be your primary source of safety information about each chemical you buy. If no MSDS is provided, search the Web for an MSDS for that chemical. Read the MSDS for every chemical, even those that present no special hazard. Print a copy of the MSDS for each chemical that presents a significant hazard, and file it in your laboratory binder. Decide conservatively which MSDS sheets to print. For example, although you probably don't need a printed copy of the MSDS for sodium bicarbonate or even ethanol, you probably should keep printed copies of the MSDS sheets for strong acids and bases, oxidizers, and poisons.

NFPA 704 FIRE DIAMOND
The *NFPA* (*National Fire Protection Association*) rates various chemicals to identify the specific types and degrees of risks they present to firefighters and other emergency personnel. Rated chemicals are assigned an *NFPA 704 fire diamond*. The fire diamond helps emergency personnel quickly and easily to identify special procedures, precautions, and equipment necessary to minimize the danger the chemical presents.

The four sections of the fire diamond specify the type of risk in four categories: health (left), flammability (top), reactivity (right), and special codes for unique hazards (bottom). The three top sections contain a number from zero to four that specifies the degree of that risk. The bottom section may contain a symbol that represents the unique danger, if any. Although it is not required, the four sections of the NFPA 704 fire diamond are usually color-coded.

Health (Blue)
4. Very short exposure may cause death or major residual injury (e.g., hydrogen cyanide gas)
3. Short exposure may cause serious temporary or residual injury (e.g., carbon tetrachloride)
2. Intense or continued but not chronic exposure may cause temporary incapacitation or possible residual injury (e.g., pyridine)
1. Exposure may cause irritation but only minor residual injury (e.g., potassium carbonate)
0. Exposure offers no hazard beyond that of ordinary combustible material (e.g., sodium chloride)

Flammability (Red)
4. Rapidly or completely vaporizes at normal temperature and pressure or is readily dispersed and will burn readily (e.g., butane gas, magnesium dust)
3. Can be ignited at normal ambient temperatures (e.g., gasoline)
2. Must be moderately heated or exposed to relatively high ambient temperatures before ignition occurs (e.g., fuel oil)
1. Must be preheated before ignition can occur (e.g., olive oil)
0. Not flammable (e.g., calcium carbonate)

Reactivity (Yellow)
4. Readily detonates or explosively decomposes at normal temperature and pressure (e.g., nitroglycerin)
3. May detonate or explosively decompose from a strong initiating force or severe shock, if heated under confinement, or in contact with water (e.g., benzoyl peroxide)
2. Undergoes violent chemical change at high temperature and pressure, reacts violently with water, or may form explosive mixture with water (e.g., phosphorus)
1. Normally stable, but may become unstable at high temperature and pressure (e.g., calcium metal)
0. Normally stable even when exposed to fire and is not reactive with water (e.g., boric acid)

Special Hazard (White)
W or WATER – Reacts with water in an unusual or dangerous manner (e.g., sodium metal)
OX or OXY – oxidizer (e.g., potassium permanganate)

SIX OF ONE...

An MSDS produced for a particular chemical by one supplier may have minor differences from the MSDS for the same chemical from another supplier, but for our purposes they can be considered interchangeable. For example, if you purchase potassium permanganate crystals from a pet supply store that doesn't provide an MSDS, you can use the potassium permanganate MSDS from J. T. Baker, Mallinckrodt, Fisher Scientific, or some other supplier. Just be sure that the MSDS refers to exactly the same chemical and in the same form and concentration. For example, an MSDS for dilute nitric acid is very different from an MSDS for fuming nitric acid.

FORM MATTERS

NFPA 704 fire diamonds and similar safety warnings may apply only to a specific form of a chemical. For example, aluminum, magnesium, or zinc metal lumps present little or no fire hazard, but those same metals in the form of dust are severe fire hazards. Don't make the mistake of judging the safety hazards of one form of a chemical based on data for another.

Although NFPA 704 defines only these two special hazards, other self-explanatory symbols are sometimes used unofficially. For example a strong acid may be flagged ACID, a strong base BASE or ALK (for alkali), and either may be flagged COR or CORR (for corrosive). More than one unique hazard may be listed in this section. For example, potassium chromate is both an oxidizer and a corrosive, so the white diamond may include both OX or OXY and COR or CORR.

Note that the NFPA 704 fire diamond is designed to warn firefighters and other emergency personnel, who may have to deal with very large quantities of a chemical in a burning building. For that reason, the NFPA 704 fire diamond may exaggerate the danger that a particular chemical represents in the very different environment of a home chem lab. A 5,000 gallon tank of acetone might explode during a fire and devastate an entire neighborhood; a 25 mL bottle of acetone might cause some excitement if it catches fire, but isn't likely to cause much damage if you have a fire extinguisher handy.

RISK PHRASES (R-PHRASES) AND SAFETY PHRASES (S-PHRASES)

Some chemicals are labeled with **Risk Phrases** (**R-phrases**). R-phrases are published in *European Union Directive 2001/59/EC*, which defines the "[n]ature of special risks attributed to dangerous substances and preparations." Although R-phrases have official standing only within the EU, they are commonly used worldwide to define the specific risks associated with particular chemicals. The following list includes all currently defined R-phrases. (Missing R-phrase numbers indicate phrases that have been deleted or replaced by another R-phrase.)

R1: Explosive when dry
R2: Risk of explosion by shock, friction, fire or other sources of ignition
R3: Extreme risk of explosion by shock, friction, fire or other sources of ignition
R4: Forms very sensitive explosive metallic compounds
R5: Heating may cause an explosion
R6: Explosive with or without contact with air

R7: May cause fire
R8: Contact with combustible material may cause fire
R9: Explosive when mixed with combustible material
R10: Flammable
R11: Highly flammable
R12: Extremely flammable
R14: Reacts violently with water
R15: Contact with water liberates extremely flammable gases
R16: Explosive when mixed with oxidizing substances
R17: Spontaneously flammable in air
R18: In use, may form flammable/explosive vapor-air mixture
R19: May form explosive peroxides
R20: Harmful by inhalation
R21: Harmful in contact with skin
R22: Harmful if swallowed
R23: Toxic by inhalation
R24: Toxic in contact with skin
R25: Toxic if swallowed
R26: Very toxic by inhalation
R27: Very toxic in contact with skin
R28: Very toxic if swallowed
R29: Contact with water liberates toxic gas
R30: Can become highly flammable in use
R31: Contact with acids liberates toxic gas
R32: Contact with acids liberates very toxic gas
R33: Danger of cumulative effects
R34: Causes burns
R35: Causes severe burns
R36: Irritating to eyes
R37: Irritating to respiratory system
R38: Irritating to skin
R39: Danger of very serious irreversible effects
R40: Limited evidence of a carcinogenic effect
R41: Risk of serious damage to eyes
R42: May cause sensitization by inhalation
R43: May cause sensitization by skin contact
R44: Risk of explosion if heated under confinement
R45: May cause cancer
R46: May cause heritable genetic damage
R48: Danger of serious damage to health by prolonged exposure
R49: May cause cancer by inhalation
R50: Very toxic to aquatic organisms
R51: Toxic to aquatic organisms
R52: Harmful to aquatic organisms
R53: May cause long-term adverse effects in the aquatic environment
R54: Toxic to flora
R55: Toxic to fauna
R56: Toxic to soil organisms
R57: Toxic to bees
R58: May cause long-term adverse effects in the environment
R59: Dangerous for the ozone layer

R60: May impair fertility
R61: May cause harm to the unborn child
R62: Possible risk of impaired fertility
R63: Possible risk of harm to the unborn child
R64: May cause harm to breast-fed babies
R65: Harmful: may cause lung damage if swallowed
R66: Repeated exposure may cause skin dryness or cracking
R67: Vapours may cause drowsiness and dizziness
R68: Possible risk of irreversible effects

R-phrases may be combined. For example, one of the R-phrases listed for copper(II) sulfate pentahydrate is R36/38. Combining R36 (Irritating to eyes) and R38 (Irritating to skin) yields a description for R36/38 of "Irritating to eyes and skin."

Whereas R-phrases define specific risks, **Safety Phrases (S-phrases)** define specific safety precautions that should be used to avoid injury. The following list includes all currently-defined S-phrases. (Missing S-phrase numbers indicate phrases that have been deleted or replaced by another S-phrase.)

S1:   Keep locked up
S2:   Keep out of the reach of children
S3:   Keep in a cool place
S4:   Keep away from living quarters
S5:   Keep contents under ... (appropriate liquid to be specified by the manufacturer)
S6:   Keep under ... (inert gas to be specified by the manufacturer)
S7:   Keep container tightly closed
S8:   Keep container dry
S9:   Keep container in a well-ventilated place
S12: Do not keep the container sealed
S13: Keep away from food, drink and animal feedstuffs
S14: Keep away from ... (incompatible materials to be indicated by the manufacturer)
S15: Keep away from heat
S16: Keep away from sources of ignition — No smoking
S17: Keep away from combustible material
S18: Handle and open container with care
S20: When using do not eat or drink
S21: When using do not smoke
S22: Do not breathe dust
S23: Do not breathe gas/fumes/vapor/spray (appropriate wording to be specified by the manufacturer)
S24: Avoid contact with skin
S25: Avoid contact with eyes
S26: In case of contact with eyes, rinse immediately with plenty of water and seek medical advice
S27: Take off immediately all contaminated clothing
S28: After contact with skin, wash immediately with plenty of ... (to be specified by the manufacturer)
S29: Do not empty into drains

S30: Never add water to this product
S33: Take precautionary measures against static discharges
S35: This material and its container must be disposed of in a safe way
S36: Wear suitable protective clothing
S37: Wear suitable gloves
S38: In case of insufficient ventilation wear suitable respiratory equipment
S39: Wear eye/face protection
S40: To clean the floor and all objects contaminated by this material use ... (to be specified by the manufacturer)
S41: In case of fire and/or explosion do not breathe fumes
S42: During fumigation/spraying wear suitable respiratory equipment (appropriate wording to be specified by the manufacturer)
S43: In case of fire use ... (indicate in the space the precise type of fire-fighting equipment. If water increases the risk add "Never use water")
S45: In case of accident or if you feel unwell seek medical advice immediately (show the label where possible)
S46: If swallowed, seek medical advice immediately and show this container or label
S47: Keep at temperature not exceeding ...°C (to be specified by the manufacturer)
S48: Keep wet with ... (appropriate material to be specified by the manufacturer)
S49: Keep only in the original container
S50: Do not mix with ... (to be specified by the manufacturer)
S51: Use only in well-ventilated areas
S52: Not recommended for interior use on large surface areas
S53: Avoid exposure — obtain special instructions before use
S56: Dispose of this material and its container at hazardous or special waste collection point
S57: Use appropriate containment to avoid environmental contamination
S59: Refer to manufacturer/supplier for information on recovery/recycling
S60: This material and its container must be disposed of as hazardous waste
S61: Avoid release to the environment. Refer to special instructions/safety data sheet
S62: If swallowed, do not induce vomiting: seek medical advice immediately and show this container or label
S63: In case of accident by inhalation: remove casualty to fresh air and keep at rest
S64: If swallowed, rinse mouth with water (only if the person is conscious)

Like R-phrases, S-phrases can be combined to better define the appropriate safety measures for a particular chemical.

HAZARD PICTOGRAMS AND LETTER SYMBOLS

Many packaged chemicals display one or more of the following pictograms and/or letter symbols to alert the user to specific hazards posed by that chemical.

These pictograms are intended to provide only general guidance. For additional information about the specific hazards involved and the steps required to handle the chemical safely, read the MSDS or other detailed safety information.

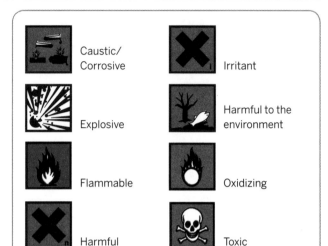

SAFE CHEMICAL HANDLING

It's important to treat your laboratory chemicals with respect and to handle them safely. But it's also important to realize that everyone routinely deals with many potentially dangerous chemicals, and that injuries occur only when those chemicals are mishandled.

For example, your bathroom medicine cabinet may contain isopropanol (flammable), ethanol (flammable), acetone nail polish remover (flammable), acetaminophen (toxic), tincture of iodine (toxic, flammable), and various prescription drugs (toxic). That can of crystal drain cleaner under the sink is almost pure sodium hydroxide (corrosive, toxic). Your automobile contains several gallons of gasoline (flammable), and its battery contains sulfuric acid (corrosive, toxic). If you use a wire brush to clean the terminals on your car battery, that powder you're scraping off is lead sulfate (toxic). You probably have a gallon of chlorine bleach (toxic, corrosive, oxidizer) in your laundry room, and the shelves in your basement shop probably hold cans of paint thinner (flammable), turpentine (flammable), and perhaps a gallon of concentrated hydrochloric (muriatic) acid (toxic, corrosive) for cleaning concrete. Your garden shed is probably full of insecticides (toxic) and that 50-pound bag of 34-0-0 fertilizer is actually pure ammonium nitrate (oxidizer, explosive). There may even be an old container of arsenic-based rat poison (toxic). Old thermometers contain mercury (toxic), as do fluorescent tubes. And so on.

The point is this: we are all exposed regularly to potentially hazardous chemicals, but awareness of the dangers and proper handling minimizes the risk of injuries. Use common sense precautions when handling chemicals. Wear gloves and protective clothing when handling chemicals, and always wear splash goggles. If you are working with powdered chemicals, wear a respirator mask. Have a fire extinguisher and first-aid kit handy.

Respect your chemicals. Do not fear them.

INCOMPATIBLE CHEMICALS
Some chemicals are incompatible with each other, in the sense that combining those chemicals may result in a violent or dangerous reaction such as intense heat, fire, explosion, or release of a toxic gas. In other cases, the reaction may not be immediately obvious. For example, by combining some chemicals you may unintentionally create a shock-sensitive explosive such as a peroxide or a fulminate, which might later detonate unpredictably.

Of course, sometimes such incompatible chemicals are intentionally combined to cause a reaction, but it's important that you avoid doing so unintentionally. If you're expecting spectacular results, you can take steps to deal with them safely. If you're not expecting a violent reaction, the results can be tragic.

Table 4-1 is a matrix that lists general incompatibilities between classes of chemicals. The matrix is not exhaustive. For example, the matrix lists organic acids, organic poisons, organic solvents, and water reactive compounds as incompatible with oxidizers, but there are many other substances that react violently with some or most oxidizers. Similarly, not every chemical in a particular category reacts badly with every chemical in an incompatible category. For example, many inorganic poisons do not react at all with many organic poisons. Still, within its limitations, this matrix provides useful guidance.

Although it is by no means exhaustive in either column, Table 4-2 lists some of the specific chemical incompatibilities that are most likely to be encountered in a home chem lab. The nature of the hazard varies. For example, acetone and hydrogen peroxide in the presence of an acid catalyst react to form the hideously dangerous acetone peroxide, an explosive that has been used by terrorists. Glycerol bursts into flame in contact with potassium permanganate. Cyanides (including complexes like ferricyanides and ferrocyanides) react with strong mineral acids to form toxic hydrogen cyanide gas. And so on.

DEADLY CHEMICALS

All of that said, there are chemicals that scare any sane chemist silly. Some chemicals are so toxic that literally one drop contacting your skin may be lethal, as may inhaling even a slight amount. (Dimethyl mercury comes to mind.)

That's why it's so important not to mix chemicals randomly or to use them other than as recommended. For example, potassium ferricyanide is used in some experiments in this book. Despite the presence of "cyanide" in the name, that chemical is of relatively low toxicity and is relatively safe to store and handle. However, if you heat potassium ferricyanide to decomposition or expose it to strong mineral acids, it produces extremely toxic hydrogen cyanide gas.

TABLE 4-1: *Chemical incompatibility matrix*

	Acids (Inorganic)	Acids (Organic)	Acids (Oxidizing)	Bases (Alkalis)	Oxidizers	Poisons (Inorganic)	Poisons (Organic)	Solvents (Organic)	Water Reactives
Acids (Inorganic)		STOP!		STOP!		STOP!	STOP!	STOP!	STOP!
Acids (Organic)	STOP!		STOP!	STOP!	STOP!	STOP!	STOP!		STOP!
Acids (Oxidizing)		STOP!		STOP!		STOP!	STOP!	STOP!	STOP!
Bases (Alkalis)	STOP!	STOP!	STOP!				STOP!	STOP!	STOP!
Oxidizers		STOP!					STOP!	STOP!	STOP!
Poisons (Inorganic)	STOP!	STOP!	STOP!				STOP!	STOP!	STOP!
Poisons (Organic)	STOP!	STOP!	STOP!	STOP!	STOP!	STOP!			
Solvents (Organic)	STOP!		STOP!	STOP!	STOP!	STOP!			
Water Reactives	STOP!	STOP!	STOP!	STOP!	STOP!	STOP!			

TABLE 4-2: *Specific chemical incompatibilities (not exhaustive)*

Chemical	Incompatible with
Acetic acid	Chromic acid, ethylene glycol, hydroxyl compounds, nitric acid, perchloric acid, permanganates, peroxides
Acetone	Concentrated nitric/sulfuric acid mixtures, hydrogen peroxide, strong bases
Ammonium nitrate	Acids, chlorates, combustible organic powders, flammable liquids, metal powders, nitrites, sulfur
Charcoal (activated)	Calcium hypochlorite, oxidizers
Chlorates	Acids, ammonium salts, combustible organic powders, metal powders, sulfur
Chlorine	Acetylene, ammonia, benzene, butadiene, butane, hydrogen, metal powders, methane, propane, sodium carbide, organic solvents
Copper	Acetylene, hydrogen peroxide
Hydrocarbons	Bromine, chlorine, chromic acid, fluorine, sodium peroxide
Ferrocyanides, ferricyanides	Acids
Hydrogen peroxide	Acetone, alcohols, aniline, chromium, copper, iron, flammable liquids, metals and metal salts, nitromethane, organic materials, nitrates
Hypochlorites	Acids, activated carbon/charcoal
Iodine	Acetylene, anhydrous or aqueous ammonia, hydrogen
Nitrates	Sulfuric acid
Nitric acid	Acetic acid, aniline, brass, chromic acid, copper, flammable liquids and gases, heavy metals, hydrocyanic acid, hydrogen sulfide
Nitrites	Acids
Oxalic acid	Mercury, silver
Potassium chlorate	Sulfuric acid and other strong acids
Potassium permanganate	Benzaldehyde, ethylene glycol, glycerol, sulfuric acid
Silver	Acetylene, ammonium compounds, fulminic acid, oxalic acid, tartaric acid
Sodium nitrite	Ammonium salts
Sulfides	Acids
Sulfuric acid	Chlorates, perchlorates, permanganates

The best ways to avoid dangerous chemical incompatibilities are to store chemicals safely and with regard to their possible interactions; to study the MSDS for each chemical you purchase, paying particular attention to listed incompatibilities; and to use the minimum quantities of chemicals necessary to complete each experiment. Finally, *NEVER MIX CHEMICALS RANDOMLY*.

STORAGE COLOR CODES

The storage color code system devised by J. T. Baker and shown in Table 4-3 has become an industry standard. This system assigns each chemical one of five storage color codes, with or without a stripe, which indicates the primary storage consideration for that chemical.

In theory, a chemical with a particular color code can safely be stored with other chemicals that have the same color code. In practice, there are many exceptions to that rule. For example, although both sulfuric acid and sodium hydroxide may be coded white for corrosive, that does not mean that they can be safely stored together. Strong acids and strong bases must be stored separately from each other.

A chemical that has a striped color code requires individual consideration, and often must be stored separately from other chemicals. For example, many suppliers color-code sodium, potassium, calcium and other alkali metals and alkaline-earth metals red with a stripe. The red indicates that the metal is flammable, and the stripe indicates that special storage requirements apply. (In this case, the metals must be kept away from liquid water, with which they react violently.) Similarly, some suppliers color-code strong mineral acids white (corrosive) and strong bases white (corrosive) with a stripe to indicate special storage considerations (in this case, that they must be stored separately from acids).

CHEMICAL ALLERGIES

Some people are allergic to substances that are generally considered to be "safe." For example, some people react very badly to the latex used in some protective gloves.

Before an allergic reaction can occur, you must be exposed at least once to the allergen, but the number of exposures required before an allergic reaction results is unpredictable. For example, I knew a man who was convinced he was immune to poison ivy. He'd handled it with bare hands dozens of times over more than 20 years, and had never had the slightest reaction to it. Until one day he did.

Contact allergies are particularly problematic in a chem lab, and for that reason you should wear gloves when handling any chemical, even one that is generally considered benign and that you have handled many times before with no adverse reaction. Otherwise, the next time you handle it you may regret not wearing gloves.

Furthermore, the color code assigned to a chemical indicates only the primary storage consideration for that chemical. For example, a chemical coded yellow (oxidizer/reactant) may also be quite toxic (blue), while a chemical coded white (corrosive) may

TABLE 4-3: *J. T. Baker storage color codes*

Storage Code	Notes
	White: Corrosive. May harm skin, eyes, and mucous membranes. Store separately from flammable and combustible materials.
	Yellow: Oxidizing and/or reactive. May react violently with air, water, or other substances. Store separately from flammable and combustible materials.
	Red: Flammable. Store separately only with other flammable reagents.
	Blue: Toxic. Hazardous to health if inhaled, ingested, or absorbed through skin. Store separately in secured area.
	Green: Presents no more than moderate hazard in any of categories above; use general chemical storage.
	Gray: Presents no more than moderate hazard in any of categories above; use general chemical storage. (Used by Fisher Scientific instead of green.)
	Orange: Presents no more than moderate hazard in any of categories above; use general chemical storage. (Obsolete color code; superseded by green.)
	Stripes: A reagent that is incompatible with other reagents of the same storage code color; store separately.

also be quite flammable (red). Different suppliers may code the same chemical differently. For example, Fisher Scientific codes concentrated nitric acid yellow (oxidizer/reactant), and other suppliers code it white (corrosive) or white with a stripe (corrosive with special storage considerations; nitric acid reacts with almost anything).

Storage codes may understate the danger. Violent reactions can occur between two chemicals for which the color codes offer little hint of danger. For example, potassium permanganate (yellow), a strong oxidizer, bursts into flames if it contacts glycerol (green). For that particular combination, yellow plus green equals extreme red. Many suppliers code such chemicals individually with a STOP icon, upraised palm icon, or similar indicator that some special storage requirement applies to that chemical.

Conversely, storage codes may exaggerate the danger, particularly for the small quantities typically stored and used in home labs. For example, some suppliers color code 5% acetic acid white (corrosive). That 5% acetic acid is exactly the same chemical and in the same concentration as the white vinegar you can buy at the supermarket.

Use the color coding on chemical bottles as a starting point rather than as an absolute guide. Study the MSDS for each chemical carefully, and pay particular attention to listed incompatibilities with other chemicals. If you are storing only small amounts of chemicals (say, 100 g or mL or less), it's probably sufficient just to segregate them on different shelves or in different cabinets. If you are storing larger quantities, pay correspondingly greater attention to proper storage.

PROPER DISPOSAL OF USED AND UNNEEDED CHEMICALS

Many home chemists dispose of all chemical wastes simply by flushing them down the drain with lots of water. That's an acceptable practice for many common laboratory chemicals, but not for chemicals that are particularly toxic or hazardous to the environment.

LOCK THEM UP

If there is even the slightest chance that children or pets may gain access to your stored chemicals, secure the storage cabinets and/or the lab itself with a sturdy lock. Keep your chemicals locked up except when you're actually using them.

Proper disposal of such hazardous chemicals is subject to an incredible maze of federal, state, and local laws and regulations. Fortunately, in practice many of these laws and regulations apply only to commercial and industrial users. For example, regulations for a particular chemical may come into play only if you are disposing of 10 kilograms or more of that chemical per month. Home chemists usually work with at most a few hundred grams of any particular chemical, so check your local hazardous waste disposal laws and regulations to determine the threshold that applies.

Before you choose a disposal strategy for your own home lab, check the web site of your local or state environmental affairs department to determine which laws and regulations apply to you. Most communities provide some means for free disposal of residential hazardous waste. Some communities hold periodic hazardous waste days, when hazardous waste can be placed separately at the curb for pickup. Others maintain hazardous waste disposal centers where you can drop off containers of hazardous waste. Take advantage of these services, and follow any labeling or other requirements they specify.

DISPOSAL OF COMMON LABORATORY CHEMICALS

Safe disposal of chemicals raises two questions that are not necessarily synonymous. First, you have to determine whether it is safe to dispose of a particular chemical by a particular means. Second, if it is safe, you have to determine whether it's legal.

Here are some general guidelines for disposing safely of common laboratory chemicals. Note that these procedures describe my own practices, which may or may not be lawful where you live. Check before you use these procedures:

- Small amounts of most flammable organic solvents (acetone, alcohols, ethers, and so on) can safely be disposed of by taking them outdoors and allowing them to evaporate. Obviously, do this in an area that is not near a flame and not accessible to children or animals. For larger amounts, place the solvent in a sealed container labeled with its contents and take it to the nearest hazardous waste disposal site.

STORAGE BINS

Heavy-duty Rubbermaid storage bins with latched lids are excellent containers to use for small amounts of chemicals. Buy several in appropriate sizes. Use a larger one for general storage of chemicals that are color-coded green. Use a smaller one color-coded white for acids, a second white one for strong bases, a yellow one for oxidizers/reactants, a red one for flammables, and a blue one for poisons.

- Mineral acids (hydrochloric, sulfuric, nitric, and so on) and most organic acids (formic, acetic, and so on) can be disposed of safely by neutralizing them with sodium bicarbonate (baking soda) until the fizzing stops. The resulting sodium salts (chloride, sulfate, nitrate, acetate, and so on) can safely be flushed down the drain with lots of water. If the acid is concentrated, pour it into ten times its volume of water to dilute it before you neutralize it. Alternatively, add a few drops of phenolphthalein solution to the dilute acid and add aqueous sodium hydroxide or sodium carbonate until the solution turns pink, indicating that the acid has been neutralized.

- Strong bases such as sodium hydroxide and potassium hydroxide can be disposed of safely by diluting them, if necessary, and then treating them with a dilute solution of hydrochloric acid. The resulting chloride salts (for example, sodium chloride—common table salt—and potassium chloride) are innocuous and can safely be flushed down the drain. To neutralize the base, add enough phenolphthalein solution to the aqueous base solution to give it a noticeable pink color, and then add dilute acid until the pink color disappears.

- Solutions that contain large amounts of free bromine or iodine are hazardous to the environment, although the small amounts that home chemists work with are generally innocuous. To safely dispose of bromine or iodine solutions, add sodium thiosulfate solution until the waste solution turns colorless. This converts the free bromine or iodine to bromide or iodide ions, which are safe to flush down the drain.

- With the exception of heavy metal ions, described in the following section, most cations are reasonably benign and can safely be flushed down the drain. "Safe" cations include aluminum, ammonium, bismuth, calcium, cerium, cesium, cobalt, gold, hydrogen, iron, lithium, magnesium, manganese, potassium, rubidium, sodium, strontium, and tin.

- Most anions are harmless, particularly in small quantities and low concentrations. "Safe" anions include acetate, bicarbonate, bisulfite, borate, bromate, bromide, carbonate, chlorate, chloride, cyanate, iodate, iodide, nitrate, nitrite, perchlorate, periodate, permanganate, phosphate, silicate, stannate, sulfate, sulfite, thiocyanate (sulfocyanate), thiosulfate, titanate, tungstate, and vanadate. Unless there is another reason not to (such as the presence of a toxic cation or anion complex), solutions of these anions can safely be flushed down the drain with copious amounts of water.

PRIVATE SEWAGE SYSTEMS

It's safe to dispose of relatively benign chemicals by flushing them down the drain only if your drain connects to a sanitary sewer system, which quickly dilutes any chemicals you flush. If you use a septic tank or other means of sewage disposal, you might want to think twice about flushing chemicals into it. It's difficult to know for sure whether a particular chemical will "poison" a septic tank, so the best practice is to avoid flushing any chemical unless you are absolutely certain that it's safe to do so. Instead, neutralize waste chemicals as described in this section and take them to a waste disposal site or other location where they can be safely disposed.

DISPOSAL OF HEAVY-METAL AND OTHER TOXIC COMPOUNDS

Although many common laboratory chemicals can be safely flushed down the drain, it is unacceptable to flush chemicals that contain heavy-metal ion species or contain other very toxic and/or persistent species. Treat any waste that contains organic compounds as hazardous unless you are certain that it is not, and treat any waste that contains heavy-metal ion species as toxic.

Fortunately, heavy-metal waste can be converted into safer forms, which can then be disposed of by taking them to a hazardous waste disposal center. Your goals in pretreating heavy-metal compounds for disposal are to convert hazardous soluble heavy-metal ions to much less hazardous insoluble solids, and to minimize the volume of solid waste that will require proper disposal. Here are the guidelines I use, although again, you must verify that these methods are legally acceptable in your own jurisdiction.

Barium

Soluble barium salts are extremely toxic and hazardous for the environment. Before disposal, treat any solution that contains soluble barium salts with an excess of a soluble sulfate salt, such as sodium sulfate. The barium ions precipitate as insoluble barium sulfate, which can be filtered, dried, and (including the filter paper) added to a hazardous waste container for later proper disposal. The filtrate contains few barium ions, and can safely be flushed down the drain.

Chromium

Soluble salts with chromium cations or chromium anion complexes (chromates, dichromates) are extremely toxic and hazardous to the environment. Chromium(III) cations can be precipitated by excess sulfate ions as insoluble chromium(III) sulfate. The much more toxic chromium(VI) cation can be reduced to chromium(III) with an excess of hydrogen peroxide and then precipitated as chromium(III) sulfate. Chromate and dichromate anions can be treated by adding a stoichiometrically equivalent amount of a soluble lead(II) salt, such as lead(II) nitrate or lead(II) acetate, which precipitates the chromium ions as insoluble lead(II) chromate. In any case, filter the precipitate and dispose of it (including the filter paper) in a hazardous waste container. The filtrate contains few chromium ions, and can safely be flushed down the drain.

Copper

Soluble copper salts are moderately toxic and moderately dangerous for the environment. Precipitate copper(I) ions by adding an excess of potassium iodide to form insoluble copper(I) iodide. Precipitate copper(II) ions by adding excess sodium carbonate or sodium hydroxide to form insoluble copper(II) carbonate or copper(II) hydroxide. Filter, and dispose of the filtrand and filter paper in a toxic waste container. The filtrate contains few copper ions, and can safely be flushed down the drain.

Lead

Soluble lead salts are extremely toxic and dangerous for the environment. Precipitate lead ions by adding an excess of sodium carbonate to form insoluble lead carbonate. Filter, and dispose of the filtrand and filter paper in a toxic waste container. The filtrate contains few lead ions, and can safely be flushed down the drain.

Silver

Soluble silver salts are toxic and dangerous for the environment. Precipitate silver ions by adding excess potassium iodide to form insoluble silver(I) iodide. Filter, and dispose of the filtrand and filter paper in a toxic waste container. The filtrate contains few silver ions, and can safely be flushed down the drain.

CHEMICALS USED IN THIS BOOK

Any home chemistry lab needs a good selection of chemicals. When I started this book, I had no real idea what I needed, so I ordered a wide variety of potentially useful chemicals. I knew that some of those chemicals would end up not being used in any of the lab sessions in the book, and that I might need more of some chemicals than I originally ordered. So I kept careful track of what chemicals I used for the various labs and in what quantities.

Table 4-4 lists the chemicals required to complete all of the laboratory sessions in this book, except those in Chapters 19, 20, 21, and 22. With the exception of a few items that must be purchased from specialty chemical suppliers, all of the chemicals are readily available from local sources such as the supermarket, drugstore, and hardware store. Table 4-5 lists the additional chemicals needed to complete the laboratory sessions in the final four chapters.

The storage, risk, and safety information in Tables 4-4 and 4-5 is neither authoritative nor definitive. For chemicals that list "none" under R(isk)-phrases and/or S(afety)-phrases, an MSDS for that chemical explicitly listed the chemical as nonhazardous and/or not requiring any safety precautions. For chemicals for which the R-phrases and/or S-phrases column is empty, no information was available. Note well: the absence of safety information does not indicate that a chemical is safe, nor does the absence of safety information indicate that no special safety measures are required. Always check a current MSDS for every chemical before you use it.

TABLE 4-4: *Recommended basic chemicals and quantities*

Chemical	Quantity	Storage Code	R(isk)-phrases	S(afety)-phrases
Acetic acid, 17.4 M (glacial)	100 mL		R10 R35	S23 S26 S45
Acetone[1]	125 mL		R11 R36 R41 R66 R67	S9 S16 S26
Alka-Seltzer tablets	6		none[21]	none[21]
Aluminum, granular, turnings, or shot	25 g		none[21]	none[21]
Ammonia, aqueous, 6 M (household)	125 mL		R20 R21 R22 R34 R36 R37 R38 R41 R50	S26 S36 S37 S39 S45 S61
Ammonium acetate[2]	25 g		R36 R37 R38	S26 S36
Ammonium chloride[3]	25 g		R22 R36	S22
Ammonium nitrate[4]	40 g		R8 R20 R21 R22 R36 R37 R38	S17 S26 S36
Calcium carbonate antacid tablets	3		R36 R37 R38	S26 S36
Celery	1 stalk		none[21]	none[21]
Charcoal, activated	25 g		R36 R37	S24 S25 S26 S36
Club soda	200 mL		none[21]	none[21]
Copper(II) sulfate[5]	250 g		R22 R36 R38 R50 R53	S22 S60 S61
Dishwashing liquid	25 mL		none[21]	none[21]
Ethanol, 70%[6]	1 L		R11 R20 R21 R22 R36 R37 R38 R40	S7 S16 S24 S25 S36 S37 S39 S45
Food coloring dyes, assorted	1		none[21]	none[21]
Gasoline	100 mL		not defined[22]	not defined[22]

TABLE 4-4 *(continued)*: Recommended basic chemicals and quantities

Chemical	Quantity	Storage Code	R(isk)-phrases	S(afety)-phrases
Glycerol (glyercin)[7]	25 mL		none[21]	S26 S36
Hydrochloric acid, concentrated[8]	250 mL		R23 R24 R25 R34 R36 R37 R38	S26 S36 S37 S39 S45
Iron filings	25 g		none[21]	none[21]
Iron shot	100 g		none[21]	none[21]
Iron or steel nail (6d to 12d)	6		none[21]	none[21]
Lead metal shot[9]	5,000 g		R23 R25	none[21]
Lemon	4		none[21]	none[21]
Lighter fluid	25 mL		not defined[22]	not defined[22]
Magnesium sulfate[10]	50 g		none[21]	S22 S24 S25
Milk, homogenized	25 mL		none[21]	none[21]
Mineral oil	25 mL		R36 R37 R38	not defined[22]
Oxalic acid[11]	25 g		R21 R22	S24 S25
Petroleum ether (ligroin)	100 mL		R11 R20 R21 R22 R45 R65	S45 S53
Phenolphthalein (powder)	1 g		R36 R37 R38	S26
Polystyrene foam, rigid	35 g		R36 R37	none[21]
Potassium hydrogen tartrate[12]	25 g		none[21]	none[21]
Potassium permanganate[13]	25 g		R8 R22	S17 S26 S36 S37 S39 S45

TABLE 4-4 *(continued):* *Recommended basic chemicals and quantities*

Chemical	Quantity	Storage Code	R(isk)-phrases	S(afety)-phrases
Sand	25 g		none[21]	none[21]
Sodium acetate[14]	25 g		none[21]	S22 S24 S25
Sodium bicarbonate	50 g		none[21]	none[21]
Sodium bisulfite	25 g		R20 R22 R36 R37 R38	S26 S36
Sodium borate (borax)	25 g		R22 R36 R37 R38 R62 R63	not defined[22]
Sodium carbonate[15]	100 g		R36 R37 R38	none[21]
Sodium chloride[16]	500 g		R36	S26 S36
Sodium hydroxide[17]	100 g		R35	S26 S37 S39 S45
Starch	25 g		none[21]	none[21]
Sucrose[18]	450 g		none[21]	none[21]
Sulfur[19]	25 g		R36 R37 R38	not defined[22]
Sulfuric acid[20]	100 mL		R23 R24 R25 R35 R36 R37 R38 R49	S23 S30 S36 S37 S39 S45
Talcum powder	25 g		none[21]	none[21]
Vegetable oil	25 mL		none[21]	none[21]
Vinegar, white distilled	250 mL		none[21]	none[21]

NOTES:

1. Acetone from the hardware store is fine.

2. You can produce ammonium acetate by neutralizing clear household ammonia with distilled white vinegar and evaporating to dryness.

3. You can produce ammonium chloride by neutralizing clear household ammonia with hardware store muriatic acid and evaporating to dryness.

4. Pure ammonium nitrate fertilizer from the lawn and garden store is fine.

5. Pure copper sulfate is sold as root killer by hardware stores.

6. Drugstore rubbing alcohol is often 70% ethanol (ethyl alcohol), but check the label to make sure that it's not isopropanol (isopropyl alcohol).

7. Glycerol under that name or as glycerin is available in drugstores.

8. Hardware stores sell hydrochloric acid as muriatic acid. Make sure that the acid concentration is listed as 30% or higher.

9. Lead metal shot is sold by sporting goods stores for use in reloading shotgun shells.

10. Pure magnesium sulfate is sold by drugstores as Epsom salts.

11. Pure oxalic acid is sold by hardware stores as wood bleach (check the label to make sure).

12. Pure potassium hydrogen tartrate is sold by grocery stores as cream of tartar.

13. Pure potassium permanganate as crystals or a solution is sold by some pet stores for aquarium use.

14. You can produce sodium acetate by neutralizing a sodium hydroxide solution with distilled white vinegar and evaporating to dryness.

15. Hydrated sodium carbonate is sold as washing soda; anhydrous sodium carbonate as soda ash.

16. You can substitute noniodized salt such as popcorn salt, kosher salt, or rock salt.

17. Some hardware stores sell pure sodium hydroxide as crystal drain cleaner (check the label to make sure that it's pure).

18. Sucrose is table sugar.

19. Lawn and garden stores sell sulfur under that name.

20. Auto supply stores sell 35% sulfuric acid as battery acid.

21. "None" means that I was able to locate risk and safety information for these substances and that no R-phrases and/or S-phrases were provided.

22. "Not defined" means that I was unable to locate published risk and/or safety information that included specific R-phrases and/or S-phrases for these substances. The absence of these R- and/or S-phrases does not indicate that no risks exist or that no safety measures should be taken. Refer to the container label for risk and safety information.

For most chemicals, Table 4-5 lists a minimum quantity of 25 g or 25 mL, even if that amount is significantly more than is actually needed for the lab sessions, because those quantities are the smallest typically offered by specialty chemical suppliers. Although some suppliers offer chemicals in smaller quantities, packaging and labeling costs are substantial even for small amounts, so it often costs little or no more to buy 25 g of a chemical than only a gram or two. Table 4-5 lists smaller recommended quantities for a few relatively expensive chemicals (such as iodine, ninhydrin, and silver nitrate) that are required in small amounts.

Compare prices among several suppliers. Quite often, a particular supplier has high prices for some chemicals and low prices for others.

ALTERNATIVE SOURCES

Many of the chemicals in Table 4-5 are available, sometimes under different names, from alternative sources such as hardware stores, lawn and garden stores, photography stores, and so on. The best source of information on alternative chemical sources is the Readily Available Chemicals list at *http://www. hyperdeath.co.uk/chemicals*.

TABLE 4-5: *Recommended supplemental chemicals and quantities*

Chemical	Quantity	Storage Code	R(isk)-phrases	S(afety)-phrases
Aluminum nitrate	25 g		R8 R36 R37 R38 R41	S17 S26 S27 S36 S37 S39
Ammonia, aqueous, 15 M (~30%)	125 mL		R20 R21 R22 R34 R36 R37 R38 R41 R50	S26 S36 S37 S39 S45 S61
Ammonium molybdate	25 g		R36 R37 R38	not defined
Ammonium oxalate	25 g		R20 R21 R25 R34	not defined
Aspirin	25 g		none	none
Barium chloride	10 g		R20 R21 R25 R36 R37 R38	S45
Barium hydroxide	10 g		R20 R21 R22 R34 R41	S26 S28
Barium nitrate	25 g		R8 R20 R25 R37	S28
Calcium nitrate	25 g		R8	not defined
Chloroform	25 mL		R20 R22 R38 R40 R48	S36 S37
Chromium(III) nitrate	50 g		R8 R20 R21 R22 R36 R37 R38	S17 S26 S27 S36 S37 S39
Cobalt(II) chloride	5 g		R25 R40 R42 R43	S22 S36 S37 S39 S45
Cobalt(II) nitrate	25 g		R8 R22 R40 R43 R50 R53	S17 S36 S37 S60 S61
Copper(II) nitrate	25 g		R8 R20 R22 R36 R37 R38	not defined
Formaldehyde, 37% to 40%	25 mL		R10 R26 R27 R28 R34 R40 R41 R43	not defined
Hydrogen peroxide, 3%	1 mL		none	S26 S36
Iodine crystals	5 g		R21 R23 R25 R34	S23 S25

TABLE 4-5 *(continued)*: *Recommended supplemental chemicals and quantities*

Chemical	Quantity	Storage Code	R(isk)-phrases	S(afety)-phrases
Iron(III) nitrate	25 g		R8 R36 R38	S26
Iron(II) sulfate	25 g		R22 R36 R37 R38	S26 S36 S37 S39
Lead nitrate	50 g		R8 R20 R21 R22 R33 R36 R37 R38 R60 R61	S17 S36 S37 S39 S45 S53
Manganese(II) sulfate	25 g		R20 R21 R22 R36 R37 R38 R40	S26 S36
Methanol	125 mL		R11 R23 R24 R25 R39	S7 S16 S24 S36 S37 S45
Nickel(II) nitrate	30 g		R8 R20 R21 R22 R36 R37 R38 R45	not defined
Ninhydrin powder	1 g		R22 R36 R37 R38	S26 S28 S36
Nitric acid, concentrated (15.8 M)	100 mL		R8 R23 R24 R25 R34 R41	S23 S26 S36 S37 S39 S45
Potassium bromide	25 g		R20 R21 R22 R36 R37 R38	S22 S26 S36
Potassium chromate	25 g		R23 R24 R25 R45 R47	not defined
Potassium hexacyanoferrate(II)	25 g		R32 R52 R53	S50B S61
Potassium hexacyanoferrate(III)	25 g		R20 R21 R22 R32	S26 S36
Potassium iodide	25 g		R36 R38 R42 R43 R61	S26 S36 S37 S39 S45
Potassium nitrate	25 g		R8 R22 R36 R37 R38	S7 S16 S17 S26 S36 S41
Potassium thiocyanate	25 g		R22 R36 R37 R38	not defined
Seawater	200 mL		none	none
Silver nitrate	5 g		R22 R34 R36 R37 R38	S26 S45

Chemical	Quantity	Storage Code	R(isk)-phrases	S(afety)-phrases
Sodium hypochlorite (laundry bleach)	125 mL		R20 R21 R22 R34 R41	S1 S2 S28 S45 S50
Sodium nitrite	25 g		R8 R25 R50	S45 S61
Sodium phosphate, tribasic	25 g		R34 R36 R38	S22 S26 S27 S36 S37 S39
Sodium sulfate	25 g		none	none
Sodium sulfite	25 g		R20 R21 R22 R36 R38 R40	S22 S26 S36
Sodium thiosulfate	25 g		R36 R37 R38	S26 S36
Strontium nitrate	25 g		R8 R36 R37 R38	S26
Sulfuric acid, 98% (18.4 M)	100 mL		R23 R24 R25 R35 R36 R37 R38 R49	S23 S30 S36 S37 S39 S45
Vitamin C tablet, 500 mg	1		none	none
Zinc metal (powder, granular, mossy)	25 g		R11	S7 S8 S43
Zinc nitrate	25 g		R8 R20 R21 R22 R34 R36 R37 R38	S17 S22 S26 S36 S37 S39

ABOUT IODINE

Just before this book went to press, the DEA reclassified elemental iodine in any quantity as a List I chemical, because iodine is sometimes used by illegal meth labs. Although it is still legal to purchase elemental iodine, many suppliers no longer offer it because of the paperwork involved.

If you can't find elemental iodine for sale (or if you just don't want your name to end up on a DEA list) you can make your own iodine crystals from potassium iodide. For instructions, visit **www.homechemlab.com/iodine.html.**

SHIPPING HAZARDOUS CHEMICALS

Some of the chemicals used in this book are classified as hazardous for shipping purposes. The rules and regulations for shipping hazardous chemicals are incredibly complex and change frequently. It's almost a full-time job to keep up with them. The specific chemical, its form (granular, powder, chunks, and so on), and the quantity in question all have a bearing on whether that chemical can be shipped by a particular method. The upshot is that it can be extremely expensive to ship even small quantities of some chemicals.

Commercial, industrial, and university laboratories deal with this problem by ordering large quantities at one time, consolidating orders, and resigning themselves to paying high transportation charges. Those aren't realistic options for most home labs. There are four ways to avoid these high shipping charges:

1. Don't use the chemical. In some cases, you can substitute another chemical that is available without hazard surcharges. Other times, you can make the chemical yourself from precursors that are not subject to shipping surcharges. (Note that some of these chemicals are listed as hazardous for a very good reason; you may want to avoid them entirely for safety's sake. On the other hand, some of them, such as acetone and alcohols, are no more hazardous than many chemicals in your medicine cabinet or garden shed.)

2. Buy the chemical locally. If you live in a large city, there are probably numerous commercial chemical supply houses within easy driving distance. Even if you live in a small town, you can find many hazardous chemicals at a hardware store (e.g., hydrochloric acid, sodium hydroxide, acetone and other ketones), drugstore (e.g., ethanol, isopropanol), pool supply store (e.g., potassium permanganate), agricultural supply store (e.g., potassium nitrate, ammonium nitrate, calcium nitrate), auto parts store (sulfuric acid), and so on.

3. Chemicals that require hazardous shipping surcharges in their solid or concentrated forms are often available as dilute solutions without surcharges. Although the price is very high per gram of chemical, if you need only a small amount of the chemical, this can be the least expensive way to get it. For example, rather than ordering solid barium nitrate and paying hazard surcharges, you might find you can order 50 mL of 0.1 M barium nitrate solution for $2.50 without hazard surcharges. That solution contains only 1.3 g of barium nitrate, but that may be all you need. This, incidentally, is also a cost-effective way of ordering any chemical that you need in very small

quantities, whether or not the solid or concentrated chemical is subject to surcharges.

4. Buy the chemical from a vendor that does not charge extra shipping for hazardous materials. (The vendor still has to pay those surcharges to the shipping company, but they simply absorb them as a cost of doing business.)

For some chemicals, notably nitric acid, option 4 is often the only economical choice. There are at least three good sources that do not charge hazardous shipping surcharges on most or all chemicals:

- ScienceKit (*http://www.sciencekit.com*) does not charge hazardous material shipping surcharges on any of the broad range of chemicals it carries. ScienceKit does not sell any chemicals to individuals, but they will sell any chemical they stock to a recognized home school (you must provide evidence that your home school is a member of an official home-schooling group) and to businesses (a copy of your business license or IRS EIN suffices as proof). As of this writing, as strange as it sounds, your luck in ordering chemicals from ScienceKit may depend on what time of day you call. If you call before 5:00 p.m. EST, you connect to their New York group, which flatly refuses to sell chemicals to anyone but public schools and universities. If you call after 5:00 p.m. EST, you connect to their California group, which will happily sell chemicals to you as long as you can prove that you're a recognized home school or business. And all you need to prove that you're a business is a copy of an Employee Identification Number (EIN), which the IRS will issue to you upon request even if you have no employees.

- Elemental Scientific (*http://www.elementalscientific.net*) sells a broad range of chemicals to individuals, most without any hazardous shipping surcharges. They do this by limiting the quantities they offer on their price list to amounts that can be shipped normally. For example, one chemical may be listed only in 1-ounce quantities, another in 1- and 4-ounce quantities, and still another in 1-, 4- and 16-ounce quantities. As long as you order only one of each chemical in a size that appears on the price list, there are no hazardous shipping surcharges. The only exceptions are chromic acid, chromium trioxide, 30% hydrogen peroxide, nitric acid, and sodium peroxide, which require a surcharge, and carbolic acid (phenol), barium chlorate, barium hydroxide, barium nitrate, barium peroxide, calcium carbide, calcium and sodium metal, ethylene dichloride, magnesium and zinc metal dust, mercuric iodide, mercuric oxide, mercurous chloride—all of which require a poison pack in addition to the hazardous shipping surcharge.

BUYING CHEMICALS THE EASY WAY

I've provided lists of the chemicals used in this book to various vendors who specialize in supply home scientists. When this book went to press, at least one of those vendors offered kits that contain some or all of the recommended chemicals in appropriate quantities, and at a discounted price compared to ordering the chemicals individually. It's likely that more vendors will offer such kits in the near future. (Note that the specific chemicals included in each kit and the quantity and grade of those chemicals may vary from vendor to vendor.)

For an updated list of these vendors and other vendors who offer chemicals to home scientists, visit *http://www. homechemlab.com/sources.html.*

- Home Science Tools (*http://www.homesciencetools.com*) sells a reasonably broad range of chemicals in small sizes, typically 30 g or 30 mL, to anyone, including individuals, without any hazardous shipping surcharges. HST doesn't offer nitric acid, but does carry such things as concentrated hydrochloric and sulfuric acids, barium compounds, and numerous oxidizers.

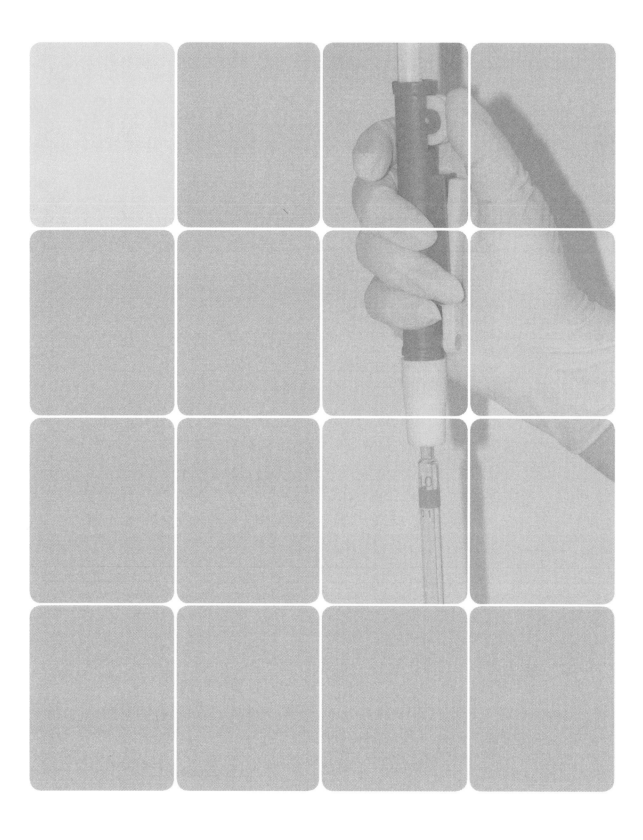

5

Mastering Laboratory Skills

This chapter describes and illustrates the laboratory skills that you need to master to complete the laboratory sessions in this book.

MEASUREMENT RESOLUTION AND SIGNIFICANT FIGURES

Measurements are a key part of chemistry, so it's important to understand them and their limitations. Some measurements, called **counts**, are exact. For example, if you have a bag of apples, you can count the apples and state definitively that the bag contains exactly eight apples. But most measurements are inexact. For example, although the label on a bag of rice may state that it contains two kilograms of rice, this is only an approximation of its mass. Depending on the resolution of the scale used to weigh the rice (and on truth-in-labeling laws), a particular bag may contain as little as 1.999 kg, 1.99 kg, or even 1.9 kg of rice, or it may contain as much 2.001 kg, 2.01 kg, or even 2.1 kg of rice. That variability (or **uncertainty**) is common to most measurements made in a chemistry lab.

The degree of uncertainty depends on the **resolution** of the measuring instrument and the degree of skill used to make the measurement. For example, if you use a ruler graduated in centimeters (cm) to measure the length of a piece of wire, you might find that the wire is between 2 cm and 3 cm long. Because the ruler is graduated in centimeters, the most you can say with certainty is that the wire is longer than 2 cm and shorter than 3 cm, because these are the two known values between which the length of the wire falls.

However, you can estimate (**interpolate**) the length of the wire to some value between the known values. For example, if the end of the wire extends almost but not quite halfway from the 2 cm marking to the 3 cm marking, you might estimate the length of the wire as 2.4 cm. That value has two **significant figures** (usually spoken as **sig figs**), which comprise the one known value and the interpolated value you added. For example, Figure 5-10 (later in this chapter) shows a burette that is graduated to (has resolution of) 0.1 mL, but allows values to be interpolated to about 0.01 mL.

> **VARIABLE SIG FIGS**
>
> It's important to understand that the number of significant figures yielded by a particular measuring instrument is not fixed. For example, if the segment of wire that we measured with the ruler graduated in centimeters was between 15 cm and 16 cm in length and we interpolated a value of 15.8 cm, that value has three sig figs rather than just two. The number of significant figures varies not just with the resolution of the measuring instrument, but also with the quantity being measured.

If you use a ruler with millimeter (mm) graduations, you can obtain a better value for the length of the wire. For example, if the wire extends past the 23 mm line but does not reach the 24 mm line, you can state with certainty that the length of the wire is between 23 mm and 24 mm. By interpolation, you may estimate that the actual length of the wire is 23.8 mm. That value has three sig figs, two known and one interpolated. By using a better instrument, you have increased the accuracy of your measurement by a factor of ten.

If you use a caliper with 0.1 mm graduations, you can obtain an even better value for the length of the wire. For example, you may find that the length of the wire falls between 23.7 mm and 23.8 mm. Once again, you may add another significant figure by interpolation and estimate the length of the wire as 23.74 mm, a value with four sig figs. But, although you now know the length of the wire with much greater accuracy than before, you still don't know the *exact* length of the wire. You can continue using better and better measuring instruments, but even the best possible instrument can still provide only an approximation of the wire's length. Some approximations are better than others, obviously, and one common way to quantify the quality of an approximation is to specify the number of significant figures.

Any calculation that uses measured data must take the uncertainly of the measuring instruments, as quantified by significant figures, into account. When you're making a calculation using measurements from more than one instrument, the least accurate instrument determines the level of certainty of the results. For example, you might want to determine a runner's average speed over a 1 km course. To get accurate results, you might use a laser rangefinder accurate to 1 m to set the start and finish lines and a stopwatch accurate to 0.01 second to time the run. The accuracy of these measuring instruments is high, and the result will have a correspondingly high number of sig figs. Conversely, if you measure the length of the course with your car's odometer, you know the distance only to within 0.1 kilometer (which can be interpolated to one further sig fig), so even using the most accurate stopwatch available cannot increase the accuracy of your result. Similarly, if you use a sundial to time the run, the high accuracy of the laser rangefinder is wasted because the elapsed time for the run is not known accurately.

That brings up another important point. Mathematical operations cannot increase the degree of accuracy. If you add, subtract, multiply, or divide values, the number of significant figures remains the same. For example, if you multiply the two values 4.03 and 1.16, the mathematical result is 4.6748, but that result ignores sig figs. Each of the two initial values is known to three sig figs, so the result can have no more than three sig figs and should therefore be recorded as 4.67. Similarly, if you multiply the values 4.0 and 1.17, the mathematical result is 4.68. But because one of the initial values had only two sig figs, you must round the result to two sig figs, or 4.7.

ANALOG VERSUS DIGITAL

Interpolation is possible only with measuring instruments that use analog readouts. Instruments that use digital readouts display only known values—assuming they are properly designed—and therefore provide no data for interpolation. For example, an electronic centigram scale provides a readout to 0.01 g, but no additional information that could be used for interpolation. Conversely, a quad-beam manual centigram scale provides known values to 0.01 g, and may allow interpolation down to the milligram level.

Alas, some digital instruments present values well beyond their actual resolutions. For example, I saw one inexpensive digital pH meter whose display specifies pH to 0.01. Unfortunately, the manual states that the actual resolution of the instrument is only ±0.2 pH, so that last decimal place is entirely imaginary.

SIGNIFICANT FIGURES IN THE REAL WORLD

Significant figures are an attempt to encode two separate pieces of information—a known value and an uncertainty—into one number, with sometimes-ambiguous results. Chemistry texts often devote a great deal of attention to significant figures, but most working scientists use a more commonsense approach. They use two separate numbers to express the two values, in the form 123.456±0.078 or 123.456(78), where 123.456 is the nominal value and 0.078 or (78) is the uncertainty. For example, if a balance indicates that the mass of a sample is 42.732 g, that mass might be recorded as 42.732±0.005 or as 42.732(5), indicating an uncertainty of 5 units in the third decimal place.

ABOUT ABOUT

Instructions in the laboratory sessions later in this book frequently say something like "transfer *about* 5 mL" of a solution or "weigh *about* 5.1 g" of a solid. So what does *about* 5 mL or *about* 5.1 g mean? The key is the number of sig figs. In the first case, anything more than 4 mL and less than 6 mL is acceptable (although you should take reasonable care to get close to 5 mL). In the second case, anything in the range of 5.0 g to 5.2 g is acceptable, although again you should take reasonable care to stay as close as possible to 5.1 g.

Often, the exact volume or mass you start with doesn't matter, but it's important that it be known accurately. For example, "weigh about 5 g of sodium hydroxide and record the mass to 0.01 g" means you should start with more or less 5 g of the chemical, but you need to determine and record the actual mass of sodium hydroxide to within 0.01 g.

HANDLING CHEMICALS PROPERLY

Aside from safety, the most important goal of proper chemical handling is to avoid contamination. To minimize the risk of contamination, use the following guidelines:

- Always store chemicals away from your work area, ideally in cupboards or covered storage containers.

- Never store chemicals in unlabeled bottles. In addition to being unsafe, the risk of contamination is high, because one chemical could easily be confused with another.

- Have only one chemical bottle open at a time.

- Measure and transfer chemicals in an area that is remote from any experimental procedures that are in progress. Otherwise, even a small accident might contaminate the entire bottle.

- Always read the label, twice, before you open the bottle. Make certain that you're actually using the chemical you think you're using.

- Before you open the bottle, wipe the exterior with a clean paper towel to remove any accumulated dust or other contamination.

- If a chemical bottle becomes grossly contaminated—for example, by leakage from another chemical bottle—either rinse the bottle thoroughly to remove all traces of the contamination or discard the chemical and obtain a new supply.

- Never return unused chemicals to their original containers. Doing so risks contaminating the entire bottle. Take only as much as you need from the original bottle, and safely discard any excess.

- Never place the bottle cap on the counter, where it may become contaminated or switched with a cap from a different chemical (which is another good reason to have only one chemical bottle open at a time). Instead, hold the cap and chemical bottle in the same hand. Depending on the size of the bottle and cap, you may find it easier to hold the cap between your thumb and forefinger, as shown in Figures 5-1 and 5-2, or between your little finger and ring finger, as shown in Figure 5-3. If the cap and bottle are too large to handle with one hand, either hold the cap in your other hand or, if absolutely necessary, place the cap flat with the inside surface up on a clean surface away from your immediate work area.

- When possible, avoid touching a solid chemical in the storage bottle with a scoop, spatula, or similar tool. Many solid chemicals are free-flowing crystals or powders. Transfer such chemicals directly to the weighing paper by rotating and gently

Safety is always the primary consideration in proper chemical handling. For detailed information about storing and handling chemicals safely, see Chapter 4.

tapping the bottle to dispense only as much as you need, as shown in Figure 5-2. Avoid letting the mouth of the bottle come into contact with the weighing paper or any other object.

- Some solid chemicals are not free-flowing or tend to form clumps. If you are using such a chemical, there is sometimes no alternative but to scoop it out of the bottle. In such cases, the best way to avoid contamination is to use a disposable scoop and discard it immediately after use. Disposable plastic spoons are excellent and inexpensive one-use scoops. You can also modify a disposable Beral pipette by cutting diagonally across the bulb portion to form a disposable scoop, as shown in Figure 5-3.

- Liquid chemicals may be transferred by using a funnel or by pouring directly from one container to another. When you transfer a liquid directly, use a clean stirring rod to prevent splashing or dripping, as shown in Figure 5-4.

- For concentrated acids and other hazardous liquid chemicals, safety trumps the risk of contamination. Rather than risk pouring such hazardous chemicals into a measuring container, uncap the bottle and use a clean (ideally new) pipette to draw out as much of the liquid as you need. Recap the bottle and transfer the liquid from the pipette directly to the measuring, mixing, or storage container or directly to the reaction vessel.

- Always recap the bottle immediately and replace the bottle in its assigned storage location.

If a chemical does become contaminated, do not attempt to salvage it. Dispose of it properly and buy a new supply.

FIGURE 5-1: *Hold the cap, rather than placing it on the bench*

FIGURE 5-2: *Tap solid chemicals directly onto the weighing paper*

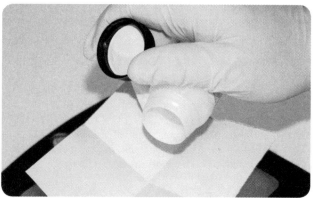

FIGURE 5-3: *Use a modified Beral pipette as a scoop*

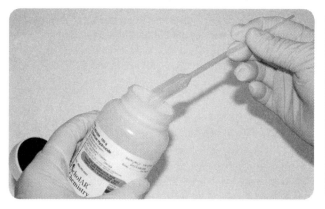

FIGURE 5-4: *Using a stirring rod to transfer a liquid*

USING A BALANCE

An accurate balance is essential for doing good science in a home lab, but even the best balance must be used and maintained properly if it is to provide reliable results. Here are some guidelines for using and maintaining your balance:

- Always read and follow the instructions supplied with your balance. Balances are delicate instruments that can be damaged easily if you use or maintain them improperly.

- Make sure that the balance is level. Many balances read inaccurately if they are even slightly out of level.

- Set up the balance on the "dry" side of your lab or workbench. Although you may sometimes use the balance to weigh liquids, it should be kept well away from the sink and other areas where large quantities of liquids are used and spills are likely.

- Know the maximum capacity of your balance and do not exceed it. Most electronic and mechanical balances are relatively well protected against small overloads, but it's easy to damage or destroy some balances by placing a gross overload on the balance pan.

- Cover the balance when it is not in use. If no dust cover was supplied with the balance, purchase a dust cover separately or use a plastic bag or a cardboard box of the appropriate size.

- Calibrate the balance periodically by using it to weigh a known mass. Many balances are supplied with calibration weights. If yours was not, purchase an inexpensive set of calibration weights that covers the range of the balance. Such sets are available from most laboratory equipment vendors.

- Never place the sample directly on the balance pan. Use a weighing paper for small samples or a weighing boat for larger or bulky samples. Although you can buy weighing papers, a 10 centimeter square of ordinary waxed paper works just as well. A plastic cup or similar container works well as a weighing boat, as shown in Figure 5-5.

- If you have an electronic balance, learn to use the tare function (rhymes with air), which zeros out the display when the balance pan already has some mass on it. For example, placing a weighing boat on the balance pan may cause the balance to read 8.47 g. Pressing the tare button (sometimes labeled 0.00, zero, Z, or T) resets the display to read 0.00 g, which means that the subsequent reading reflects only the mass of the sample that you add to the weighing boat.

MEASURING LIQUIDS BY MASS

It's often convenient to measure liquids by mass rather than by volume. For example, if you need 0.1 mole of hydrochloric acid for a reaction, you could calculate the volume of concentrated (37%, 12 M) hydrochloric acid solution that contains 0.1 mole of the acid and measure that volume of the concentrated acid using a graduated cylinder. Alternatively, because you know that one gram of 37% hydrochloric acid contains 0.37 gram of the acid and that the gram-molecular mass of hydrochloric acid is about 36.5 g/mol, it's easy to calculate the amount you need by mass. You need 0.1 mole of acid, or 3.65 g. The concentrated acid contains 0.37 g of acid per gram of solution, so you need (3.65 g)/(0.37 g/mL) = 9.85 g of concentrated acid solution.

To obtain that mass of concentrated hydrochloric acid solution, place a small beaker or similar container on the balance pan, tare the balance to read 0.00 g with the beaker in place, and use a pipette to transfer 9.85 g of concentrated hydrochloric acid solution to the beaker.

FIGURE 5-5:
Using a weighing boat to measure the mass of a bulky solid sample

MEASURING LIQUIDS BY VOLUME

Liquids are usually measured volumetrically, using glassware that is graduated to indicate the volume that it contains. In general, volumetric accuracy varies inversely with the capacity of the glassware. For example, a 100 mL graduated cylinder may be accurate to ± 1.0 mL, and a 10 mL graduated cylinder may be accurate to ± 0.1 mL. Similarly, volumetric glassware with a small bore is more accurate than glassware with a larger bore. For example, a beaker or flask may be accurate to ± 5%, a graduated cylinder to ± 1%, and a pipette to ± 0.1%. (Remember that these percentages refer to full capacity, so if you use a 10 mL graduated cylinder to measure 1.0 mL of liquid, for example, your measurement is accurate to 0.1 mL, or 10%, rather than the 1% full-scale accuracy.)

Unfortunately, the surface of a liquid in a glass container is not flat, because the glass attracts or repels the liquid. This attraction or repulsion causes the surface of the liquid to assume a curved shape called a *meniscus*. Water and aqueous solutions are attracted to glass and therefore exhibit a concave meniscus, with the center of the column of fluid lower than the edges. Figure 5-10 shows an example of a concave meniscus. (Mercury and a few other fluids are repelled by glass, and therefore form a convex meniscus, with the center of the column of fluid higher than the edges, but convex menisci are seldom seen in a home lab.)

For accurate volumetric measurements, you must take the meniscus into account. Take the measurement with the meniscus at eye level, recording the reading at the center (bottom) of the meniscus. One advantage of plasticware for volumetric measurements is that it neither attracts nor repels water or aqueous solutions, so there is no meniscus to complicate readings.

USING A VOLUMETRIC FLASK

Using a volumetric flask allows you to make up fixed quantities of solutions with very high accuracy. For example, a 100 mL volumetric flask allows you to make up exactly (within stated error) 100 mL of a solution, and a 500 mL volumetric flask allows you to make up exactly 500 mL of a solution. To make up solutions with a volumetric flask, take the following steps:

1. Fill the flask half to three-quarters full with distilled or deionized water.
2. Carefully weigh or measure the solid or liquid solute and transfer it to the volumetric flask using a funnel. Make sure to do a *quantitative transfer*, which is a fancy way of saying you should ensure that all of the chemical is transferred to the flask. For example, tap the weighing paper against the funnel to make sure none of the chemical adheres to the weighing paper, and rinse the inside of the funnel two or three times with several mL of distilled water from a wash bottle.
3. Using distilled water from a wash bottle, fill the flask until the liquid level reaches the neck of the flask, within a few centimeters of the index line on the neck.
4. Stopper the flask and swirl or invert it until the chemical dissolves completely.
5. When the solute appears to be fully dissolved, invert the flask several times to ensure that the solution is reasonably homogeneous. Allow a moment or two for the solution adhering to the upper part of the flask to drain into the body of the flask.
6. Rinse the stopper, directing the rinse water into the flask to make sure that any chemical adhering to the stopper is not lost.
7. Carefully fill the flask until the bottom of the meniscus just touches the index line. Use an eyedropper or Beral pipette to add the last mL or so of liquid.
8. Stopper the flask and again invert it repeatedly until the solution is thoroughly mixed. Allow the solution to drain from the neck, and verify that the bottom of the meniscus is just touching the index line.
9. Transfer the solution to a labeled storage bottle or other container.

Dissolving some chemicals such as concentrated acids and strong bases is extremely exothermic (heat-producing), sometimes producing enough heat to cause a volumetric flask made of ordinary glass to crack. Although most volumetric

> **DR. MARY CHERVENAK COMMENTS:**
>
> I was curious whether I could find another fluid, other than mercury, that formed a convex meniscus. Some molten metals and ink solutions form convex meniscuses/menisci.

FIGURE 5-6:

Make up highly exothermic solutions in a separate container

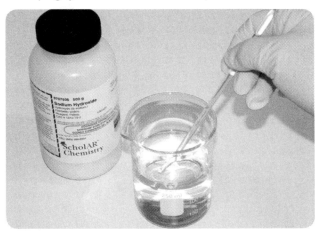

FIGURE 5-7:

Lubricate the ground glass stopper

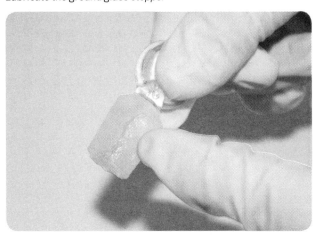

flasks are made of Pyrex or similar glass, there's no reason to risk your volumetric flask. For an accurate measurement, you need to allow the solution to cool to room temperature anyway, so you might as well make up the solution in a separate container, as shown in Figure 5-6, and do a quantitative transfer to the volumetric flask once the solution has cooled.

To do so, fill the volumetric flask with distilled water to about 75% of its capacity (depending on the solubility of the chemical) and transfer that water to a beaker. Dissolve the chemical in the beaker, allow the solution to cool to room temperature, and use a funnel to pour the solution into the volumetric flask. Rinse the beaker several times with a few mL of distilled water, and transfer the rinse water to the flask, rinsing the funnel as you do so. Add distilled water to the volumetric flask until the bottom of the meniscus matches the index line on the flask.

Inexpensive volumetric flasks (and some professional-grade models) come with a plastic snap-on cap. Most professional-grade volumetric flasks come with a ground-glass stopper. If your volumetric flask has a ground-glass stopper, be very careful using it. If you don't lubricate the stopper properly, it may weld itself to the flask, rendering the flask useless. This is particularly likely to happen when you are making up concentrated solutions of sodium hydroxide, potassium hydroxide, or other strong bases that can etch glass, but it can happen with any solution. To avoid problems, always apply a small amount of silicone-based lubricant to the ground-glass part of the stopper and spread it evenly, as shown in Figure 5-7. Apply just enough lubricant to cause the ground glass to clear, and no more. When you finish using the volumetric flask, wash thoroughly and set it aside to dry. Store the stopper separately.

USING A PIPETTE

A Mohr pipette, which is also called a serological pipette or a graduated pipette, has graduations that you can use to measure and dispense small quantities of liquids accurately. For example, a 1 mL Mohr pipette has 0.01 mL graduations and can be interpolated to 0.001 mL, and a 10 mL Mohr pipette has 0.1 mL graduations and can be interpolated to 0.01 mL. Pipettes are indispensable for dealing accurately with small quantities of liquids.

To transfer an accurately measured quantity of fluid with a Mohr pipette, take the following steps:

1. Obtain a suitable amount of the liquid to be measured in a beaker. Pipetting directly from the reagent bottle risks contaminating the entire contents of the bottle.
2. Draw a small amount of the liquid into the pipette and use it to rinse the inside of the pipette thoroughly. Discard the rinse liquid in the appropriate container.
3. Use a pipette pump or pipette bulb, as shown in Figure 5-8, to draw liquid from the container until the meniscus is well above the zero index graduation on the pipette.
4. Remove the pipette pump or pipette bulb, and quickly place your index finger over the top end of the pipette.
5. Wipe the outside of the pipette with a paper towel or rinse it with distilled water into the sink or a waste container. Make sure that the partial drop suspended from the tip of the pipette is removed.
6. Reduce the pressure of your index finger enough to allow the liquid to drain slowly into the beaker or a waste container, watching the level closely until it falls to the 0.00 index mark. Make sure to read the meniscus at eye level.
7. By altering the pressure of your fingertip, transfer the desired volume of the liquid to the reaction vessel and discard the excess.

FIGURE 5-8:
*Use a pipette pump or pipette bulb
to transfer liquids safely and accurately.*

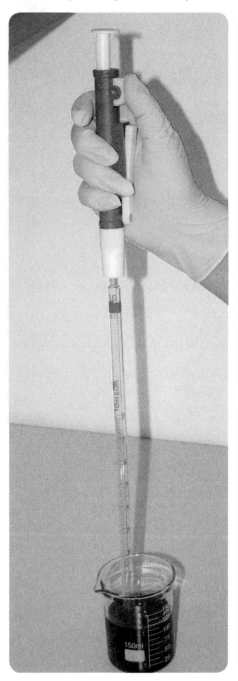

CALIBRATING A VOLUMETRIC FLASK WITH A BALANCE AND A THERMOMETER

Although small volumes of liquids can be measured accurately with a pipette or similar instrument, it's much more difficult to measure larger volumes volumetrically. For example, a 100 mL volumetric flask may be accurate only to 0.5 mL, plus or minus. But if the mass of that volumetric flask when filled with water is less than the maximum capacity of your balance, you can calibrate that flask by mass, using the known density of distilled water at various temperatures, listed in Table 5-1. To do so, place the empty flask (less stopper) on the balance pan and tare the balance. Fill the flask with distilled water until the water level is a few millimeters under the index line on the flask and replace the flask on the balance pan. Add water dropwise to the flask until the bottom of the meniscus just touches the index line, and then note the mass of the water.

For example, if you're calibrating a 100 mL volumetric flask at 20°C, the mass of the water should be 99.82 g. The actual mass is likely to be somewhat different, particularly if you're using a student-grade volumetric flask. You can calculate the actual capacity of the flask by dividing the actual mass of the water by the density of water at the ambient temperature. For example, if the actual mass of the water is 100.08 g, dividing that mass by 0.9982071 tells you that the actual capacity of the flask is about 100.26 mL. Record that actual capacity and use it rather than 100.00 mL for calculations when you use the flask to make up solutions of known concentrations.

MOHR PIPETTES VERSUS VOLUMETRIC PIPETTES

A Mohr pipette is finely graduated from zero to its maximum capacity, and so can be used to measure and dispense any arbitrary amount of liquid up to its maximum capacity. A volumetric pipette, which is instantly recognizable by the bulge in its center, is designed to contain a specific amount of liquid when it is filled to the index line, like a volumetric flask but more accurate. The mechanics of using either of these types of pipettes are the same, but volumetric pipettes are seldom needed in home chemistry labs.

Remember that most Mohr pipettes are calibrated To Deliver (TD or EX) and most volumetric pipettes are calibrated To Contain (TC or IN). The designation determines how you treat the last drop in the pipette. With a TD pipette, the last drop is not part of the measured volume, and so remains in the pipette. With a TC pipette, also called a blow-out pipette, for obvious reasons, that last drop is part of the measured volume and so must be transferred to the receiving vessel.

TABLE 5-1:

Densities of distilled water at various temperatures

Temperature (°C)	Density (g/mL)
15	0.9991026
16	0.9989460
17	0.9987779
18	0.9985986
19	0.9984082
20	0.9982071
21	0.9979955
22	0.9977735
23	0.9975415
24	0.9972995
25	0.9970479

CLEANING A PIPETTE

The very small bore makes it difficult to clean a pipette thoroughly. Formal laboratories either use disposable one-use pipettes or have a pipette cleaner that repeatedly draws cleaning solution through pipettes to clean them. You can accomplish the same thing manually by using your pipette pump or pipette bulb. To clean a pipette, take the following steps:

1. The first rule is to clean a pipette immediately after you use it, or at least rinse it thoroughly. If you allow the pipette to dry, the solute may crystallize out inside the bore of the pipette, making it difficult or impossible to clean properly.
2. To begin, rinse the pipette inside and out with a thin stream of warm tap water.
3. Put a few drops of dishwashing liquid in a beaker and fill the beaker with warm tap water.
4. Use your pipette pump or pipette bulb to draw the sudsy water through the pipette repeatedly, making sure to clean the outside surface as well.
5. Rinse the pipette thoroughly with tap water, inside and out.

DON'T SUCK

The cardinal rule for using a pipette is to never, *ever* pipette by mouth. If you do, the day will come when you suck something noxious into your mouth. Guaranteed. Just don't do it.

6. Fill a beaker with distilled water, and use your pipette pump or pipette bulb to draw the distilled water repeatedly through the pipette.
7. Use the pipette pump or pipette bulb to blow out as much of the final distilled water rinse as possible, and then set the pipette vertically in a rack to drain and dry.

CALIBRATING A DISPOSABLE PLASTIC PIPETTE

Disposable one-piece soft plastic pipettes, usually called Beral pipettes, have many uses around the lab. I buy them in bags of 100 or 500 at a time. At a few cents each, they're cheap enough to treat as one-use items. But they're also accurate enough to substitute for a Mohr pipette when you need to measure small quantities of liquids.

Although some Beral pipettes are roughly calibrated, typically to 0.5 mL, that's insufficient for critical work. Fortunately, Beral pipettes are very consistent from one to another, and all of them deliver consistent and repeatable drop sizes. For example, if you know that a particular type of Beral pipette consistently delivers 40 drops per mL, you also know that one drop from that pipette is 0.025 mL. If you need, say, 0.50 mL of a solution, you can use that Beral pipette to transfer 20 drops.

Different models of disposable plastic pipettes are rated to deliver anything from 20 drops/mL to 50 drops/mL, but those values are nominal. Just as you can accurately calibrate a volumetric flask with a balance, you can do the same to calibrate Beral pipettes. To do so, place a small beaker or similar container on the balance pan and tare the balance to read 0.00 g. Fill the Beral pipette completely with distilled water, and then count the drops as you empty all but the last few drops from the Beral pipette into the beaker. (Make sure to deliver drops consistently. Hold the pipette vertically. Don't squirt drops into the beaker, but allow each drop to form slowly and fall from gravity alone.)

For example, one Beral pipette that I tested was rated to deliver 23 drops/mL. I dispensed 75 drops of water and then set the pipette aside. The balance told me the mass of those 75 drops of distilled water was 3.26 g. At room temperature, the density of distilled water can be taken as 1.00 g/mL for the small quantity I used, so 3.26 g equals 3.26 mL. Dividing 75 drops by 3.26 mL gives about 23.01 drops/mL, which is certainly close enough.

FIGURE 5-9: *Calibrating a disposable pipette*

END POINT VERSUS EQUIVALENCE POINT

The **end point** of a titration is the point at which the titration is complete, typically when an added indicator such as phenolphthalein changes color. (The end point may also be determined by other methods, such as the conductivity of the analyte reaching a minimum or the pH of the analyte reaching 7.00.) The **equivalence point** is closely related to but not necessarily identical with the end point. The equivalence point is the point at which the number of moles (or equivalents) of titrant exactly equals the number of moles (or equivalents) of analyte.

Ideally, the end point should exactly equal the equivalence point, but in the real world they are usually slightly different. For example, you may titrate a hydrochloric acid analyte with sodium hydroxide titrant, using phenolphthalein as an indicator. Phenolphthalein is colorless in acid solutions, and pink in base solutions, but no color change occurs until the pH of the solution reaches about 8.2, well into the basic range.

I wasn't so lucky with another pipette, which was rated to deliver 43 drops/mL but in fact delivered closer to 44 drops/mL, at least using my dropping technique. Still, once the actual delivery rate is known, the nominal delivery rate is immaterial. Because I've calibrated them, I can use those pipettes to deliver accurately known small amounts of liquid reliably.

The number of drops per mL depends on the viscosity of the liquid. For example, if you calibrate the number of drops per mL using distilled water, that number is not valid for a more viscous liquid such as olive oil or concentrated sulfuric acid, nor for a less viscous liquid such as acetone or methanol. Fortunately, the viscosity of the dilute aqueous solutions commonly used in labs does not significantly differ from the viscosity of pure water.

USING A BURETTE

A burette is used to transfer solutions while measuring the volume transferred with high accuracy. Although burettes are sometimes used to measure and transfer nominal fixed volumes (called *aliquots*) of a solution, for example, by transferring 5.0 mL of a solution to each of several beakers that are to be used in a subsequent procedure, that function is ordinarily better done using a volumetric or Mohr pipette.

By far the most common use of a burette is for *titration*, a procedure in which a solution (called the *titrant*) whose concentration is known very accurately is dispensed by a burette and reacted with a known volume of another solution of unknown concentration (called the *analyte*). By measuring the amount of

titrant needed to neutralize the analyte, you can determine the concentration of the analyte accurately.

Here is the proper way to use a burette:

1. Rinse the inside of a clean burette thoroughly with the solution it will contain. Allow the solution to run out through the stopcock. Drain the burette completely. Repeat the rinse at least once.
2. Make sure that the outside of the burette is clean and dry, and then mount it securely to a laboratory stand using a burette clamp of the proper size.
3. Fill the burette to above the zero mark, using a graduated cylinder, small beaker, or other container. Use a funnel if necessary to prevent spillage.
4. Run some solution through the stopcock to fill the burette tip completely, making sure that there are no air bubbles and that the level of the solution falls to or below the zero mark. (It's not necessary to hit the 0.00 mark exactly.)
5. Record the starting volume. When you complete the titration, you'll subtract this starting volume from the final volume to determine the amount of solution you've added. Read the value by placing your eye at the level of the solution and reading the value at the bottom of the meniscus, as shown in Figure 5-10 (which indicates a reading of about 34.37 mL).

When you complete the titration, empty any remaining solution into a waste container (using the proper disposal method for

FIGURE 5-10: *Reading the meniscus on a burette*

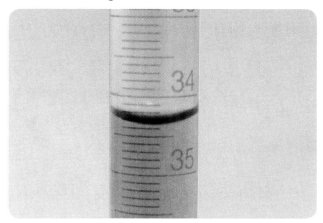

waste solutions) and remove the burette from the clamp. Rinse the burette thoroughly inside and out with tap water, making sure to run the rinse water through the tip. If the burette uses a pinchcock, remove the pinchcock and rubber tubing. If it uses a ground glass stopcock, remove the stopcock. If it uses a Teflon stopcock, release the tension on the stopcock and/or remove the stopcock, depending on the design of the burette. Rinse the burette, tip, and stopcock assemblies thoroughly with tap water and then with distilled water and place them aside to dry thoroughly.

PERFORMING TITRATIONS WITHOUT A BURETTE

If you don't have a burette, you can use one of the methods described in the following sections to perform titrations with differing levels of convenience and accuracy.

TITRATE USING A GRADUATED CYLINDER

If your only volumetric glassware is a graduated cylinder, you can use it to perform quick-and-dirty titrations with fair accuracy. To titrate using your graduated cylinder, take the following steps:

1. If possible, calculate the approximate amount of titrant you expect to be required to complete the titration.
2. Fill the graduated cylinder with titrant until the titrant reaches the top graduation mark as closely as possible. Use a Beral pipette to add the final few drops to bring the level of the meniscus as close as possible to the line.
3. Use the Beral pipette to withdraw about 2 mL of titrant, and set the pipette aside. Invert the pipette to make sure that none of the titrant is lost.
4. Slowly trickle titrant from the graduated cylinder into the titration vessel with swirling until you have nearly reached the equivalence point. You're approaching the equivalence point when the indicator changes color where the titrant is being added, but that color change disappears as you swirl the titration vessel.

CONSERVING TITRANT

If you are certain that the titration will require much less than the full capacity of the burette, you can conserve your titrating solution by partially filling the burette. Make sure that you fill it far enough to allow the tip to be filled and air bubbles to be eliminated. For example, if you are using a 50 mL burette and have calculated that the titration should require about 20 mL of titrant, fill the burette to about 30 mL and run a few mL through the tip to eliminate air bubbles. Leave yourself some slack. If you run out of titrant before you've neutralized the analyte, you'll either have to repeat the titration from scratch or add more titrant to the burette, which reduces the accuracy of your titration.

ADDING PARTIAL DROPS

As you near the end point of a titration, you can add partial drops by manipulating the stopcock carefully. As you release each partial drop, use a wash bottle filled with distilled water to flush the partial drop from the tip of the burette directly into the solution being titrated.

PRELIMINARY TITRATIONS

If you have no idea where the end point of a titration will occur, you can save a lot of time by doing a fast trial run instead of slowly adding titrant from the beginning of the titration. To do so, run titrant quickly into the titration vessel until the indicator begins to change color. If you're lucky, the titration will not have reached its end point, and you can begin adding titrant more slowly until you reach the end point. If you overshoot the mark while adding titrant quickly, note the reading on the burette and use something less than that as the starting point for the real titration. For example, if adding 22.0 mL of titrant was just a bit too much, when you start the real titration you can quickly run 20.0 mL of titrant into the titrating vessel and then add titrant drop-by-drop until you reach the end point.

5. When you are near the equivalence point, stop adding titrant from the graduated cylinder, and begin adding titrant drop-by-drop from the Beral pipette. Continue adding titrant dropwise from the Beral pipette until you reach the equivalence point, which is indicated by a color change that persists for at least 10 or 15 seconds as you continue to swirl the titration vessel.

TESTING BURETTE ACCURACY

Some burettes, particularly inexpensive "student" models, are not linear, because the interior walls are not straight and parallel. That is, the accuracy may differ according to which section of the burette you use. For example, the burette may give accurate readings near the top of the tube, inaccurate readings in the middle of the tube, and accurate readings near the bottom of the tube.

Pure water weighs 1.00 g/mL, so you can use your balance to calibrate your burette to determine its accuracy over its full range. To do so, fill the burette as closely as possible to the 0.00 mL (top) graduation line with distilled water, using an eyedropper or Beral pipette to add the final drops necessary. Weigh a beaker that has at least the capacity of your burette, and record the mass to 0.01 g. Transfer 5.00 mL of water to the beaker, reweigh it, and record the mass. Transfer 5.0 mL of water and reweigh the beaker repeatedly until you have emptied the burette. (You can speed this process up by leaving the beaker on the balance pan and setting up the burette above it.) When you finish, subtract each of the mass readings from the next mass reading to determine how much mass corresponded to each 5.00 mL transfer. Ideally, each 5.00 mL transfer should increase the mass by 5.00 g, but if the burette isn't linear, the mass increase may not always correspond to the volume added.

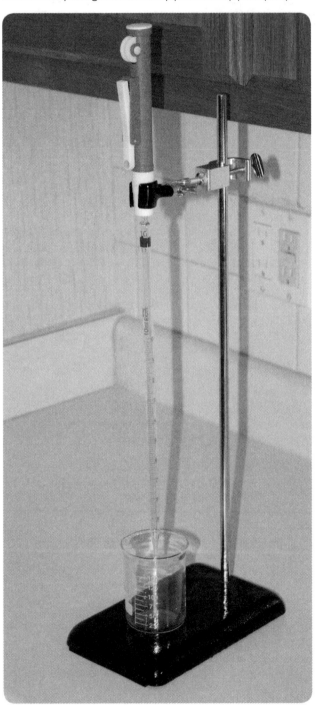

6. Transfer the remaining titrant from the Beral pipette back into the graduated cylinder. Read the final volume on the graduated cylinder and determine the volume used by subtraction.

TITRATE USING A MOHR PIPETTE

You can use a Mohr pipette with a pipette pump or pipette bulb, as shown in Figure 5-11, to perform titrations with high accuracy. There are a few tricks to getting good results when titrating with a pipette:

- Use a high-capacity Mohr pipette. Most Mohr pipettes contain from 0.2 mL to 2.0 mL, but 10.0 mL models are readily available. (One inexpensive source is *http://www.indigo.com*.) I use a 10.0 mL model with 0.1 mL graduations that allow interpolation to 0.02 mL or better.

- Estimate as closely as possible the volume of titrant that will be needed, and make sure it falls well within the capacity of the pipette.

- Practice using the pipette pump or pipette bulb until you get the hang of releasing one drop at a time.

- For finer dispensing control, you can use a short length of flexible tubing, a pinchcock clamp, and a short piece of glass tubing drawn to a point to convert the pipette to an actual burette. If you use this method, fill the pipette/burette by releasing the pinchcock clamp and using a pipette bulb or pipette pump to draw titrant up into the pipette/burette through the tip. When the pipette/burette is full, clamp the tubing and remove the pipette bulb or pump.

Perform titrations using this setup just as you would if you were using an actual burette.Titration by mass difference takes advantage of the fact that a laboratory balance can measure masses very accurately. If your balance has at least centigram (0.01 g) resolution and you have a volumetric flask, your results with this method will be at least as accurate as those you obtain with a burette.

Determining the mass of an accurately known volume of titrant allows us to calculate the mass per unit volume (or density) of the titrant with high accuracy. If we subsequently run a titration starting with a known initial mass of titrant, we can determine the mass of the titrant remaining when the titration is complete. By subtracting the final mass of titrant from the initial mass, we can determine what mass of titrant was required to complete the titration. Knowing the density of the titrant, we can convert that mass to volume by dividing the mass of titrant required by the density of the titrant. Table 5-2 shows example values for a titration I completed using the mass difference method.

To titrate by mass difference, take the following steps, modifying them as necessary to account for the capacity of your balance and the volumetric glassware you have available.

1. Weigh an empty 100 mL graduated cylinder or (better) a 100 mL volumetric flask. Record the mass as accurately as possible.
2. Carefully fill the graduated cylinder or volumetric flask to the 100.0 mL line. Reweigh the container and record the mass as accurately as possible.
3. Subtract the mass of the container when empty from the mass of the container when filled and record that value as the mass of 100.0 mL of the titrant.
4. Transfer sufficient titrant to complete the titration to a small beaker or similar container. Add a Beral pipette to the beaker, determine the combined mass of the beaker, titrant, and Beral pipette as shown in Figure 5-12, and record the combined mass as accurately as possible.

ABSOLUTE ERROR VERSUS RELATIVE ERROR

Typical graduated cylinders are accurate to about 1% at full scale, which means that a 100 mL graduated cylinder provides results accurate to about 1.0 mL and a 10 mL graduated cylinder provides results accurate to about 0.1 mL. These errors are absolute, so if your titration uses only 10.0 mL from the 100 mL graduated cylinder, for example, the absolute error is still ± 1.0 mL, which translates to a 10% error of the smaller volume.

You can minimize this source of error by using a graduated cylinder with a maximum capacity that is near the volume of titrant you expect to require. For example, if you expect to need about 8 mL of titrant, start with a 10 mL graduated cylinder filled to capacity. A typical 10 mL graduated cylinder is accurate to within 0.1 mL, so your results are an order of magnitude more accurate if you use the smaller graduated cylinder.

5. Complete the titration and return the Beral pipette to the beaker.
6. Again determine the combined mass of the beaker, remaining titrant, and Beral pipette, and record that combined mass as accurately as possible.
7. Subtract the final combined mass from the initial combined mass to determine how many grams of titrant were required to complete the titration.
8. Convert grams of titrant to mL of titrant by multiplying the number of grams of titrant required to complete the titration by 100 and then dividing that result by mass of 100.0 mL of titrant to determine the number of mL of titrant that were needed to complete the titration.

TABLE 5-2: *Example values for a titration by mass difference*

Item	Value
A. Mass of empty 100 mL volumetric flask	69.94 g
B. Mass of filled 100 mL volumetric flask	189.47 g
C. Mass of 100.0 mL of titrant (B – A)	119.53 g
D. Mass of beaker, titrant, and Beral pipette (initial)	135.11 g
E. Mass of beaker, titrant, and Beral pipette (final)	106.87 g
F. Mass of titrant used (D – E)	28.24 g
G. Volume of titrant used (E·100/C)	23.63 mL

FIGURE 5-12: *Determining the combined mass of beaker, titrant, and Beral pipette*

FILTRATION

Filtration is a method used to separate a solid, called the *filtrand* or *residue*, from a liquid, called the *filtrate* or *supernatant fluid*. To filter a solid-liquid mixture, you pour it into a funnel that contains a barrier—typically a piece of filter paper or a similar membrane—that passes the liquid and stops the solid. The liquid is caught by a beaker, flask, or other receiving vessel and the solid remains on the filter. Depending on the process, you may retain the solid and discard the liquid or vice versa. In some processes, the solid and liquid are both retained for further processing.

To use gravity filtration to separate a solid-liquid mixture, take the following steps:

1. Set up your filtering funnel, which can be any standard funnel. You can rest the funnel inside a support ring on a ring stand with a beaker or flask underneath it as the receiving vessel, or simply rest the funnel on top of an Erlenmeyer flask.

2. If you are doing a quantitative procedure (where the mass of the filtrand matters), weigh a piece of filter paper and record its mass to 0.01 g or better, depending on the resolution of your balance.

3. Fold the circular filter paper in half, creasing it at the fold.

4. Fold the filter paper in half again to form pie-shaped quarters.

5. Open up the folded filter paper, and lay it flat on the work surface.

6. Fold each quarter in half to form eighths, with all of the folds pointing up. Make sure all folds come to the same point at the center of the paper.

7. Turn over the filter paper, so that the folds you've just made are pointed down, and fold each eighth into sixteenths, with the folds pointing up. Again, make sure that all folds come to the same point at the center of the paper.

8. Expand the fan-folded filter paper into a cone, as shown in Figure 5-13, and place it in the funnel.

9. Place the receiving vessel in position under the funnel. If you're using a beaker, put the stem of the funnel against the side of the beaker to prevent splashing. If you're supporting the funnel on top of an Erlenmeyer flask or similar small-mouth container and the funnel has a smooth exterior, place a folded piece of paper or similar small object between the funnel and the mouth of the container, as shown in Figure 5-14. Otherwise, the funnel may form a seal with the mouth of the container, preventing the liquid from running through the funnel.

10. Swirl or stir the solid-liquid mixture that is to be filtered to make sure that the solid is suspended in the liquid, and pour the mixture into the funnel, making sure that the level of the liquid does not rise higher than the top of the filter paper cone. If necessary, use a stirring rod to transfer the mixture without splashing or loss.

11. Wait for the liquid to drain into the receiving vessel. If necessary, continue pouring more of the original mixture into the funnel until all of it has been transferred.

12. Using distilled water from a wash bottle, rinse the original container and the stirring rod several times with a few mL of water each time and pour the rinse water into

GRAVITY FILTRATION VERSUS VACUUM FILTRATION

In this book, we'll use only *gravity filtration*, in which the force of gravity alone is used to separate the liquid from the solid. Academic and industrial labs frequently use *vacuum filtration*, in which a vacuum is used to draw the liquid through the filter. Vacuum filtration has several advantages, not least of which is that it speeds up filtration significantly. Gravity filtration may require anything from about a minute to several hours to finish, depending on the amount to be filtered, the permeability of the filter, the size range of the solid particles, and other factors. Vacuum filtration typically takes only a few seconds, as the liquid is quickly drawn through the filter. Another advantage of vacuum filtration is that less-permeable filters can be used, which catch finer particles but slow down filtration dramatically. Finally, vacuum filtration yields a drier filtrand, because it draws more of the liquid out of the solid. In fact, vacuum filtration is routinely continued long after all of the liquid has been drawn through the funnel, because drawing additional air through the solid sample evaporates additional liquid, drying the sample completely.

If you have a spare $50 or so, you can add vacuum filtration capability to your home laboratory. That buys you a filter flask—essentially a heavily constructed Erlenmeyer flask with a sidearm to which you attach the vacuum source—a Büchner funnel and stopper to fit the filter flask, and a manually operated hand vacuum pump. I have such a setup in my home lab, but for most people that $50 could be better spent elsewhere.

the funnel. Make sure that all of the solid and liquid is transferred into the funnel.

13. When all of the liquid has drained through the filter paper, rinse the solid filtrand several times with a few mL of water to make sure that all of the soluble material has been rinsed into the receiving vessel.

14. If the filtrand is waste material, remove the funnel and dispose of the filter paper and filtrand properly.

15. If the filtrand is your product, carefully remove the filter paper and filtrand from the funnel and transfer them to a Petri dish or similar container for drying. Once the filter paper and filtrand are thoroughly dry, you can determine their combined mass and subtract the mass of the filter paper to determine the mass of the filtrand.

16. Wash and dry the funnel and any other vessels you used and put them away.

FIGURE 5-13:
Making a filter paper cone

FIGURE 5-14:
Proper filtration procedure

SEPARATIONS

In a laboratory sense, *separation* means physically dividing two or more immiscible liquid layers, usually an aqueous layer and an organic layer. A separation most commonly follows an *extraction*, in which a substance dissolved or suspended in one solvent is extracted from that solvent by agitating the solution with a quantity of a second, immiscible solvent. If the solute is more soluble in the second solvent than in the first, it is preferentially extracted into the second solvent. Because the two liquids are immiscible, they separate upon standing into two physical layers, which can subsequently be divided and isolated from each other.

For example, in one of the laboratory sessions later in this book, we produce an aqueous solution that contains elemental iodine. Iodine is relatively insoluble in water, but is very soluble in many organic solvents. Adding a quantity of an organic solvent to the reaction vessel and agitating the mixture allows the iodine to migrate from the aqueous solution to the organic solvent. Upon standing, the solvent layers separate, with the less dense organic layer—which now contains nearly all of the iodine—floating on top of the aqueous layer, which still contains nearly all of the other reaction products. By drawing off and later evaporating the organic layer, we can isolate nearly all of the iodine in relatively pure form.

In formal laboratories, most separations are done with a separatory flask, usually called a sep flask. If you don't own a sep flask, you can still do a good separation using inexpensive standard glassware and a disposable plastic pipette. In the following example, assume that you begin with two layers, each of about 100 mL, in a 250 mL flask or beaker. If the actual quantities differ significantly, you can modify this procedure accordingly.

1. First, make absolutely certain that you know which layer contains your product and which layer is waste. Many novice chemists have embarrassed themselves by pouring the product layer down the drain and saving the waste layer. In this example, we'll assume that the top layer is the product layer.

2. After allowing the two layers to separate completely, carefully pour the top layer into your 100 mL graduated cylinder. Make sure to get all of the top layer, and as little as possible of the bottom layer—at most, a few mL.

3. Allow the layers to separate completely again, and then pour as much as possible of the top layer into a receiving beaker or flask, making sure that none of the bottom layer is transferred. At this point, a few mL at most should remain in the 100 mL graduated cylinder.

4. Pour the entire remaining contents of the 100 mL graduated cylinder into your 10 mL graduated cylinder and allow the layers to separate.

5. Use the disposable plastic pipette carefully to draw up as much as possible of the top layer, as shown in Figure 5-15, and transfer that liquid to the receiving beaker or flask.

AVOID SUPERHEATING

As strange as it sounds, it's possible for a liquid to be heated above its boiling point without boiling. A liquid in that unstable state is *superheated*, and is extremely dangerous. A superheated liquid can begin boiling spontaneously and explosively, ejecting large amounts of boiling liquid from the container. Microwave ovens are notorious for superheating liquids, but it is also possible to superheat a liquid with an alcohol or gas burner, particularly if the heat is focused on a small part of the bottom of the container.

To avoid superheating, add a *boiling chip* to the liquid before you begin heating it, particularly if you are heating it in a microwave oven. A boiling chip is simply a pebble of limestone or a similar porous material that prevents superheating from occurring by providing a locus for boiling to begin. If you are boiling a liquid in a beaker or flask over an alcohol or gas burner, use a ceramic wire gauze to spread the heat and place a stirring rod in the beaker or flask so that the tip of the stirring rod is in contact with the bottom of the container at the point where the heat is most intense, as shown in Figure 5-16. Like a boiling chip, the stirring rod provides a locus for boiling to begin, preventing superheating from occurring. Avoid superheating when you heat a test tube by directing the flame toward the middle of the solution rather than the bottom, as shown in Figure 5-17, and keep the tube moving constantly to distribute the heat.

FIGURE 5-15:
Using a Beral pipette to separate layers

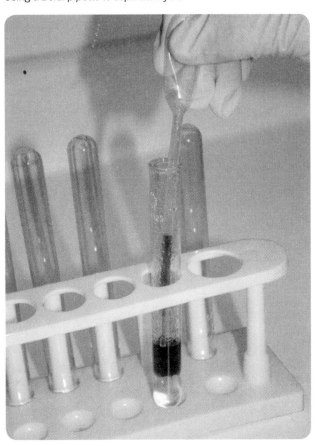

USING HEAT SOURCES

Heat sources are essential in any chemistry laboratory, but you risk being burned any time you use a heat source. Heat sources that use open flame add the risk of fire. Use the following guidelines for safe use of heat sources.

- To minimize the danger of fire, use a hotplate rather than an alcohol lamp or gas burner whenever possible.

- When you use flame, always have a fire extinguisher handy. Know how to use it.

- Before you light a burner, make absolutely certain that there are no flammable substances nearby, including burner fuel that is not in the burner.

- Never use open flame to heat a container that contains a flammable substance.

- When you finish using a heat source, turn it off or extinguish it immediately.

- Remember that hot glass looks exactly like cold glass. Always use tongs or insulated gloves to handle glassware unless you are certain that it has cooled.

FIGURE 5-16: *Using a stirring rod to avoid superheating*

FIGURE 5-17: *Heating a test tube properly*

- Know how to treat minor burns, and always have a first-aid kit immediately available.

- Know how to identify burns that require professional treatment. (Any burn that is more severe than reddened skin and very minor blistering should be examined by a physician.)

USING AN ALCOHOL LAMP

For more than 100 years, alcohol lamps have been the traditional heat source in home chemistry labs. Alcohol lamps are inexpensive, easy to use, and provide a gentle heat suitable for heating test tubes and other small containers. Follow these guidelines when using an alcohol lamp:

- Use only alcohol in the lamp—either ethanol (ethyl alcohol) or isopropanol (isopropyl alcohol) from the drug store. The 70% solutions can be used, but the 91% or higher solutions provide a cleaner, hotter flame. Never use acetone, gasoline, kerosene, or other flammable solvents.

- Make sure that the alcohol lamp has cooled completely before you refill it.

- After a refill, allow the wick to become completely saturated with alcohol before you light the lamp. Otherwise, you're burning the wick instead of alcohol.

- Adjust the exposed length of the wick to control the size of the flame. Exposing more wick makes the flame larger.

- When the wick becomes severely frayed, trim it off. Always keep a spare wick or two on hand.

- Extinguish the lamp by placing the cap (also called the snuffer) over the wick.

- Store the lamp with the cap in place to prevent evaporation.

- If you need a higher temperature than the lamp provides, use a blowpipe to force more oxygen into the flame, allowing it to burn hotter. Place the tip of the pipe near the hottest part of the flame (near the tip of the flame) and blow gently and steadily through the blowpipe, directing the tip of the flame at the object you want to heat. For most purposes, a length of ordinary glass tubing makes a good blowpipe. If you need an even hotter (but smaller) flame, draw the glass tubing (see "Working with Glass Tubing," later in this chapter) to provide a smaller tip.

USING A GAS BURNER

Gas burners provide a much hotter, more intense flame than alcohol lamps. Because they deliver more heat, gas burners are better for heating larger volumes of liquids (although a hot plate is usually an even better choice). Because the flame is very hot—how hot exactly depends on the type of gas used and how the burner is adjusted—gas burners are the best choice when you need to heat a solid sample to a very high temperature.

Formal laboratories use Bunsen or Tyrell burners connected directly to natural gas taps, which is not practical for a home laboratory. Fortunately, there are good and inexpensive alternatives to having natural gas installed in your home lab. One of the most convenient is a portable butane burner, shown in Figure 5-18.

FIGURE 5-18:
Adjusting the flame on a portable butane gas burner

These burners have internal fuel storage, and are filled from inexpensive disposable butane canisters sold by drugstores and tobacconists. Most run for an hour or more at their highest setting on a full tank of fuel. Because they are not tethered to a natural gas tap, you can move these burners around the lab as needed. Flame spreaders and other accessories are often available for such burners, either included with the burner or as options.

The only drawback to these portable butane burners is their relatively high price—$35 to $50 or so. One reasonable substitute is a propane torch from the hardware store. These torches, sold under the Bernz-o-Matic and other tradenames, are inexpensive; use cheap, disposable propane cannisters; and have a wide variety of accessories available, including flame spreaders and wire stands. The flame is adjustable from a tiny point to one larger than you should ever need in a home lab. Although propane contains less heat per unit volume than butane, propane burns at almost exactly the same temperature as butane—just under 2,000°C—so a propane torch can substitute for a butane burner in any application.

Follow these guidelines when using a gas burner:

- Never heat glass directly with the burner flame, particularly the tip of the flame. The gas burner flame is hot enough to damage even Pyrex glass. Use a ceramic-filled wire gauze between the flame and the glassware.

- Do not use a gas burner to heat a test tube or other small container directly. The hot gas flame can rapidly superheat a small part of the solution, which may instantaneously flash to steam, forcefully ejecting very hot liquid from the container. This dangerous phenomenon, called *bumping*, can be avoided by using a less-intense source of heat (or a flame spreader on the gas burner) and by using a boiling chip or stirring rod to prevent superheating. The possibility of bumping also means that it's imperative to keep the mouth of the container pointed away from you and anyone else present.

- Allow the burner to cool completely before you refill the reservoir or replace the gas cylinder.

- Follow the instructions in the manual to adjust the burner or torch to provide the most efficient flame. Some burners allow you to adjust both gas flow and air flow, and it's important to have both properly adjusted.

- Clean the burner regularly, as recommended by the manual.

EVAPORATING AND DRYING
It's frequently necessary to remove water from samples that you have prepared. For example, if your product is an aqueous solution of a salt, you may need to determine the mass of the dry salt to determine your percent yield. Or you may have a precipitate on filter paper that is still damp with solvent. In either case, you need to remove all of the liquid before you can determine an accurate mass for the solid.

The best container for drying liquid samples is a porcelain evaporating dish, although you can substitute a Pyrex Petri dish, Pyrex saucer, or similar container that exposes the surface of the solution over as large an area as possible to speed evaporation. To dry a solid sample on filter paper, place the filter paper in an evaporating dish, Petri dish, or saucer.

Follow these steps to dry a sample:

1. Weigh the empty container or filter paper before you begin, and record its mass. That way, once the sample has been dried, you can determine its mass accurately by weighing the combined mass of the sample in the container or filter paper and then subtracting the mass of the empty container or filter paper. In particular, if the sample is on filter paper, make sure to weigh the filter paper and record its mass before using it, because it's often impossible to remove all of the product from the filter paper.

2. If your product is a dilute solution, you need to remove a great deal of water, ideally as quickly as possible. To do so, transfer the solution to a porcelain evaporating dish and boil the solution gently to evaporate most of the water. The idea is to remove most of the water, but not to boil it to dryness. (Heating a dry solid strongly may cause it to decompose.) If the evaporating dish is not large enough to contain all of the solution at once, continue adding more of the original solution to the evaporating dish as the liquid level falls.

3. When most of the water has been removed (or if you start with a product that is only moist, such as a solid precipitate in filter paper), move the container to a drying oven or place it under a heat lamp, as shown in Figure 5-19, and heat the sample until the last traces of water are gone. If you are using a drying oven, set the temperature to at least 150°C to ensure that all of the water vaporizes. If you do not have a heat lamp, you can use an ordinary gooseneck lamp with an incandescent bulb placed as near as possible to the sample. (Take care not to splash water on the hot bulb, or it will shatter.)

4. Depending on the amount of water remaining, you may need to heat the sample for anywhere from a few minutes to several hours. If the sample crusts over, use the tip of a clean stirring rod to break up the mass as much as possible to make sure that there are no puddles of liquid concealed by the encrustation. Continue heating until all of the water appears to have been driven off.

FIGURE 5-19: *Using a lamp to dry a sample*

5. When the sample appears to be completely dry, allow it to cool, weigh it, and record the mass.
6. Return the sample to the oven or replace it under the heat lamp, and heat it strongly for at least another 15 minutes.
7. Reweigh the sample, record the mass, and compare that mass with the mass you previously recorded. If the new mass is lower, not all of the water was extracted from the sample when you made the previous weighing.
8. Repeat steps 5 and 6 until the mass no longer changes. At that point, you can be sure that the sample is completely dry.

WORKING WITH GLASS TUBING

Glass tubing is used in conjunction with flexible rubber or plastic tubing and holed stoppers to route liquids and gases between containers. Glass tubing is available in various outside and inside diameters, wall thicknesses, types of glass, and lengths. The most important characteristic is the outside diameter, which should be chosen to fit your stoppers and flexible tubing. Tubing with an outside diameter (OD) of 5 mm is most common, but verify the proper OD before you order.

Most glass tubing is made of ordinary flint glass, which softens at the relatively low temperature provided by an alcohol lamp. By heating the tubing until it softens, you can bend it, stretch it, and otherwise manipulate it into the shapes needed to construct distillation setups, gas-generating bottles, and similar apparatus.

For high-temperature use, glass tubing is available made from Pyrex or similar heat-resistant borosilicate glasses. Such high-temperature tubing is seldom needed, if ever, in a home lab, but it costs little more than flint-glass tubing and is often stocked by laboratory supply vendors. If you find that heating a piece of glass tubing with an alcohol lamp doesn't soften it, that glass tubing is almost certainly made of heat-resistant glass. You can still manipulate and shape such tubing, but you'll need the hotter flame of a gas burner to do so.

CUTTING GLASS TUBING

Glass tubing is supplied in standard lengths, typically 6" (~150 mm), 12" (~300 mm), and longer. You'll often need shorter lengths, which are easy enough to cut from the stock tubing. To do so, take the following steps:

1. Use the edge of a steel file to score the glass once, perpendicular to its length, as shown in Figure 5-20. Do not saw away at the glass or attempt to cut deeply into it. Just a slight notch suffices.
2. Place both thumbs near the notch, as shown in Figure 5-21, with the notch on the other side of the tubing from your thumbs, and apply gentle pressure until the tubing snaps in two. Wear eye protection, and use heavy gloves or wrap the tubing in a towel to prevent cuts.
3. The two fresh ends of the tubing are quite sharp, and need to be fire-polished before use. To fire polish the tubing, hold the sharp end in the flame of an alcohol lamp or gas burner, as shown in Figure 5-22, and rotate the tubing slowly until the sharp end is smoothed. Don't overdo the fire-polishing or you'll melt the end of the tubing enough to block it. An alcohol lamp isn't hot enough to fire-polish Pyrex tubing, which requires a gas burner.

BENDING AND DRAWING GLASS TUBING

Bending and drawing glass tubing is easy enough, but requires a bit of practice. (*Drawing* is the process of stretching tubing to reduce its diameter—for example, to make a nozzle for a wash bottle.) To bend glass tubing, perform the following steps:

1. Hold the tubing horizontally in the hottest (blue) part of the flame of an alcohol lamp or gas burner, as shown in Figure 5-23. Heat the tubing at the point you want to make the bend and for a centimeter or so on either side. (If you use a gas burner, a flame spreader is very helpful in distributing the heat evenly.) Rotate the tubing continuously to make sure that it's heated around its full circumference.

FIGURE 5-20:
Use the edge of a file to notch the glass tubing

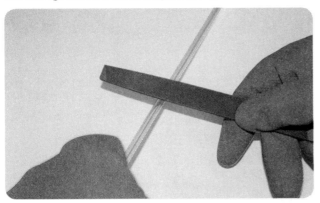

FIGURE 5-21:
Place both thumbs near the notch and snap the glass tubing

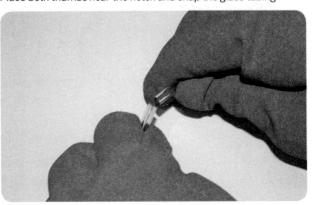

FIGURE 5-22: *Rotate the end of the glass tubing in the flame to smooth the edges*

FIGURE 5-23: *Heat the section of the glass tubing to be bent by rotating it in the flame*

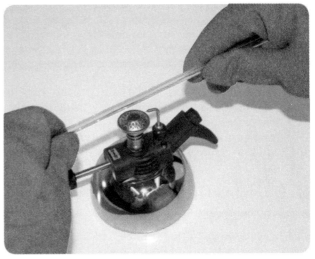

2. As you heat the tubing, apply extremely gentle bending pressure on it continuously. When you feel that the tubing is near the point that it can be bent, stop applying pressure. Continue rotating and heating the tubing for few seconds longer. (If you heat the tubing to the point where it begins to sag under its own weight, you've gone much too far.)

3. Remove the tubing from the flame and allow it to cool for a couple of seconds, keeping it straight.

4. Using one smooth motion, quickly bend the tubing gently to the desired angle. Hold the tubing in position for several more seconds until it hardens and then set it aside on a heat-resistant surface to cool completely.

A good bend is one in which the tubing assumes the desired angle smoothly with no internal crimping or constriction. If you haven't heated the tubing sufficiently, the bend will be uneven. If you've overheated the tubing, the bend may be too sharp and there may be a constriction or blockage at the most extreme part of the bend.

Drawing glass tubing requires the same initial steps as bending it. To draw tubing, heat the tubing until it is pliable and then pull the two ends straight away from each other until the heated section of the tubing is drawn down to the desired size. Cut the tubing in the middle of the drawn section and fire-polish the ends. Figure 5-24 shows a section of tubing that has been drawn to a fine point.

FIGURE 5-24: *A section of glass tubing that has been drawn*

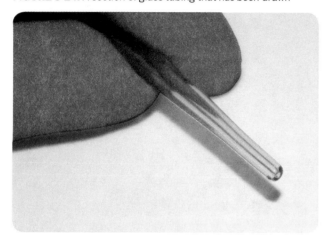

FIGURE 5-25: *Lubricate the glass tubing with glycerol or oil and then rotate it as you slide it into the stopper*

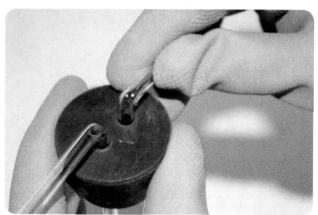

INSERTING GLASS TUBING INTO CORKS AND STOPPERS SAFELY

Trying to force glass tubing into a cork or stopper is one of the most frequent causes of injury in home labs. If you apply pressure even slightly off-axis, the glass tubing may snap and impale you. Fortunately, it's easy to avoid such accidents. Proceed as follows:

1. First, make absolutely certain that the glass tubing is of the proper size to fit the hole in the stopper. If you try to force 5 mm tubing into a stopper drilled for 4 mm tubing, for example, things are sure to end badly.

2. Lubricate the tubing with glycerol, mineral oil, or a similar substance. (Many chemists use water, but I recommend something slipperier.)

3. Using heavy gloves or a towel to protect your fingers, align the tubing with the hole in the stopper, and slide it into the hole, rotating the tubing constantly as you do so. Apply only as much pressure as needed to cause the tubing to slide into the stopper, and apply that pressure as close as possible to the face of the stopper, as shown in Figure 5-25.

CLEANING GLASSWARE

Dirty glassware is one sure sign of a sloppy chemist. Use the following guidelines to keep your glassware in like-new condition.

- Always do at least a preliminary cleaning as soon as possible after you finish using the glassware. Contamination that rinses away easily when fresh may harden overnight into a deposit that's difficult or impossible to remove.

- Before you clean glassware, always examine it for cracks, starring, chipping, or other damage. Dispose of damaged glassware by rinsing it to remove as much contamination as possible, wrapping it in several layers of newspaper, and discarding it with the household trash.

- For small-bore items such as glass tubing, pipettes, and so on, force or draw clean tap water through the glassware several times to rinse out most of the contamination. Then draw clean sudsy water through the glassware several times, followed by drawing several rinses of tap water through the glassware, and finally by drawing distilled or deionized water through the glassware several times. Use a rubber bulb or zero-residue canned "air" (*not* your mouth) to blow out any remaining rinse water and set the glassware aside to dry. No further cleaning is needed.

- For glassware with accessible inside surfaces—such as beakers, flasks, and test tubes—as soon as you finish using the glassware, let it cool (if applicable), dispose of the contents properly, rinse the glassware thoroughly under running hot tap water, and then sit it aside inverted to dry. Make sure that it's in an area devoted to dirty glassware or is otherwise marked as requiring thorough cleaning.

- If the glassware is very dirty—such as a test tube or flask that contains a precipitate that won't rinse away—use a brush and hot sudsy water to remove as much of the contamination as possible. If the glassware still looks dirty, submerge it in a sink or plastic tub that contains hot sudsy water, allow it to soak for several hours or overnight, and again use a brush and hot sudsy water to scrub it out. If the glassware still looks dirty, try soaking it overnight in a 1 M solution of hydrochloric acid (hardware store muriatic acid diluted one part acid to ten parts water is fine for this purpose) and then scrubbing it out again. If the glassware then appears clean, rinse it with tap water and set it aside to drain until you are ready to do a final cleaning. Otherwise, discard the glassware.

- Glassware that has undergone preliminary cleaning may appear clean to the eye, but it's not clean enough for lab use. A good preliminary cleaning removes nearly all of the contamination, but a final cleaning is necessary to ensure the glassware is really clean. Begin by filling a sink or tub with hot sudsy water. Using an appropriate brush or brushes, scrub the entire surface of the glassware, inside and out. Rinse the glassware thoroughly under hot running tap water and then examine it to ensure it is pristine. Invert the glassware on a drying rack and allow it to drain completely. Finally, rinse the glassware, particularly the inside, with distilled water and invert it on the drying rack to allow the final rinse to drain. Although it no doubt horrifies our friends who do quantitative analyses, I add one drop (literally) of dishwashing liquid (actually, I use Kodak Photo-Flo) per liter of distilled rinse water. The dishwashing liquid or Photo-Flo breaks the surface tension and allows the remaining water to sheet cleanly off the glassware rather than beading, leaving the glassware dry to the touch almost immediately.

- Once the glassware is completely dry, return it to its storage location.

> **DR. MARY CHERVENAK COMMENTS:**
>
> Actually, I've never used hydrochloric acid to clean glassware. If the glassware is extremely dirty, I soak it in a **base bath** (concentrated ethanolic potassium hydroxide). Once the soak is complete, rinse the glassware thoroughly with water and dry with acetone.

INDUSTRIAL STRENGTH CLEANING

When I worked in chemistry labs back in the early 1970s, the standard method for cleaning glassware was to use a solution of chromium trioxide (or potassium dichromate) in concentrated sulfuric acid. On the plus side, this solution works wonders. Glassware that goes in filthy comes out pristine and sparkling. The only minor drawbacks are that chromic-sulfuric acid solution is very expensive, extremely toxic, a known carcinogen, hideously corrosive, emits deadly chromyl chloride fumes, may explode unpredictably, and is almost impossible to dispose of safely. But, boy, does it ever clean glass squeaky clean.

Fortunately, there are other cleaning solutions that are almost as good and somewhat safer to use. One of the best is a concentrated solution of sodium hydroxide or potassium hydroxide in 91% isopropanol or 95% ethanol. Although this solution is extremely corrosive (and requires using full protective gear including goggles and face shield) and very flammable, it is inexpensive, does not contain heavy metal ions, and is easy to dispose of safely. It's strong enough to dissolve glass, literally, so you must either make it up as needed or store it in a nonglass container that is rated for storage of strong alkalis. On the few occasions I use it—usually to clean a very dirty piece of expensive glassware—I mix up a fresh batch just large enough to do what needs to be done. When the job is finished, I neutralize the alkali with dilute acid and flush the neutralized solution down the drain with lots of water.

Note that *I do not recommend using this cleaning solution*, but mention it only for completeness.

WHEN TO CLEAN AND WHEN TO DISCARD

Don't get too attached to your glassware. It's usually better to discard seriously dirty glassware and replace it with new, particularly if the glassware is an inexpensive item such as a test tube, pipette, beaker, or Erlenmeyer flask. For more expensive glassware, such as a volumetric flask or burette, more heroic measures may be justified. (Of course, you shouldn't be doing things with expensive glassware that cause it to become seriously dirty anyway; most expensive glassware is intended only for volumetric use, not for use as a reaction vessel.) If scrubbing the glassware with soap and water doesn't work and soaking it in dilute hydrochloric acid doesn't work, it's best just to discard it. If you do decide to risk using the alcohol/hydroxide solution and it doesn't work, it's time to replace the glassware.

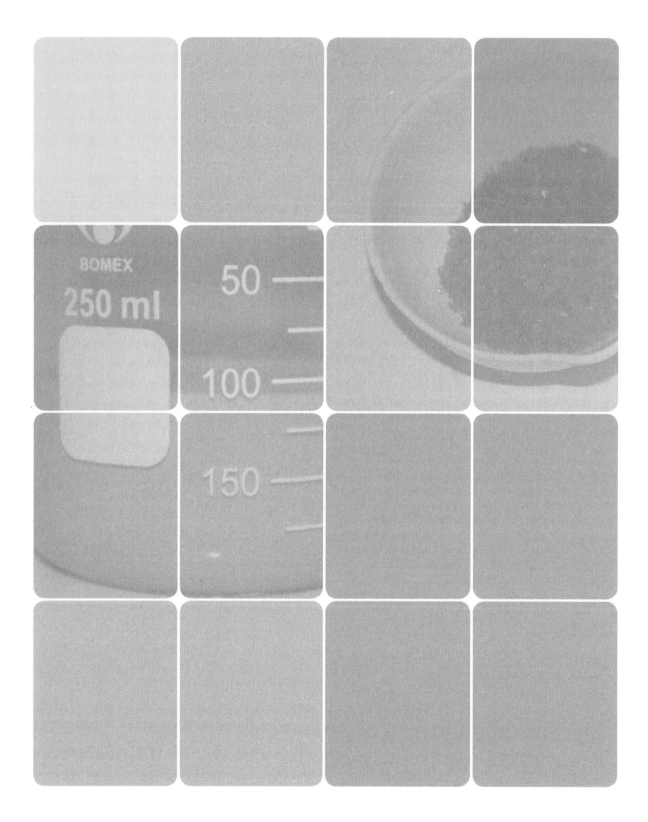

Laboratory: Separating Mixtures

6

A ***mixture*** is a substance that comprises two or more elements and/or compounds that are physically intermingled but that have not reacted chemically to form new substances. A mixture may be a solid, liquid, gas, or some combination of those states.

We are surrounded by mixtures, both in the chemistry lab and in everyday life. The air we breathe is a mixture of nitrogen, oxygen, and small amounts of other gases. Our soft drinks are complex mixtures of water, sugar, carbon dioxide, and various organic compounds that provide the color and flavor. The foods we eat are complex mixtures of organic and inorganic compounds.

Mixtures are often created intentionally because a particular mixture possesses special desirable characteristics. For example, stainless steels are mixtures of iron, chromium, carbon, nickel, manganese, and other elements in specific proportions, chosen to optimize such characteristics as resistance to corrosion, hardness, tensile strength, color, and luster. Similarly, concretes are complex mixtures of components chosen to minimize cost while optimizing strength, durability, resistance to road salts, permeability to water, and other factors, depending on the purpose for which the concrete will be used.

Because the components of a mixture have not reacted chemically, it is possible to separate the mixture into its component substances by using purely physical means. Chemists have devised numerous methods for separating compounds based on differential physical characteristics, including ***differential solubility***, ***distillation***, ***recrystallization***, ***solvent extraction***, and ***chromatography***. In this chapter, we'll examine these common methods of separating mixtures.

LABORATORY 6.1:
DIFFERENTIAL SOLUBILITY: SEPARATE SAND AND SUCROSE

Differential solubility was one of the earliest methods developed for separating mixtures. Differential solubility depends on the fact that different substances have different solubility in different solvents. In this lab, we'll examine the simplest example of differential solubility: separating a mixture of two compounds, one of which is freely soluble in water and one of which is insoluble in water.

EVERYDAY SEPARATIONS

Mixtures are the starting points for many of the materials we use every day.

- Refineries use distillation and other separation methods to process crude oil—a complex mixture of hydrocarbons—into gasoline, diesel fuel, and hundreds of other useful organic compounds.

- Industrial chemists synthesize thousands of valuable compounds, from pharmaceuticals and plastics to dyes, which must be extracted from complex mixtures of by-products. These compounds are often purified using one or more of the separation methods that we'll examine in this chapter.

- Mining companies extract, crush, and treat tons of rock to obtain only a few grams of a precious metal ore. For example, typical gold ore contains less than 50 grams of gold per 1,000 kilograms of ore (0.005%).

REQUIRED EQUIPMENT AND SUPPLIES

- ☐ goggles, gloves, and protective clothing
- ☐ balance and weighing papers
- ☐ hotplate
- ☐ microwave or conventional oven
- ☐ beaker, 150 mL (2)
- ☐ watch glass
- ☐ filter funnel
- ☐ filter paper
- ☐ stirring rod
- ☐ wash bottle (water)
- ☐ dry sand (10.0 g)
- ☐ sucrose (10.0 g)
- ☐ water

SUBSTITUTIONS AND MODIFICATIONS

- You may substitute any convenient heat-resistant containers of similar size for the beakers.

- You may substitute an ordinary kitchen funnel for the filter funnel, and a coffee filter for the filter paper.

- If the sand is not completely dry, spread it in a thin layer and dry it in a microwave or conventional oven.

- You may substitute ordinary table sugar for the sucrose.

PROCEDURE

1. If you have not already done so, put on your splash goggles, gloves, and protective clothing.
2. Weigh about 10.0 g of dry sand and about 10.0 g of sucrose and add both of them to a beaker. Record the masses on lines A and B of Table 6-1.
3. Stir the mixture or swirl the beaker until the sucrose and sand are thoroughly mixed.
4. Add about 25 mL of water to the beaker, and stir thoroughly to dissolve the sucrose.
5. Weigh a piece of filter paper and record the mass in Table 6-1.
6. Weigh the second beaker and record the mass in Table 6-1.
7. Set up your filter funnel over the second beaker, using the filter paper you weighed in the preceding step.
8. Swirl the contents of the first beaker to keep as much sand as possible suspended in the liquid, and pour the liquid through the filter funnel.
9. Use the wash bottle to rinse the beaker with a few mL of water and pour that water with any remaining sand through the filter funnel, collecting the *filtrate* (the liquid that passes through the filter paper; the solid remaining in the filter paper is called the *filtrand*) in a second beaker. Repeat if necessary until all of the sand has been transferred from the first beaker to the filter funnel.
10. Remove the filter paper from the funnel, being careful not to lose any of the sand. Place the filter paper on a watch glass, as shown in Figure 6-1, and heat it gently under a heat lamp or in a microwave or conventional oven until all moisture is driven off.
11. Weigh the filter paper and record the mass of the filter paper plus sand in Table 6-1. Calculate the mass of the sand.
12. Place the second beaker on the hot plate and bring it to a gentle boil. Continue boiling the solution until nearly all of the water has been vaporized. As the volume of the solution is reduced, the sugar begins to crystallize. Use the stirring rod to break up any large masses of crystals as they form. Reduce heat and continue heating gently until the sucrose remaining in the beaker appears dry. Be careful not to overheat the sucrose and char it.
13. After allowing it to cool completely, weigh the second beaker and record the mass of the beaker plus sucrose in Table 6-1. Calculate the mass of the sucrose.

CAUTIONS

Although the chemicals used in this laboratory session are not hazardous, it is a matter of good laboratory practice to wear splash goggles, gloves, and protective clothing at all times. Be careful when heating the mixture with the hotplate.

DR. PAUL JONES COMMENTS:

Make observations of the physical appearance of the sand and sugar. Does it change as you go through the process? This is a good rule of thumb to consider during any physical or chemical manipulation.

FIGURE 6-1: *The filter paper captures the sand, allowing the sucrose solution to pass into the receiving beaker*

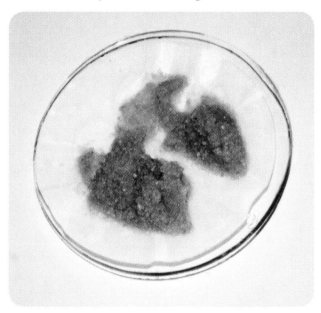

DISPOSAL: All of the waste material from this laboratory can be disposed of with household waste.

TABLE 6-1: *Differential solubility—observed and calculated data*

Item	Mass
A. Sand, initial	_____.___ g
B. Sucrose, initial	_____.___ g
C. Filter paper	_____.___ g
D. Second beaker	_____.___ g
E. Filter paper + sand	_____.___ g
F. Second beaker + sucrose	_____.___ g
G. Sand, final (E − C)	_____.___ g
H. Sucrose, final (F − D)	_____.___ g

REVIEW QUESTIONS

Q1: After this procedure, you should end up with separate piles of sand and sucrose with masses that are the same as the initial masses. How closely do your experimental results correspond to the expected results? Propose an explanation for why the final masses of the sand and/or sucrose might be higher or lower than the starting masses.

_____

_____

_____

_____

_____

Q2: You are mixing concrete, but the only sand available is heavily contaminated with salt, which weakens concrete. You have plenty of water available, but no means of filtering. How might you remove the salt from the sand?

_____

_____

_____

_____

Q3: You are presented with a mixture of two compounds, both of which are freely soluble in water. What might you do to separate these compounds using differential solubility?

_____

_____

LABORATORY 6.2:
DISTILLATION: PURIFY ETHANOL

Distillation is the oldest method used for separating mixtures of liquids. Distillation uses the fact that different liquids have different boiling points. When a mixture of liquids is heated, the liquid with the lower (or lowest) boiling point vaporizes first. That vapor is routed through a **condenser**, which cools the vapor and causes it to condense as a liquid; the liquid is then collected in a receiving vessel. As the original liquid mixture continues to be heated, eventually some or all of the lower-boiling liquid is driven off, leaving only the higher-boiling liquid or liquids in the distillation vessel.

I say "some or all" because distillation is an imperfect method for separating mixtures of liquids that form **azeotropes**. An azeotrope, also called a **constant boiling mixture**, is a mixture of two or more liquids at a specific ratio, whose composition cannot be altered by simple distillation. Every azeotrope has a characteristic boiling point, which may be lower (a **positive azeotrope** or **minimum-boiling mixture**) or higher (a **negative azeotrope** or **maximum-boiling mixture**) than the boiling points of the individual liquids that make up the azeotrope.

For example, ethanol forms a positive azeotrope with water. The boiling point of a mixture of 95.6% ethanol (by weight) with 4.4% water is 78.1°C, which is lower than the boiling point of pure water (100°C) or pure ethanol (78.4°C). Because the azeotropic mixture boils at a lower temperature, it's impossible to use simple distillation to produce ethanol at concentrations higher than 95.6%. (More concentrated ethanol solutions can be produced by using drying agents such as anhydrous calcium chloride that physically absorb the water from a 95.6% solution of ethanol. These solutions must be stored and handled carefully; otherwise, they absorb water vapor from the air until they reach the 95.6% azeotropic concentration.)

Ethanol also forms azeotropes with many other liquids, including some that are poisonous or taste bad. This allows production of **denatured ethanol**, which is toxic, cannot be drunk, and can therefore be sold cheaply without cannibalizing sales of (and taxes on) much more expensive potable ethanol, such as vodka and other distilled beverages.

REQUIRED EQUIPMENT AND SUPPLIES

- ☐ goggles, gloves, and protective clothing
- ☐ balance
- ☐ volumetric flask, 100 mL
- ☐ distillation apparatus (see Substitutions and Modifications)
- ☐ beaker, 250 mL
- ☐ hotplate or other flameless heat source
- ☐ boiling chips
- ☐ ethanol, 70% (200 mL)

SUBSTITUTIONS AND MODIFICATIONS

- You may substitute a 100 mL graduated cylinder for the 100 mL volumetric flask, with some loss of accuracy.

- If you don't have a ready-made distillation apparatus, you can create one from standard equipment. I used the apparatus shown in Figure 6-2, made with two Erlenmeyer flasks, one- and two-hole stoppers to fit them, some short lengths of glass tubing, some plastic tubing, and a cold water bath. Use a two-hole stopper in the receiving flask to prevent pressure from building up in the apparatus. The cold water bath keeps the receiving flask cold, which causes the ethanol vapor to condense in the receiving flask as soon as it arrives. Depending on the capacity of your distillation apparatus, the maximum capacity of your balance, the capacities of your volumetric flask or other volumetric glassware, and the initial concentration of the ethanol, you may need to modify the starting quantity of ethanol.

- Ethanol is sold in drugstores, sometimes by name, and sometimes as ethyl alcohol, ethyl rubbing alcohol, or simply rubbing alcohol. (If the bottle is labeled as rubbing alcohol, make sure it's ethanol rather than isopropanol.)

(continues ...)

SUBSTITUTIONS AND MODIFICATIONS *(continued)*

- Most drugstore ethanol is 70% by volume (about 64.7% by weight), but 90% and 95% concentrations are also common. If you start with more concentrated ethanol, dilute it with water to 70% before you begin the distillation.

FIGURE 6-2: *My distillation apparatus*

In this laboratory, we'll use distillation to increase the concentration of an ethanol solution. At 25°C, pure water has a density of 0.99704 g/mL and pure ethanol a density of 0.78736 g/mL. Solutions of ethanol and water have densities between these figures. If you add the densities of the pure liquids and divide by two, you get 0.89220 g/mL, which you might assume is the density of a 50/50 ethanol-water mixture. As it turns out, that's not true. Ethanol and water do not mix volumetrically; that is, if you mix 100 mL of pure ethanol with 100 mL of pure water, you do not get 200 mL of solution, for the same reason that dissolving 100 mL of sucrose in 100 mL of water does not yield 200 mL of solution.

Nonetheless, it's possible to determine the concentration of ethanol by measuring the density of the solution and comparing that value to ethanol-water density tables available in the CRC handbook and similar publications. We'll measure the density of the starting solution and the resulting distillate and compare

those values to published values to determine the ethanol concentrations of the two solutions.

Figure 6-2 shows the distillation apparatus that I used for this laboratory session. Note that both flasks are securely clamped to prevent tipping, and that the stopper in the receiving flask (right) has an open hole to vent the flask and prevent pressure buildup. The receiving flask sits in a large container of ice water. As the alcohol vapor passes through the tubing and into the receiving flask, it condenses immediately when it contacts the cold air in the receiving flask.

PROCEDURE

1. If you have not already done so, put on your splash goggles, gloves, and protective clothing.
2. Weigh the volumetric flask on your balance and record the mass in on line A of Table 6-2.
3. Transfer 100.00 mL of ethanol to the volumetric flask and reweigh the flask. Record the mass on line B of Table 6-2.
4. Subtract line A from line B and record the result (the mass of the 100 mL of ethanol) on line C. Divide line C by 100 to give the density of the ethanol in g/mL and record this value on line D.
5. Empty the ethanol from the volumetric flask into the distillation vessel and add about 100 mL more ethanol. Add one or two boiling chips to the flask.
6. Reassemble the distillation apparatus, using the 250 mL beaker as the receiving vessel.
7. Turn on the hotplate and bring the ethanol to a gentle boil. As the ethanol begins to boil, the temperature indicated by the thermometer in the distillation apparatus will begin to climb rapidly until it stabilizes at the temperature of the ethanol-water vapor.
8. Continue the distillation until more than 100 mL of distillate has collected in the receiving vessel. (Make

sure that the indicated temperature does not climb above the stable value; if it does, that indicates that the ethanol has been eliminated from the distillation vessel and what you're transferring to the receiving vessel is pure water. If that occurs, start again using more of the original ethanol mixture.)

9. Allow the distillate to cool to room temperature.
10. Transfer 100.0 mL of the distillate to the volumetric flask, weigh the flask and contents, and record the mass on line E of Table 6-2.
11. Subtract line A from line E to determine the mass of the distillate, and record this value on line F of Table 6-2.
12. Divide line F by 100 to give the density of the distillate in g/mL and record this on line G.

> **DISPOSAL:** You can retain the distilled ethanol and use it as fuel for your alcohol lamp or for any other purpose that requires denatured ethanol where concentration is not critical, or you can safely pour it down the drain. If you retain the distillate, label it properly.

TABLE 6-2: *Distillation of ethanol—observed and calculated data*

Item	Data
A. Volumetric flask	_____.__ g
B. Volumetric flask with 100 mL 70% ethanol	_____.__ g
C. Mass of 100 mL 70% ethanol (B – A)	_____.__ g
D. Density of 70% ethanol (C/100)	_____.__ g/mL
E. Volumetric flask with 100 mL distilled ethanol	_____.__ g
F. Mass of 100 mL distilled ethanol (E – A)	_____.__ g
G. Density of distilled ethanol (F/100)	_____.__ g/mL

OPTIONAL ACTIVITIES

If you have enough time and the required materials, consider performing these optional activities:

• Repeat the distillation, transferring the original distillate to the distillation vessel for redistillation. The volume of the second distillate will be smaller than the volume produced during the first distillation, so you'll either need to distill considerably more than 100 mL originally or else use a smaller volumetric flask to determine the density of the second distillate. Alternatively, you can use a graduated cylinder to determine the density of the second distillate, with some loss of accuracy.

• Run the distillation a third time, and determine the density (and ethanol concentration) of the third distillate.

HIGH BOILING POINT AZEOTROPES

Hydrochloric acid is one familiar example of a negative (high-boiling) azeotrope. Pure hydrogen chloride has a boiling point of -84°C and pure water a boiling point of 100°C. A solution of 20.2% hydrogen chloride (by weight) in water has a boiling point of 110°C, higher than the boiling point of either component. This means that boiling a solution of hydrochloric acid of any concentration eventually produces a solution of exactly 20.2% hydrogen chloride by weight. If the starting solution is more dilute, water is driven off until the solution reaches 20.2% concentration. If the starting solution is more concentrated, hydrogen chloride gas is driven off until the solution reaches 20.2% concentration.

REVIEW QUESTIONS

Q1: Would you expect the density of the distilled ethanol solution to be lower or higher than the density of the original ethanol solution? Why?

Q2: Using values for the densities of various concentrations of ethanol and water that you obtain from the *CRC Handbook of Chemistry and Physics* (CRC Press) or a reliable online source, estimate the ethanol concentrations in the original solution and the distillate.

Q3: A drugstore offers denatured ethanol in concentrations of 70%, 95%, and 99% by weight. The 70% and 95% solutions are relatively inexpensive, but the 99% solution is very costly. Why?

Q4: Distilling wine produces not colorless pure ethanol, as you might expect, but brandy, which is deeply colored and contains complex flavors. Why?

LABORATORY 6.3:

RECRYSTALLIZATION: PURIFY COPPER SULFATE

In this lab, we'll use a procedure called *recrystallization* to purify crude copper sulfate. Crude copper sulfate is a mixture of copper sulfate with various impurities that may include copper carbonate, copper oxides, and other copper compounds. You can obtain crude copper sulfate at a hardware or lawn and garden store, where it is sold as a root killer or for pond treatment.

Many chemicals besides copper sulfate are available inexpensively in impure forms, such as technical and practical grades. Although these chemicals may be insufficiently pure for general lab use, many of them can be purified to the equivalent of laboratory grade or even reagent grade by using recrystallization.

Successful recrystallization depends upon two factors. First, crystals are the purest form of a chemical, because there is no room for impurities in the crystalline lattice. As crystals grow in a solution of an impure chemical, the impurities remain in solution. Second, some chemicals are much more soluble in hot water than in cold. For example, the solubility of copper sulfate pentahydrate, the chemical that we'll purify in this lab, is 203.3g/100mL at 100°C, versus only 31.6g/100mL at 0°C.

In other words, a saturated solution of copper sulfate at 100°C contains about six times more copper sulfate than the same amount of solution at 0°C. If we saturate boiling water with impure copper sulfate and then cool that solution to 0°C, about 5/6 of the copper sulfate crystallizes in pure form, leaving the impurities in solution (along with about 1/6 of the original copper sulfate).

Once the crystals have formed, they can be separated from the supernatant liquid by filtration or by decanting off the liquid. In either case, some of the contaminated liquid remains mixed with the purified copper sulfate crystals. You remove this last bit of contamination by rinsing the crystals with a small amount of ice-cold water or another solvent that is miscible with water, such as acetone, leaving only purified copper sulfate crystals.

We'll use acetone for the final rinse, because copper sulfate is almost insoluble in acetone. If we used ice-cold water, some of our purified copper sulfate would dissolve in the rinse water, lowering our final yield. Using acetone flushes away impurities

without dissolving the copper sulfate crystals. Volatile organic solvents like acetone also evaporate faster and more completely than water.

REQUIRED EQUIPMENT AND SUPPLIES

- ☐ goggles, gloves, and protective clothing
- ☐ balance and weighing papers
- ☐ hotplate
- ☐ beaker, 250 mL (2)
- ☐ 100 mL graduated cylinder
- ☐ stirring rod
- ☐ filter funnel and support
- ☐ filter paper
- ☐ beaker tongs
- ☐ thermometer
- ☐ copper sulfate pentahydrate (250 g crude crystals)
- ☐ acetone (a few mL)
- ☐ sodium carbonate heptahydrate (50 g)
- ☐ ice bath

SUPERSATURATED SOLUTIONS

A *supersaturated solution* is one that contains more solute per volume of solution than the solution would ordinarily contain at that temperature. Supersaturation normally occurs when the temperature of a saturated solution is gradually reduced. Rather than precipitating out, the excess solute remains in solution until some event occurs that causes the excess solute to precipitate suddenly. That event may be as minor as dropping a tiny crystal of the solute into the solution or even simply tapping the container. This is called *nucleation*. Crystals produced from a supersaturated solution are not necessarily pure, because the rapid crystallization may trap impurities within the crystal structure.

SUBSTITUTIONS AND MODIFICATIONS

• If your hotplate has heating coil rather than a flat heating surface, use a large tin can lid or a burner cover between the coil and beaker rather than putting the beaker in direct contact with the burner.

• The quantity of copper sulfate used in this laboratory assumes that you really do want to purify several ounces of copper sulfate to use in the later labs in this book. If you want only to demonstrate the principle of recrystallization, you can reduce the quantity of copper sulfate (and water) and the size of the beakers proportionately.

• You can substitute 50 g of crude washing soda for the 50 g of sodium carbonate heptahydrate.

• You can substitute 23 g of anhydrous sodium carbonate for the 50 g of sodium carbonate heptahydrate.

• Rather than using an ice bath, you can cool the solution in a freezer. If you do that, make sure the solution is in a sealed container to make sure it can't contaminate any of the food items in the freezer. Label the solution to make absolutely certain it cannot be confused with food items. (This should go without saying; labeling containers is good lab technique, and you should never store any product in an unlabeled container.)

PROCEDURE

1. If you have not already done so, put on your splash goggles, gloves, and protective clothing.
2. Place a weighing paper on the balance pan and tare the balance to read 0.00 g. Add the crude copper sulfate pentahydrate until the balance indicates about 250 g. Record the mass of the crude copper sulfate to 0.01 g on line A of Table 6-3. If the maximum capacity of your balance is less than 250 g, weigh multiple samples until you accumulate about 250 g total. Transfer the copper sulfate to the second 250 mL beaker.
3. Set your balance to read 0.00 g and weigh a piece of filter paper. Record the mass of the filter paper to 0.01 g in Table 6-3, and use a pencil to write the mass on the filter paper itself.
4. Fan-fold (flute) the massed filter paper and set it aside for later use.
5. Set up your filter funnel with another piece of fan-folded filter paper over one of the 250 mL beakers.
6. Use the 100 mL graduated cylinder to measure 150 mL of hot tap water, and add it to the copper sulfate in the 250 mL beaker.

7. Place the beaker on the hotplate and bring the water to a gentle boil. Stir the solution until all of the crude copper sulfate has dissolved. Any solid matter that refuses to dissolve is an insoluble impurity and can be ignored.

DR. PAUL JONES COMMENTS:

I wouldn't say that it can be "ignored." As described in the following step, any insoluble residue should be trapped via hot filtration and discarded. Don't add more water to try to dissolve it. Also, hot filtration works best if you heat the receiving vessel up prior to filtration. Use a little boiling water to heat the filter paper and funnel. This reduces crystallization on the paper.

8. Using the beaker tongs, pour the hot copper sulfate solution through the filter paper as quickly as possible. Some copper sulfate may recrystallize when it contacts the cooler filter paper, funnel, or receiving beaker, but it should go back into solution as you continue to pour the hot solution through the funnel.
9. As the solution cools, copper sulfate will begin crystallizing. Slow cooling produces larger crystals, and fast cooling smaller crystals. Smaller crystals are easier to handle and weigh, so we want to produce crystals as small as possible. Our goal, then, is to cool the solution as fast as possible from the boiling point to near the freezing point. To do that, place the beaker of hot solution in an ice bath (a larger container full of crushed ice) or a freezer.

DR. PAUL JONES COMMENTS:

Yes, but it's much cooler to grow big crystals by letting the solution sit to cool slowly without stirring, and then you get to watch them grow. You could even shoot a time-lapse video.

10. As the solution cools, stir it periodically with the thermometer. While you're waiting for the solution to cool to 0°C, remove the used filter paper from the filter funnel, rinse the funnel clean, and put in the fan-folded filter paper that you weighed earlier. Also rinse the original 250 mL beaker clean and place it under the funnel.
11. When the solution has cooled to 0°C, use the stirring rod to free any crystals that have grown on the sides of the beaker. Swirl the contents of the beaker gently to keep the crystals suspended in the solution, and pour the solution through the filter funnel.
12. Use a few mL of acetone to rinse any remaining crystals from the beaker and pour the acetone through the filter

funnel, trying to make sure it wets all of the crystals in the filter paper.

13. Carefully remove the filter paper and set the purified copper sulfate crystals aside to dry.

14. When the filter paper and copper sulfate is completely dry, weigh it and record the mass of the filter paper and copper sulfate crystals in Table 6-3.

15. Transfer the purified copper sulfate crystals to a labeled bottle, and retain them for use in later laboratory sessions.

SOLUBILITY VERSUS SOLVENT TEMPERATURE

Most solid water-soluble compounds are more soluble in hot water than in cold. Gases and a few solid compounds are more soluble in cold water than in hot water, a phenomenon called *retrograde solubility*. For those solid compounds, it's possible to do a recrystallization in reverse. Instead of making a saturated solution of the compound in hot water and then cooling the solution to cause crystals to form, you make a saturated solution of the compound in cold water, and then heat the solution to cause crystals to form.

The change in solubility with solvent temperature differs from compound to compound. Some compounds, such as the copper sulfate we use in this lab, are much more soluble in hot water than in cold. But the solubility of many compounds is little affected by solvent temperature. For example, the solubility of sodium chloride in water is 355 g/L at 0°C, increasing only about 10% to 390 g/L at 100°C. Recrystallization is inefficient for purifying such compounds, because most of the compound remains in solution regardless of the temperature.

WHEN CRUDE ISN'T

Ironically, the "crude" copper sulfate crystals we bought turned out to be quite pure. The label included an assay that listed "copper sulfate — 99.0%," which is nearly reagent grade purity. Dr. Paul Jones, one of my technical reviewers, is a chemistry professor. I showed him the two-pound bottle of copper sulfate I'd bought at a local big-box DIY store. He examined the label, commented on its high purity, and asked, "How much did this cost?" When I told him I paid about $7.50 for the two-pound bottle and added that at the same time I'd also paid about $5.00 for a two-pound bottle of sodium hydroxide with an assay of 100.0%, Paul commented, "I may have to start buying some of my chemicals there." Indeed.

FIGURE 6-3:
Recrystallized copper sulfate

FIGURE 6-4:
Precipitated copper carbonate

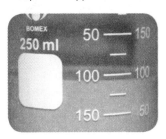

At this point, we have (assuming no losses) about 200 g of pure copper sulfate crystals on the filter paper and about 50 g of crude copper sulfate dissolved in about 150 mL of solution. Rather than waste that 50 g of crude copper sulfate, we'll use it to produce some crude copper carbonate, which we'll use in a later laboratory session.

Copper sulfate is fairly soluble in water, as are sodium carbonate and sodium sulfate, but copper carbonate is extremely insoluble in water. We'll take advantage of these differential solubilities to produce, separate, and purify copper carbonate. By reacting the waste copper sulfate solution with a solution of sodium carbonate, we precipitate nearly all of the copper ions as the insoluble carbonate salt, leaving sodium sulfate in solution.

1. Weigh about 50 g of sodium carbonate heptahydrate and transfer it to the empty 250 mL beaker.

2. Use the graduated cylinder to measure about 100 mL of warm tap water and transfer it to the beaker with the sodium carbonate. Stir until the sodium carbonate dissolves.

3. Pour the 100 mL of sodium carbonate solution into the beaker of copper sulfate solution, with stirring.

4. Put a fresh piece of fan-folded filter paper into the filter funnel and place the empty beaker beneath the funnel. Swirl the contents of the beaker to keep the precipitate suspended in the solution and pour the solution through the filter paper.

5. Rinse the precipitate three times with about 50 mL of water each time. This step removes nearly all of the sodium sulfate from the precipitate, as well as any other soluble salts formed by trace contaminants in the original waste solution.

6. Carefully remove the filter paper and set the copper carbonate aside to dry. If you have sufficient acetone, you can do a final rinse with 25 mL to 50 mL of acetone to remove nearly all of the water and allow the copper carbonate to dry faster.

7. Transfer the copper carbonate to a labeled bottle, and retain it for use in a later laboratory session.

TABLE 6-3: *Recrystallization of copper sulfate—observed and calculated data*

Item	Data
A. Crude copper sulfate	_____.__ g
B. Filter paper	_____.__ g
C. Filter paper plus purified copper sulfate	_____.__ g
D. Purified copper sulfate (C – B)	_____.__ g

DR. MARY CHERVENAK COMMENTS:

Paul has also started purchasing sodium bicarbonate from the grocery store, because it's much, much cheaper and just as pure (if not purer, as it's routinely used in baking) than the containers that he buys from his usual laboratory chemical suppliers!

DISPOSAL: The filtrate from step 19 contains only sodium sulfate and a small amount of copper. You can flush it down the sink with plenty of water. The used filter paper can be disposed of with household waste.

REVIEW QUESTIONS

Q1: The solubility of copper sulfate pentahydrate is 2033g/L at 100°C, which means that 305 g should dissolve in 150 mL of water at 100°C. Why do you suppose we used only 250 g of crude copper sulfate rather than 305 g?

_____

_____

_____

_____

Q2: We obtained about an 80% yield (roughly 200 g of purified copper sulfate from 250 g of crude copper sulfate). We might have improved the percent yield somewhat by dissolving more than 250 g of crude copper sulfate initially, but at best that method would still result in about a sixth of the crude copper sulfate going to waste. What method might we use to obtain much higher percent yields?

_____

_____

_____

_____

Q3: A chemist requires copper sulfate of at least 99.9% purity for a particular procedure. She recrystallizes crude copper sulfate as we have done, but analysis shows that her purified copper sulfate is only 99.4% pure. What might she do to obtain copper sulfate of 99.9% or higher purity?

_____

_____

_____

_____

LABORATORY 6.4:
SOLVENT EXTRACTION

Solvent extraction, also called *liquid-liquid extraction* or *partitioning*, is a procedure used to separate compounds based on their solubility in two immiscible liquids—usually water and an organic solvent. During solvent extraction, one or more of the solutes in one of the liquid phases migrates to the other liquid phase. The two liquid phases are then physically separated and the desired product is isolated from the phase that contains it.

Solvent extraction is one of the most commonly used laboratory purification methods, particularly in organic chemistry labs. Solvent extractions done in chemistry labs are usually small-scale, batch-mode operations using a *separatory funnel*. Solvent extraction is also widely used in industrial operations. Some industrial applications use batch-mode extraction, albeit usually on a much larger scale than laboratory solvent extractions. Other industrial applications use continuous-mode solvent extractions, often on a gigantic scale, where the two solvents are continuously added to and removed from a large reaction vessel.

In some solvent extractions, the desired product migrates from the original liquid phase to the second liquid phase. In other solvent extractions, impurities migrate from the original liquid phase to the second liquid phase, leaving the product in the original liquid phase. The phase that contains the desired product is called the *product layer*. The phase that contains impurities, excess reactants, and other undesirable compounds is called the *waste layer*. Many first-year organic chemistry students come to grief by discarding what they think is the waste layer, only to learn later that they actually discarded the product layer.

In this laboratory, we'll use solvent extraction to isolate the iodine present in Lugol's solution, an aqueous solution of iodine and potassium iodide.

REQUIRED EQUIPMENT AND SUPPLIES

- ☐ goggles, gloves, and protective clothing
- ☐ test tube with solid stopper
- ☐ test tube rack
- ☐ eyedropper
- ☐ watch glass
- ☐ Lugol's solution (1 mL)
- ☐ lighter fluid (~2 mL)
- ☐ tap water

SUBSTITUTIONS AND MODIFICATIONS

- You may substitute a Petri dish, small beaker, or saucer for the watch glass.

- You may substitute tincture of iodine for the Lugol's solution. Tincture of iodine is a solution of iodine in ethanol, so using it means that you're doing a simple transfer of iodine from one layer to another rather than a separation of the iodine and potassium iodide present in Lugol's solution. Still, the principle can be illustrated just as well with tincture of iodine.

- You may substitute gasoline, paint thinner, or any other organic solvent that is not miscible with water for the lighter fluid.

DR. PAUL JONES COMMENTS:

Maybe you're relying on them learning the hard way, which to be honest really is the only way to learn this lesson, but I like to explicitly tell students not to throw *anything* away until they have gotten the product they want/expect. But, yes, no one listens. Alas, the life of a teacher. They think you're an idiot as long as you're in their presence and only after they're miles and years away do they realize you weren't a boob.

PROCEDURE

1. If you have not already done so, put on your splash goggles, gloves, and protective clothing.
2. Fill a test tube about halfway with water.
3. Add about 1 mL (~20 drops) of Lugol's solution to the test tube. Stopper the tube and agitate it until the contents are thoroughly mixed. Record anything unusual about the appearance of the aqueous solution on line A of Table 6-4.
4. Use the eyedropper to carefully add about 1 mL of lighter fluid to the test tube by allowing it to run down the inner surface of the test tube, as shown in Figure 6-5. (If you substitute another solvent for the lighter fluid, make sure that it forms a distinct layer on top of the aqueous solution of iodine. If it doesn't, you'll have to start again, using a solvent that does not mix with water.) Record the appearance of the organic solvent layer on line B of Table 6-4.
5. Stopper the test tube and agitate it vigorously for 15 seconds. Record the appearance of the mixed layers immediately after you complete the agitation on line C of Table 6-4.
6. Place the test tube in the rack and allow it to settle for 30 seconds or so. Record the appearance of the aqueous and organic layers on lines D and E of Table 6-4.
7. After the two layers have separated completely, use the eyedropper to draw off the top (organic) layer as completely as possible. Transfer the organic layer to the watch glass and set it aside to allow the solvent to evaporate.
8. Add about 1 mL of lighter fluid to the aqueous solution. Stopper the tube, agitate it, and wait for the layers to separate. Record the appearance of the aqueous and organic layers on lines F and G of Table 6-4.

CAUTIONS

Iodine is toxic and irritating, and stains skins and clothing. Stains can be removed with a dilute solution of sodium thiosulfate. Lighter fluid is flammable. Wear splash goggles, gloves, and protective clothing.

OPTIONAL ACTIVITIES

If you have time and the required materials, consider performing these optional activities:

- Repeat the experiment, beginning with two test tubes, each of which contains 10 mL of water and 1 mL of Lugol's solution. Add 1 mL of lighter fluid to the first test tube and add 5 mL of lighter fluid to the second. Agitate the tubes. After settling, compare the appearances of the aqueous layers to determine whether using additional lighter fluid causes more of the iodine to be extracted from the aqueous layer.

- Repeat the experiment, substituting three or four different food colorings for the iodine (one color per tube).

DISPOSAL: The aqueous waste can be flushed down the drain with plenty of water. The few milligrams of solid iodine can be flushed down the drain or disposed of with solid household waste.

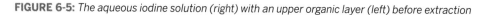

FIGURE 6-5: *The aqueous iodine solution (right) with an upper organic layer (left) before extraction*

TABLE 6-4: *Solvent extraction of iodine—observed data*

Layer	Appearance
A. Aqueous iodine solution	
B. Organic layer (before agitation)	
C. Mixed layers (after agitation)	
D. Aqueous layer (after settling)	
E. Organic layer (after settling)	
F. Aqueous layer (after second extraction)	
G. Organic layer (after second extraction)	

REVIEW QUESTIONS

Q1: The colors of the two layers after agitating and settling make it clear that most of the iodine was extracted from the aqueous layer into the organic layer, but that does not explain the color change from orange in the aqueous layer to magenta in the organic layer. Use the Internet to research the cause for this color change.

Q2: You are running a solvent extraction to isolate the products of a synthesis, using an aqueous layer and an organic layer. After agitation and settling, the two layers are of very similar appearance. You know that the organic layer contains your product and the aqueous layer is the waste layer, but you are not 100% certain which layer is which. What simple test might you perform to determine which is the organic layer?

LABORATORY 6.5:

CHROMATOGRAPHY: TWO-PHASE SEPARATION OF MIXTURES

Chromatography (from the Greek for "color-writing") is a method for separating mixtures. In chromatography, a mixture (called the *analyte* or *sample*) dissolved in a *mobile phase* (sometimes called the solvent or *carrier*) passes through a *stationary phase* (sometimes called the *substrate*). During the passage, various components of the analyte carried by the mobile phase selectively adhere to the stationary phase with greater or less affinity, separating the components physically across the stationary phase.

The Russian botanist Mikhail Semyonovich Tsvet invented chromatography in 1900, and used it to separate chlorophyll and other plant pigments. Tsvet named his new procedure "chromatography" because his chromatograms were literally colorful. Nowadays, chromatography is often used to separate colorless compounds, so the original rationale for the name no longer applies.

The simplest chromatography method—the one we explore in this laboratory—is called *paper chromatography*. Unlike other chromatography methods, paper chromatography requires no expensive equipment or special materials. In paper chromatography, the mobile phase is water or another solvent, and the stationary phase is paper. The solvent is absorbed by the paper, and dissolves a spot of the analyte. As the solvent is drawn up the paper by capillary action, the various dissolved components of the mixture are deposited on the paper at different distances from the original spot.

Other chromatography methods—including thin-layer chromatography (TLC), gas chromatography, and liquid chromatography—are often used in laboratories and industrial processes. These other methods have various advantages compared with paper chromatography, including faster throughput, sharper separations, higher sensitivity, and smaller required analyte amounts. Most chromatography done in laboratories is *analytical chromatography*, which requires only small amounts of analyte and is used to separate mixtures for subsequent instrumental or wet analyses and to follow reactions to completion or equilibration. Conversely, industrial chromatography is *preparative chromatography*, which uses

REQUIRED EQUIPMENT AND SUPPLIES

- ☐ goggles, gloves, and protective clothing
- ☐ chromatography jar and cover (3)
- ☐ chromatography paper strips (12)
- ☐ chromatography paper squares (6)
- ☐ black and brown permanent and water-soluble marking pens
- ☐ pencil
- ☐ ruler (mm scale)
- ☐ paper clips (12)
- ☐ stiff wire (3 segments)
- ☐ water
- ☐ acetone (as required)
- ☐ petroleum ether (as required)

SUBSTITUTIONS AND MODIFICATIONS

- You may substitute any wide-mouth containers of appropriate size for the chromatography jars. Ideally, use transparent jars that are large enough for a 15 cm paper square. Use containers made of glass or another material that is impervious to acetone and petroleum ether. To minimize problems caused by evaporation of the solvent, cover the containers while they are in use.

- Although chromatography paper gives the best results, you may substitute filter paper if chromatography paper is unavailable. Whatever paper you use, you'll need strips about as long as your chromatography jars are tall. The squares should be as large as possible, but small enough to fit the chromatography jar without contacting the side.

- You may substitute marking pens in colors other than brown and black. I chose those colors because the inks used in them are usually a mix of two or more different dyes. *(continues ...)*

SUBSTITUTIONS AND MODIFICATIONS (continued)

- A coat hanger trimmed to length works well for the stiff wire. You'll need three segments, one for each of the chromatography jars, each longer than the diameter of the jar. These wires will be used with the paper clips to suspend the chromatography strips and squares inside the jars.

- You may substitute methyl ethyl ketone (MEK) for acetone.

- You may substitute any other nonpolar solvent, such as lighter fluid or toluene, for petroleum ether.

CAUTIONS

Most of the solvents used in this laboratory are flammable. Use extreme care, avoid open flame, and have a fire extinguisher handy. Wear splash goggles, gloves, and protective clothing.

DR. PAUL JONES COMMENTS:

You can generally correct mid-experiment if you started at too low of a polarity; it is difficult to make the correction the other direction on the fly.

large (sometimes huge) amounts of analyte, and is used to purify compounds on a commercial scale.

In paper chromatography, it's important to choose an appropriate solvent. **Polar solvents** are more efficient carriers for polar analytes, such as ionic compounds. **Nonpolar solvents** are more efficient carriers for molecular (covalent) compounds. Some examples of common solvents in decreasing order of polarity are:

- water
- amides (e.g., N,N-dimethylformamide)
- alcohols (e.g., ethanol, isopropanol)
- ketones (e.g., acetone, methyl ethyl ketone)
- esters (e.g., ethyl acetate, amyl acetate)
- chlorocarbons (e.g., dichloromethane, carbon tetrachloride, chloroform)
- ethers (diethyl ether)
- aromatics (benzene, toluene)
- alkanes (hexanes, heptanes, petroleum ether)

Mixed solvents are used in paper chromatography to separate mixtures that contain both polar and nonpolar compounds, or to increase separation of mixtures of components that have similar behavior with a single solvent. The rule of thumb for mixed solvents is that a small amount of a polar solvent mixed with a large amount of a nonpolar solvent tends to behave as a polar solvent.

The key metric for paper chromatography is called the **retardation factor** (R_f), which is the ratio of distance the analyte moves from the initial point to the distance the solvent front moves from the initial point. In other words, R_f = (migration distance of the analyte)/(migration distance of the solvent).

For example, if a spot of analyte migrates 2 cm from the initial point during the time it take the solvent front to migrate 5 cm from the initial point, the R_f for that compound with that particular

solvent and that particular substrate is 0.40 (2 cm/5 cm). The R_f of a particular compound is a dimensionless number that is fixed for any particular combination of solvent and substrate, but may vary dramatically for other combinations of solvent and substrate. (Many texts incorrectly define R_f as **retention factor** (k), which is a related but separate concept.)

In this lab, we'll examine many aspects of paper chromatography, including ionic and molecular analytes, polar and nonpolar solvents, and two-dimensional chromatography.

PROCEDURE

This lab has two parts. In Part I, we'll use basic paper chromatography to examine the effect of polar and nonpolar solvents on the dyes contained in permanent and water-soluble marking pens. In Part II, we'll use two-dimensional chromatography to achieve better separation of these dyes.

PART I: BASIC CHROMATOGRAPHY

1. If you have not already done so, put on your splash goggles, gloves, and protective clothing.
2. Label each chromatography jar with the solvent that it will contain. Transfer enough solvent into each chromatography jar to provide about 2.5 cm of liquid in the bottom of the jar. Cover the jars. Ideally, we want the air in each jar to become saturated with the vapor from the solvent contained in that jar.
3. Prepare 12 chromatography strips, each about 15 mm to 25 mm wide and about as long as your chromatography jars are tall. Draw a fine pencil line across each strip about 4 cm from one end. At the opposite end, label the strips A through L.
4. Prepare the 12 labeled chromatography strips by spotting each of them with the water-soluble and permanent black and brown marking pens in the center of the pencil line. The goal is to create spots that are as small and as

concentrated as possible. Choose one of the pens and touch it to the center of the pencil line on strip A. Allow the paper to absorb ink until there is a spot 2 to 3 mm in diameter. Remove the pen, and allow the ink spot to dry completely while you spot strips B and C with the same pen (one strip for each of the three solvents). Return to strip A and place the tip of the marker pen in the center of the existing ink spot. Allow the paper to absorb more ink without significantly increasing the size of the spot. Retreat strips B and C similarly, and repeat this process until each strip has been treated at least five times. Repeat this procedure using the second marking pen for strips D, E, and F, the third marking pen for strips G, H, and I, and the fourth marking pen for strips J, K, and L. At the end of this procedure, you have 12 strips, 3 marked with each pen: the black water-soluble pen, the brown water-soluble pen, the black permanent pen, and the brown permanent pen. Record the analyte (pen type) used for each strip in Table 6-5.

5. Attach a paper clip to the labeled end of each strip.

6. Thread the stiff wire through the paper clips for strips A, D, G, and J (one strip of each marking pen type). Suspend those four strips in the first chromatography jar. The bottom of each strip should be immersed about 2 cm into the solvent. Make sure that the spot on each strip remains well above the level of the solvent. Repeat this procedure to suspend strips B, E, H, and K in the second chromatography jar, and strips C, F, I, and L in the third chromatography jar.

7. Observe the rate at which each solvent is drawn up the paper by capillary action. Different solvents and different types of paper progress at different rates. Typically, it might take from five minutes to one hour for the solvent front to approach the top of the strip. If the solvent front stops climbing before it reaches the top of the strip, that means that the capillary action and evaporation rate have reached equilibrium. Stop the process for strips using that solvent when the solvent front reaches that level. Otherwise, allow the solvent front to approach (but not reach) the top of the strip.

8. When the solvent front approaches the top of a set of strips (or stops climbing), immediately remove those strips from the chromatography jar. Use the pencil to make a small mark at the highest point that the solvent front reached on each strip. Suspend the strips and allow them to dry thoroughly.

9. For each strip, measure the distance between the pencil line and the maximum height reached by the solvent front. Record that value on the strip itself and in Table 6-5. Also measure the distance between the pencil line and any significant color spots on the strip—measure from the line to the middle of the spot—recording their descriptions, distances, and R_f values in Table 6-5.

DON'T OVERDEVELOP

As you are waiting for the strips to develop, you can start work on Part II. Keep a close eye on the strips, though. If the solvent front reaches the top of the strip, that strip is ruined, because an accurate R_f value can no longer be calculated.

Figure 6-6 shows five example chromatograms. None of these show particularly good separation of the components, but all illustrate some important issues.

From the left, the first two chromatograms are of black and blue permanent markers, using acetone as the solvent. (Neither dye was at all soluble in water.) No separation is evident, which almost certainly means that these pens use a single dye. (It's just possible, although very unlikely, that two or more dyes were used that had identical characteristics with this substrate and solvent, in which case there would also be no separation.) Clearly, the blue dye bonds very tightly to the acetone solvent, and quite loosely to the paper substrate, which accounts for its very sharply delineated spot. Conversely, the black dye, although readily soluble in acetone, clearly also bonds more tightly to the paper substrate, accounting for the strung-out spot it produces.

The third and fourth strips were spotted with green food coloring dye, which is clearly a mixture of dyes. The third strip used the nonpolar acetone as a solvent. The blue component of the green food coloring dye obviously binds tightly to acetone and reasonably tightly to the paper substrate. The yellow component

FIGURE 6-6: *These chromatograms show the separation of the ink and food coloring components with polar (water) and nonpolar (acetone) solvents*

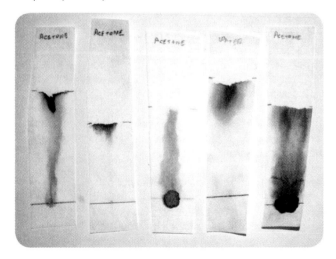

is just emerging from the spot, showing that, although it is at least somewhat soluble in acetone, it binds to the acetone solvent and the paper substrate about equally. The fourth strip was also spotted with green food coloring dye, but uses water, which is extremely polar, as the solvent. Both the blue and yellow components of the dye bind tightly to water, and were therefore carried far up the strip. Note that some separation is evident, with the blue component concentrated toward the top of the solvent front and the yellow component lagging slightly.

Finally, the fifth strip was spotted with a mixture of green and red food coloring dye, and developed using acetone. Clearly, the red component of the mixed dyes bonds very tightly to acetone, and is concentrated and relatively sharply delineated near the solvent front. The blue component lags somewhat, and is more spread out. As was true with the third strip, the yellow component has barely made any progress, and is visible as a brownish area near the starting point, where the yellow dye is mixed with some remaining red and blue dye.

TABLE 6-5: *Basic chromatography—observed data*

Strip	Analyte	Solvent	Solvent front	Notes
Example	Brown permanent pen	Acetone	9.15 cm	Red component at 3.80 cm (R_f 0.42); green at 6.75 cm (R_f 0.74)
A.				
B.				
C.				
D.				
E.				
F.				
G.				
H.				
I.				
J.				
K.				
L.				

PART II: TWO-DIMENSIONAL CHROMATOGRAPHY

If an analyte contains several different compounds, it sometimes happens that a particular solvent achieves better separation for some of the compounds than for others. Another solvent might provide better separation of those problem compounds, but poorer separation of the compounds that were well separated by the first solvent. One way to separate such analyte mixtures is to use two passes with different solvents, reorienting the substrate by 90° between the runs. The solvent used in the first pass achieves good separation of some of the components. Those components that are not well separated by the first solvent are separated by the second solvent. Because the new solvent front moves at 90° relative to the first solvent front, the resulting chromatogram is two-dimensional. Such two-dimensional chromatograms yield characteristic (and often colorful) patterns even with complex mixtures of analytes.

1. If you have not already done so, put on your splash goggles, gloves, and protective clothing.
2. Prepare six squares of chromatography paper and label them A through F.
3. Draw a pencil line parallel to and about 4 cm from one edge of each square, and label that line 1. Draw a second perpendicular pencil line about 4 cm from an adjoining edge, and label that line 2.
4. Spot each square at the intersection of the lines, as described in Part I. Spot squares A, B, and C with both the brown and black permanent marking pens, at least five times each. Spot squares D, E, and F with both the brown and black water-soluble marking pens, at least five times each. Record the analyte for each square in Table 6-6.
5. Using the stiff wire and two paper clips, suspend square A in the first chromatography jar, square B in the second jar, and square C in the third jar.
6. When the solvent front approaches the top of each square, remove the square from the chromatography jar, mark the highest point reached by the solvent front, and suspend the square to dry.
7. While the first set of squares is drying, suspend square D in the first chromatography jar, square E in the second jar, and square F in the third jar. When the solvent front approaches the top of each square, remove the square from the chromatography jar, mark the highest point reached by the solvent front, and suspend the square to dry.
8. When the first set of squares is dry, rotate square A 90° to put the second pencil line parallel with the bottom of the chromatography jar and suspend square A in the second chromatography jar. Repeat this procedure to put square B in the third jar and square C in the first jar. When the solvent front approaches the top of each square, remove the square from the chromatography jar, mark the highest point reached by the solvent front, and suspend the square to dry.

TABLE 6-6: *Two-dimensional chromatography—observed data*

Square	Analyte	Solvent 1	Solvent front 1	Solvent 2	Solvent front 2
Example	*Permanent pens*	*Acetone*	*8.85 cm*	*Petroleum ether*	*9.20 cm*
A.					
B.					
C.					
D.					
E.					
F.					

9. When the second set of squares is dry, rotate square D 90° to put the second pencil line parallel with the bottom of the chromatography jar and suspend square D in the second chromatography jar. Repeat this procedure to put square E in the third jar and square F in the first jar. When the solvent front approaches the top of each square, remove the square from the chromatography jar, mark the highest point reached by the solvent front, and suspend the square to dry.

DR. PAUL JONES COMMENTS:

You might also try chromatography of an uncolored substance. Dissolve aspirin in acetone (or water) and apply to paper, run it up the paper and let the paper dry. Of course, you then have a problem. The aspirin is colorless, so how can you see where it is on the paper? One common method used to make invisible substances visible on the chromatography paper is iodine fuming. Place a few crystals of iodine on the bottom of a large beaker or similar container. Put the chromatogram in the container, and cover the container with a watchglass or plastic wrap. The iodine vaporizes and reacts with the aspirin (and many other analytes) to form a visible mark on the paper. At room temperature, this process may take several hours. You can speed things up by heating the bottom of the beaker gently with an alcohol lamp or by partially submerging the beaker in warm water to vaporize the iodine quickly. If you're going to run a lot of chromatograms, it makes sense to keep an iodine-saturated container for this purpose. That way you don't have to wait.

DISPOSAL: Aqueous waste can be flushed down the drain with plenty of water. Unwanted chromatograms can be disposed of with household waste. Acetone, petroleum ether, and other organic solvents can be allowed to evaporate.

OPTIONAL ACTIVITIES

If you have time and the required materials, consider performing these optional activities:

- Repeat the experiment, using polar and nonpolar solvents, with an extract of the coloring matter contained in spring or summer leaves. (Extract the analytes by macerating the leaves manually or in a blender and soaking them in small amounts of various solvents. Evaporate as much as possible of the solvent to concentrate the analytes.) In the autumn, after the leaves have turned, collect samples from the same types of trees. Extract the analytes and run chromatograms with the same solvents that you used for the spring or summer leaves. Determine whether the yellow, red, and orange colors of autumn leaves are caused by substances present only in autumn leaves or if those substances are also present in spring or summer leaves and are masked by the bright green of the chlorophyll.

- Make your own TLC (thin-layer chromatography) plates using microscope slides. Use a thin layer of egg white as the adhesive. Try different substrates, such as cornstarch, powdered silica (from silica gel packets), talcum powder, and so on. Test your TLC plates by using the same analytes and solvents that you used with paper to determine whether you can achieve better separations.

REVIEW QUESTIONS

Q1: What is the purpose of chromatography?

Q2: By what mechanism does chromatography separate the components of a mixture?

Q3: Why did we use a pencil rather than a pen to label and mark the chromatography paper?

Q4: Why should the solvent front be marked immediately after the chromatograph develops?

Q5: Define the term "R_f value."

Q6: How is an R_f value calculated?

Q7: On a particular chromatogram, the solvent front moved 9.5 cm from the point of application and one of the components moved 6.2 cm from the point of application. What is the R_f value for that component?

Q8: Two of the components of an analyte mixture have R_f values of 0.25 and 0.83. Which of these components is more soluble in the mobile phase? How do you know?

_____

_____

_____

Q9: You have run a chromatogram on an analyte mixture in which one of the components remains unidentified. You have calculated an R_f value for that unknown analyte and are comparing that value against a table of known R_f values. In addition to the R_f value itself, what other information do you need about the known value to make your comparison valid? Why do you need this additional information?

_____

_____

_____

_____

_____

LABORATORY 6.6:
DETERMINE THE FORMULA OF A HYDRATE

Many ionic compounds exist in two or more forms. The **anhydrous** form of the compound contains only molecules of the compound itself. The **hydrated** form or forms of the compound contains molecules of the compound and one or more molecules of water loosely bound to each molecule of the compound. These water molecules are referred to as **water of hydration** or **water of crystallization**, and are incorporated into the crystalline lattice as the compound crystallizes from an aqueous solution.

Because these water molecules assume defined positions within the crystalline lattice, the proportion of water molecules to compound molecules is fixed and specific. For example, copper sulfate exists as an anhydrous compound ($CuSO_4$) and in hydrated form as the pentahydrate ($CuSO_4 \cdot 5H_2O$). Copper sulfate does not exist in the tetrahydrate ($CuSO_4 \cdot 4H_2O$) or hexahydrate ($CuSO_4 \cdot 6H_2O$) forms, because the physical geometry of the crystalline lattice does not permit four or six water molecules to associate with one copper sulfate molecule. The number of molecules of water in a hydrate is usually an integer, but not always. For example, some hydrates exist in the form $X_2 \cdot 5H_2O$, where each molecule of the compound X is associated with a fractional number (in this, case 2.5) molecules of water.

Some compounds, including copper sulfate, have only one stable hydrated form. (Monohydrate and trihydrate forms of copper sulfate are known, but are difficult to prepare and tend to spontaneously convert to the more stable anhydrous or pentahydrate forms by absorbing or giving up water molecules.) Other compounds have two or more common hydrated forms. For example, sodium carbonate exists in anhydrous form (Na_2CO_3), monohydrate form ($Na_2CO_3 \cdot 1H_2O$), heptahydrate form ($Na_2CO_3 \cdot 7H_2O$), and decahydrate form ($Na_2CO_3 \cdot 10H_2O$).

Many anhydrous compounds are **hygroscopic**, which means they absorb water vapor from the air and are gradually converted to a hydrated form. Such compounds, such as calcium chloride ($CaCl_2$), are often used as drying agents. (Some of these compounds absorb so much water vapor from the air that they actually dissolve in the absorbed water, a property called

REQUIRED EQUIPMENT AND SUPPLIES

☐ goggles, gloves, and protective clothing

☐ balance and weighing papers

☐ crucible with cover and tongs

☐ gas burner

☐ ring stand, ring, and clay triangle

☐ copper sulfate pentahydrate (~5 g)

SUBSTITUTIONS AND MODIFICATIONS

• You may substitute a Pyrex custard cup and saucer or similar heat-resistant items for the crucible and cover, which also eliminates the need for the ring stand, ring, and clay triangle. Make sure that the saucer allows venting of the water vapor driven off during heating. You can place a small nail or similar heat-resistant object between the lip of the custard cup and the saucer to provide a vent. A Pyrex vessel with a small spout (like a liquid measuring cup) will vent nicely as well. If you use a Pyrex container with a gas burner, use ceramic wire gauze between the flame and the container.

• You may substitute a test tube for the crucible and use a correspondingly smaller amount of copper sulfate pentahydrate. When you heat the test tube, hold it nearly horizontal to keep the hydrate spread out in as thin a layer as possible. With a test tube, you'll find that the water vapor that's driven off tends to condense in the upper, cooler part of the test tube. You can eliminate this problem by also heating the upper part of the test tube to vaporize the water.

• If you use a Pyrex custard cup or similar substitute for the crucible, you can also substitute a hotplate for the gas burner. If you do that, don't place the custard cup directly on the burner coil. Use a large tin can lid or burner to isolate the custard cup from the burner element.

• You may substitute any other available hydrate for the copper sulfate pentahydrate. Magnesium sulfate heptahydrate ($MgSO_4 \cdot 7H_2O$) is inexpensive and widely available in drugstores as Epsom salts.

deliquescence.) Conversely, the water molecules in some hydrated compounds are so loosely bound that the compound spontaneously loses some or all of its water of hydration if left in a dry environment, a property called **efflorescence.** Some compounds may be either hygroscopic or efflorescent, depending on the temperature and humidity of the environment. For example, anhydrous copper sulfate exposed to a humid atmosphere gradually absorbs water vapor and is converted to the pentahydrate form, and copper sulfate pentahydrate exposed to warm, dry air gradually loses water and is converted to the anhydrous form.

Because the water of crystallization in a hydrate can be driven off by heating the hydrate, a hydrate is actually a mixture of an anhydrous salt with water rather than a separate compound. (Recall that a mixture is a substance that can be separated into its component parts by physical means, such as heating, as opposed to a substance that can be separated into its component parts only by using chemical means.) Because the number of water molecules in a hydrate are in fixed proportion to the number of molecules of the compound, it's possible to determine that fixed proportion by weighing a sample of a hydrate, heating the compound to drive off the water of crystallization, weighing the resulting anhydrous compound, and using the mass difference between the hydrated and anhydrous forms to calculate the relationship.

In this laboratory, we'll heat hydrated copper sulfate to drive off the water of crystallization and use the mass differential to determine how many molecules of water are associated with each molecule of hydrated copper sulfate. We could have used any number of common hydrates, but we chose copper sulfate because the hydrated and anhydrous forms have distinctly different appearances. In hydrated form, copper sulfate forms brilliant blue crystals; in anhydrous form, copper sulfate is a white powder (see Figure 6-7).

PROCEDURE

1. If you have not already done so, put on your splash goggles, gloves, and protective clothing.
2. Set up your ring stand, support ring, clay triangle, and burner. Heat the crucible and cover gently for a minute or two to vaporize any moisture.
3. Remove the heat and allow the crucible and cover to cool to room temperature, which may require 10 or 15 minutes.
4. Weigh the crucible and cover, and record their mass to 0.01 g on line A of Table 6-7.
5. Transfer about 5.0 g of copper sulfate pentahydrate to the crucible. (The copper sulfate pentahydrate should be in the form of fine crystals. If it is in the form of large lumps, use your mortar and pestle to crush it into finer crystals.)
6. Reweigh the crucible, cover, and contents and record the mass to 0.01 g on line B of Table 6-7. Subtract the mass of the empty crucible and lid from the mass of the crucible with the copper sulfate and record the initial mass of the copper sulfate pentahydrate on line C of Table 6-7.
7. Place the crucible and cover on the heat source and begin heating them gently. As the crucible warms up, increase the heat gradually until it is at its highest setting. Continue heating the crucible for at least 15 minutes.
8. Remove the heat and allow the crucible, cover, and contents to cool to room temperature, which may require 10 or 15 minutes.
9. After you are sure the crucible has cooled, reweigh crucible, lid, and contents. Record the mass to 0.01 g on line D of Table 6-7. Subtract the initial mass of the crucible and lid (line A) from this value, and record the mass of the anhydrous copper sulfate to 0.01 g on line E of Table 6-7. Subtract the mass of the anhydrous copper sulfate (line E) from the mass of the copper sulfate pentahydrate (line C), and record the mass loss to 0.01 g on line F of Table 6-7.

CAUTIONS

This laboratory uses strong heat. Use extreme care with the heat source and hot objects. A hot crucible looks exactly like a cold crucible. Wear splash goggles, gloves, and protective clothing.

DISPOSAL: Retain the anhydrous copper sulfate for use in later experiments. You can store it in a sealed bottle, or allow it to remain exposed to air until it rehydrates to form blue crystals.

FIGURE 6-7: *Copper sulfate samples before (left) and after heating (right) to remove water of crystallization*

TABLE 6-7: *Formula of a hydrate—observed and calculated data*

Item	Data
A. Mass of crucible and cover	_____.__ g
B. Initial mass (crucible, cover, and hydrate salt)	_____.__ g
C. Mass of hydrate salt (B – A)	_____.__ g
D. Final mass (crucible, cover, and anhydrous salt)	_____.__ g
E. Mass of anhydrous salt (D – A)	_____.__ g
F. Mass loss (C – E)	_____.__ g

OPTIONAL ACTIVITIES

If you have time and the required materials, consider performing these optional activities:

- Repeat the experiment, substituting magnesium sulfate heptahydrate (Epsom salts) for the copper sulfate pentahydrate.
- Repeat the experiment, substituting hydrated sodium carbonate (washing soda) for the copper sulfate pentahydrate.

REVIEW QUESTIONS

Q1: How does anhydrous copper sulfate differ in appearance from hydrated copper sulfate?

Q2: The formula weight of hydrated copper sulfate is 249.7 grams per mole (g/mol). How many moles of hydrated copper sulfate were present in your sample (line C of Table 6-7)?

Q3: The formula weight of anhydrous copper sulfate is 159.6 g/mol. How many moles of anhydrous copper sulfate were present in your sample (line E of Table 6-7)?

_____

_____

_____

_____

Q4: Would you expect the number of moles of hydrated copper sulfate and the number of moles of anhydrous copper sulfate to be the same? Why or why not?

_____

_____

_____

_____

Q5: Assuming that all of the mass loss (line F of Table 6-7) represented water of crystallization, how many moles of water are present in hydrated copper sulfate for each mole of copper sulfate? (Hint: the formula weight of water is 18.015 g/mol. Divide the mass loss by the formula weight of water to calculate the number of moles of water present in the hydrated sample. Divide that number by the number of moles of copper sulfate to determine the proportion of water molecules to copper sulfate molecules and round the result to the nearest whole number.)

_____

_____

_____

_____

Q6: How closely did your experimental results correspond to the actual value of five molecules of water per molecule of copper sulfate? Propose several explanations for any variation.

_____

_____

_____

_____

_____

_____

_____

Laboratory:
Solubility and Solutions

7

Chemists routinely make up and use solutions. A *solution* is a homogeneous mixture of two or more substances. In a solution, the *solute* is dissolved in another substance called the *solvent*. For example, if you add a teaspoon of sugar to a cup of coffee, sugar is the solute and coffee is the solvent.

The solute and solvent may be in any phase. One common form of solution is a solid solute (such as salt or sugar) dissolved in a liquid solvent (such as water). A solution in which water is the solvent is called an *aqueous solution*. Many other solvents are commonly used in chemistry labs, including ethanol and other alcohols, acetone, benzene, and other organic liquids. Some solutions are mixtures of one liquid dissolved in another liquid. For example, the isopropyl rubbing alcohol found in drugstores contains about 30% water (the solute, in this case) dissolved in about 70% isopropanol (the solvent). A solution may also be made by dissolving a gas in a liquid. For example, aqueous ammonia (often incorrectly called ammonium hydroxide) is a solution of ammonia gas dissolved in water, and hydrochloric acid is a solution of gaseous hydrogen chloride dissolved in water.

Although the most common types of solutions used in chemistry labs are solid-liquid, liquid-liquid, and gas-liquid solutions, there are other types of solutions. Mixtures of gases can also be thought of as solutions. Our atmosphere can accurately be described as a solution of about 21% oxygen gas in about 79% nitrogen gas, with minor amounts of other gases present. Similarly, two or more solids can form a solution. For example, stainless steel is a solution in which iron is the solvent and small amounts of chromium and other metals are the solutes.

SOLVENT OR SOLUTE?

Conventionally, in a solid-liquid mixture, the liquid is always considered to be the solvent and the solid the solute. Some solids, however, are so soluble in a given solvent that a very concentrated solution may contain more of the solid by mass and by volume than it contains of the liquid. In such cases, it's not unreasonable to consider the solid as the solvent and the liquid as the solute. In a liquid-liquid solution, the liquid present in the greater amount is considered the solvent.

The most important characteristic of a solution is its concentration. Concentration is sometimes referred to in general terms. A *dilute solution* is one that contains a relatively small amount of solute per volume of solvent. A *concentrated solution* is one that contains a relatively large amount of solute per volume of solvent. A *saturated solution* is one that contains the maximum amount of solute that the volume of solvent is capable of dissolving at a specified temperature. (A *supersaturated solution* contains more solute than the solvent is capable of dissolving; see Laboratory 6.3.)

For many applications, it's important to have a specific value for concentration. Chemists specify concentration in numerous ways, using the mass, volume, and/or number of moles of solute and solvent. Some ways, such as parts per million (ppm) are used primarily in specialized applications, such as trace metal analysis or environmental science. Others are obsolescent or seldom used.

Here are the primary methods chemists use to specify concentrations:

Molarity

Molarity, abbreviated *mol/L* or *M*, specifies the number of moles of solute per liter of *solution* (not per liter of *solvent*). Molarity is the most commonly used way to specify concentration. Technically, the word "molarity" and the unit symbol M are obsolete. The official replacements are *amount-of-substance concentration* and *mol/dm³*, neither of which

---

DR. MARY CHERVENAK COMMENTS:

Expressing concentration as ppm or wt/wt percentage is common in industry. As an example, biocides and preservatives are added to paint, adhesives, wash water, cosmetics, oil, metal working fluids, and sizing baths in ppm. Additionally, the components that form a synthetic latex or plastics are measured in ppm. The pharmaceutical industry is, of course, different.

DILUTE/CONCENTRATED VERSUS WEAK/STRONG

It's a mistake to refer to a dilute solution as "weak" or a concentrated solution as "strong." The words weak and strong have different meanings to a chemist. For example, a concentrated solution of acetic acid (a weak acid) is still a weak acid solution, and a dilute solution of hydrochloric acid (a strong acid) is still a strong acid solution.

---

anyone actually uses. The advantage of using molarity to specify concentration is that it makes it very easy to work with mole relationships. The disadvantages are that it is much more difficult to measure volumes accurately than it is to measure masses accurately, which makes it difficult to prepare solutions of exact molarities, and that the molarity of a solution changes with temperature because the mass of solute remains the same while the volume of the solution changes with temperature.

Molality

Molality, abbreviated *mol/kg* or *m*, specifies the number of moles of solute per kilogram of *solvent* (not per kilogram of *solution*). The primary advantage of using molality to specify concentration is that, unlike its volume, the mass of the solvent does not change with changes of temperature or pressure, so molality remains constant under changing environment conditions. Molality is used primarily in tasks that involve the colligative properties of solutions, which are covered in the following chapter. For the dilute aqueous solutions typically used in laboratories, molarity and molality are nearly the same, because nearly all of the mass of these solutions is accounted for by the solvent (water) and the mass of water at room temperature is almost exactly one kilogram per liter.

Normality

Normality, abbreviated *N*, specifies the number of *gram equivalents* of solute per liter of *solution* (not per liter of *solvent*). The concept of gram equivalents takes into account the dissolution of ionic salts in solution. For example, a 1.0 M solution of calcium chloride ($CaCl_2$) can be made by dissolving one mole (111.0 g) of anhydrous calcium chloride in water and bringing the volume up to 1.0 liter. The calcium chloride dissociates in solution into one mole of calcium ions and two moles of chloride ions. That solution is 1.0 N with respect to calcium ions, but 2.0 N with respect to chloride ions. Accordingly, it is nonsensical to label such a solution of calcium chloride as 1.0 N or 2.0 N, unless you specify the ion species to which the normality refers.

Acids and bases are often labeled with their normalities, the assumption being that normality for an acid always refers to the hydronium (H_3O^+) ion concentration and the normality for a base to the hydroxide (OH^-) ion concentration. For monoprotic acids such as hydrochloric acid (HCl) and nitric acid (HNO_3), normality and molarity are the same, because these acids dissociate in solution to yield only one mole of hydronium ions per mole of acid. For diprotic acids such as sulfuric acid (H_2SO_4), the normality is twice the molarity, because one mole of these acids dissociates in solution to form two moles of hydronium ions. For triprotic acids such as phosphoric acid (H_3PO_4), normality is three times molarity, because in solution these acids yield three hydronium ions per acid molecule. Similarly, dibasic molecules such as barium hydroxide, $Ba(OH)_2$, dissociate in solution to yield two moles of hydroxide per mole of compound, so for solutions of these compounds the normality is twice the molarity.

Mass percentage

Mass percentage, also called *weight-weight percentage* or *w/w*, specifies the mass of the solute as a percentage of the total mass of the solution. For example, dissolving 20.0 g of sucrose in 80.0 g of water yields 100 g of a 20% w/w sucrose solution. Mass percentage is often used to specify concentration for concentrated acids. For example, a bottle of reagent-grade hydrochloric acid may specify the contents as 37% HCl, which means that 100 g of that solution contains 37 g of dissolved HCl. (Such solutions are often also labeled with the density of the solution, which allows them to be measured volumetrically rather than requiring weighing to transfer an accurate amount of solute.)

Mass-volume percentage

Mass-volume percentage, also called *weight-volume percentage* or *w/v*, specifies the mass of the solute as a percentage of the total volume of the solution. For example, dissolving 5.0 g of iodine in ethanol and making up the final volume to 100.0 mL yields 100 mL of a 5% w/v iodine solution. Mass-volume percentage is often used to specify concentration for indicators such as phenolphthalein and other reagents that are typically used dropwise.

Volume-volume percentage

Volume-volume percentage, abbreviated *v/v*, specifies the volume of the solute as a percentage of the total volume of the solution, and is often used when mixing two liquids. For example, a 40% v/v solution of ethanol in water can be made by measuring 40 mL of 100% ethanol and adding water until the final volume is 100 mL. I phrased that last sentence very carefully, because volumes are not necessarily additive. For example, adding 60.0 mL of water to 40.0 mL of ethanol yields something less than 100.0 mL of solution.

Keep significant digits in mind when you make up and label solutions. The necessary accuracy of concentration depends on the purpose of the solution.

Bench solutions

Bench solutions specify concentration to only one, two, or three significant figures (e.g., 1 M, 1.0 M, or 1.02 M). For example, you might make up 500 mL of of a 1 M bench solution of dilute hydrochloric acid using a graduated cylinder to measure the concentrated acid and a large beaker to make up the solution to about 500 mL. The advantage of bench solutions is that they can be prepared quickly and can be used when the exact concentration doesn't matter.

A MOLE OF EGGS?

A mole is a specific number of objects, much like a dozen or a gross, but with a much larger numeric value. While it's convenient to group eggs by the dozen or small parts by the gross, atoms and molecules are so unimaginably tiny that we group them in much larger numbers. Rather than 12 or 144 objects, one mole (abbreviated "mol") of a substance contains $6.0223 \cdot 10^{23}$ (Avogadro's number) of the elementary entities (atoms or molecules) that make up that substance.

One mole of an element or compound is, for all practical purposes, that quantity whose mass in grams has the same numeric value as the atomic mass or molecular mass of the substance. For example, because the atomic mass of iron is 55.847 unified atomic mass units (u, also called Daltons, Da), one mole of elemental iron has a mass of 55.847 grams. Similarly, because the molecular mass of molecular oxygen is 31.9988, one mole of oxygen molecules has a mass of 31.9988 grams. (Conversely, one mole of atomic oxygen, whose atomic mass is 15.9994, has a mass of 15.9994 grams.)

Moles are often more convenient for chemists to use than masses, because moles represent specific numbers of atoms or molecules. For example, one atom of sodium (Na) reacts with one atom of chlorine (Cl) to form one molecule of sodium chloride (NaCl). Because one mole of sodium contains the same number of atoms as one mole of chlorine, one mole of sodium reacts with one mole of chlorine to form one mole of sodium chloride. Or, stated differently, 22.989768 grams (one mole) of sodium reacts with 35.4527 grams (one mole) of chlorine to form 58.442468 grams (one mole) of sodium chloride. Using moles rather than masses makes the proportions of reactants and products clear.

Standard solutions

Standard solutions are used in quantitative analyses and are made up to whatever level of accuracy is required. Typically, standard solutions are accurate to four or five significant figures (e.g., 1.002 M or 1.0008 M) and are *standardized* against a *reference standard* (a solution of which the concentration is known very accurately).

For example, to make up 1 L of a nominal 1.0000 M solution of sodium carbonate requires one mole (105.99 g) of sodium carbonate. You begin by weighing an amount of sodium carbonate as close to 105.99 g as possible, but the exact amount is less important than recording the actual mass to within 0.01 g. You might, for example, end up with 106.48 g of sodium carbonate in the balance pan (or 104.37 g) but the important thing is that you know the mass to within 0.01 g. After recording the actual mass, you transfer the sodium carbonate to a 1 L volumetric flask that contains perhaps 800 mL of distilled water and swirl or shake the flask to dissolve the solute. You then rinse any small amount of sodium carbonate that remains in the weighing boat into the flask, fill the flask to within a centimeter of the reference line, and mix the contents again by swirling or shaking the flask. Finally, you fill the flask exactly to the reference line using a dropper.

At this point, you have 1 L of sodium carbonate solution with the concentration known to a high degree of accuracy. Depending on how much sodium carbonate you actually used, that concentration may be (for example) 1.0046 M, and the storage container can be labeled accordingly. To verify the exact concentration, you can titrate the solution against a reference standard.

Stock solutions

Stock solutions are concentrated solutions of stable chemicals, often saturated or nearly saturated, that are normally diluted before use. Stock solutions may be purchased or made up. The most common examples of purchased stock solutions are concentrated acids (such as acetic, hydrochloric, nitric, and sulfuric acids), some bases (such as aqueous ammonia), and other common liquid laboratory chemicals, such as concentrated hydrogen peroxide and formalin.

Many chemists keep on hand a variety of made-up stock solutions of solid chemicals that they use frequently. Stock solutions minimize required storage space and make it easy and convenient to produce more dilute bench solutions simply by diluting the stock solution in some proportion. For example, a 12 M stock solution of hydrochloric acid can easily be diluted with an equal part of water to provide 6 M HCl, with three parts of water to provide 3 M HCl, or with 11 parts of water to provide 1 M HCl. Stock solutions can also be standardized and diluted more accurately to provide more dilute standard solutions.

When you're making up solutions, it's important to consider the *solubility* of the solute, which specifies the maximum amount of solute that dissolves in the solvent at a particular temperature. For example, you might decide to make up 100 mL of a 1.00 M potassium permanganate stock solution. You look up the formula weight of potassium permanganate and find that it is 158.04 g/mol, which means you need 15.8 g to make up 100 mL of 1.00 M solution. So far, so good.

But if you start weighing and dissolving without further checking, you're going to have an uh-oh moment. Why? Because, according to the **CRC Handbook**, at 20°C, 100 mL of water dissolves only 6.38 g of potassium permanganate and the other 9.42 g of potassium permanganate will remain undissolved in the bottom of your 100 mL volumetric flask. Oops. Time for a bit more figuring.

If the solubility of potassium permanganate in water at 20°C is 63.8 g per liter and the formula weight of potassium permanganate is 158.04 g/mol, that means that a saturated solution of potassium permanganate is [63.8 g] / [158.04 g/mol] = 0.40+ M. Because you don't want potassium permanganate crystallizing out of the stock solution if the temperature in your lab happens to fall below 20°C, make up your potassium permanganate stock solution to 0.3 M by dissolving 4.74 g of potassium permanganate and making the solution up to 100 mL.

MISCIBILITY

Some liquid solutes and solvents can form solutions in any proportion. For example, acetone is freely soluble in water. (Another way of saying this is that water is freely soluble in acetone.) That means a water-acetone solution may contain anything from just under 100% water to just under 100% acetone. Such liquids are referred to as *miscible*.

EVERYDAY SOLUTIONS

We're surrounded by solutions, quite literally, because the atmosphere itself is a solution. You can find many other examples of commonplace solutions just by looking around you:

- An average home contains hundreds of solutions, from beverages, vinegar, lemon juice, maple syrup, and vanilla extract in the kitchen to the bleach and fabric softener in the laundry room and the medicines and personal care products in the bathroom.

- When you turn the key to start your car, the power needed to crank the engine is supplied by a solution of sulfuric acid in the car battery. When the engine starts, it's fueled by gasoline, which is a complex solution of hydrocarbons and additives.

- When you record a television program to a rewritable DVD, your DVD recorder stores that program by melting small pits into the complex solid solution of rare-earth metals that makes up the recording surface of the disc.

- Our bodies are made up largely of water, which serves as the solvent for the many complex solutions that serve various functions in our bodies. For example, our blood plasma is a complex solution of proteins, salts, glucose and other carbohydrates, amino acids, hormones, vitamins, gases, and metabolic waste products. Cytosol, the fluid inside our cells, is a similarly complex solution.

STABLE VERSUS UNSTABLE

Before you make up a stock solution—particularly a standardized stock solution—check to make sure that the chemical is stable in solution. Some chemicals break down in solution, and others react with the container or air. For example, sodium hydroxide solutions absorb carbon dioxide from the air, which converts some of the sodium hydroxide to sodium carbonate. Sodium hydroxide solutions also dissolve glass containers, further changing the concentration and introducing contaminants. The best practice is to make up solutions of such chemicals only as needed. If you must make up stock solutions of unstable chemicals, at least make the minimum amounts possible and store them in inert plastic accordion bottles, such as those sold by photography supply stores. Opaque, screw-cap Nalgene bottles are also good for storage.

LABORATORY 7.1:

MAKE UP A MOLAR SOLUTION OF A SOLID CHEMICAL

In this laboratory, we make up 100 mL of a stock solution of copper (II) sulfate, which is used in many of the other labs in this book. Although we won't standardize the solution, we will make every effort to achieve an accurate concentration by weighing masses carefully and measuring volumes carefully.

The first question we need to answer is what the concentration of the stock solution should be. Ideally, we want the concentration to be as high as possible without quite being saturated. (We don't want copper sulfate crystallizing out of solution if one day the lab happens to be a bit cooler than usual or if some of the solvent evaporates.) So, the first thing we need to determine is the molarity of a saturated solution of copper sulfate at room temperature. From that, we can decide what the molarity of our stock solution should be, and calculate how much copper sulfate is required to make up 100 mL of stock solution at that molarity. We proceed as follows:

1. Looking up copper sulfate pentahydrate in a reference book, we find that its its solubility at 20°C is 317 g/L and its formula weight is 249.7 g/mol.
2. Dividing 317 g/L by 249.7 g/mol tells us that a saturated solution of copper sulfate contains about 1.27 mol/L, which is 1.27 M.
3. To give us a safety margin against the solution crystallizing at lower temperatures, we decided to make our stock solution 1.00 M, or 1.00 mol/L.
4. Because we're making up 100 mL (0.1 L) of solution, we need 0.1 moles of copper sulfate pentahydrate.
5. The formula weight of copper sulfate pentahydrate is 249.7 g/mol, so we need 24.97 g of copper sulfate pentahydrate for our 100 mL of solution.

The assay on the bottle of copper sulfate pentahydrate lists the contents as 99.0% copper sulfate pentahydrate, which means that this substance contains 1.0% of something other than copper sulfate pentahydrate (probably mostly metallic copper, copper oxide, and other insoluble copper salts). Dividing 24.97 g by 0.99 (99%) tells us that we need to weigh out 25.22 g of the substance in the bottle to get 24.97 g of copper sulfate pentahydrate. With the calculations complete, it's time to head into the lab and actually make up the stock solution of copper sulfate pentahydrate.

REQUIRED EQUIPMENT AND SUPPLIES

☐ goggles, gloves, and protective clothing

☐ balance and weighing papers

☐ volumetric flask, 100 mL

☐ graduated cylinder, 100 mL

☐ beaker, 150 mL

☐ funnel

☐ eye dropper

☐ wash bottle (distilled or deionized water)

☐ labeled storage bottle

☐ copper sulphate pentahydrate (25.22 g)

SUBSTITUTIONS AND MODIFICATIONS

- You may substitute a 100 mL graduated cylinder for the 100 mL volumetric flask, with some loss in accuracy.

- You may substitute any convenient mixing container of the appropriate size for the 150 mL beaker.

- You may substitute 16.12 grams of copper sulfate anhydrate (the anhydrous form) for the 25.22 g of the pentahydrate salt. (Note that this substitution assumes 99.0% purity.)

- If you are using the copper sulfate pentahydrate that you recrystallized yourself (see Lab 6.3), I suggest that you assume 99% purity.

- If you are using technical-grade copper sulfate pentahydrate that has no assay listed, I suggest you assume 96% purity and recalculate the required mass accordingly. This is one of the hazards of using technical-grade compounds; you're never certain of the actual purity.

PROCEDURE

1. If you have not already done so, put on your splash goggles, gloves, and protective clothing.
2. Place a weighing paper or boat on the balance and tare the balance to read 0.00 g.
3. Transfer copper sulfate pentahydrate crystals to the weighing paper until the balance reads as close as possible to 25.22 g. Read the mass to 0.01 g and record it. (It's not important to have exactly 25.22 g of copper sulfate on the weighing paper, but it is important to know the mass as exactly as possible.)
4. Use the graduated cylinder to measure about 85 mL of distilled water, and transfer it to the beaker.
5. Transfer the copper sulfate pentahydrate from the weighing paper to the beaker, making sure that all of the crystals are added to the beaker.
6. Swirl the beaker gently until all of the copper sulfate has dissolved. This may take a few minutes, because this solution is nearly saturated.
7. Place the funnel in the mouth of the volumetric flask, and carefully pour the copper sulfate solution into the volumetric flask.
8. Use the wash bottle to rinse the inside of the beaker with 2 or 3 mL of water, and add the rinse water to the volumetric flask, rinsing the funnel as you do so. (Be careful not to use too much rinse water; the volumetric flask already contains about 85 mL of solution, and we can't go over 100 mL.)
9. Repeat the rinse with another 2 or 3 mL of water. This procedure is called a quantitative transfer, and the goal is to make sure that all of the copper sulfate has been transferred to the volumetric flask.
10. Remove the funnel from the mouth of the volumetric flask, and use the dropper to add water dropwise until the level of solution in the flask reaches the reference line.
11. Insert the stopper in the volumetric flask and invert the flask several times to mix the solution thoroughly.
12. Using the funnel, transfer the solution from the volumetric flask to the labeled storage bottle.
13. Using the exact mass you recorded in step 3, calculate the molarity to the correct number of significant figures and record that molarity on the label. Also, record the date on the label.
14. Rinse the beaker, funnel, and volumetric flask and stopper with tap water and then with distilled water, and set them aside to dry.

CAUTIONS

Copper sulfate is moderately toxic by ingestion, inhalation, or skin contact. Wear splash goggles, gloves, and protective clothing at all times.

RUNNING THE NUMBERS

In this case, the calculation is trivial; we're making up a one-tenth of a liter of 1 M solution, so we obviously require 0.1 moles of the solute. For other volumes and molarities, the calculations may be a bit more complicated. For example, to determine how much solute is required to make up 25 mL of a 0.75 M solution of copper sulfate, you might calculate as follows:

(25 mL) · (1 L/1000 mL) · (0.75 mol/L) · (249.7 g/mol) = 4.68 g

Milliliters (mL) cancel out between multiplicands one and two, liters (L) cancel out between multiplicands two and three, and moles cancel out between multiplicands three and four, leaving only grams (g).

DISPOSAL: The weighing paper can be disposed of with household waste.

OPTIONAL ACTIVITIES

If you have the time and the required materials, consider performing these optional activities:

- Table 7-1 lists compounds for which stock solutions are handy to have on hand in any lab. Among them, these compounds represent the most common cation and anion species we work with in this book. For each compound, look up the formula and formula weight (don't forget to consider water of hydration) and enter it in the table. Look up the solubility of the compound and use it to determine an appropriate molarity for a stock solution. Calculate the mass of the compound that is required to make up 100 mL of the stock solution.

- For as many of these compounds as you have access to (and time available), mix up stock solutions. Record the actual mass that you use and determine the actual molarity of each of your stock solutions using the correct number of significant digits.

Table 7-1:

Make up a molar solution of a solid chemical: observed and calculated data

Compound	Formula	Formula Weight	Nominal M	Volume	
cupric sulfate pentahydrate	$CuSO_4 \cdot 5H_2O$	249.68 g/mol	1.0 M	100 mL	
aluminum nitrate		_____.___ g/mol	___.__ M	100 mL	
ammonium acetate		_____.___ g/mol	___.__ M	100 mL	
ammonium chloride		_____.___ g/mol	___.__ M	100 mL	
ammonium molybdate		_____.___ g/mol	___.__ M	100 mL	
ammonium nitrate		_____.___ g/mol	___.__ M	100 mL	
ammonium oxalate		_____.___ g/mol	___.__ M	100 mL	
barium chloride		_____.___ g/mol	___.__ M	100 mL	
barium nitrate		_____.___ g/mol	___.__ M	100 mL	
calcium nitrate		_____.___ g/mol	___.__ M	100 mL	
chromium (III) nitrate		_____.___ g/mol	___.__ M	100 mL	
cobalt (II) nitrate		_____.___ g/mol	___.__ M	100 mL	
copper (II) nitrate		_____.___ g/mol	___.__ M	100 mL	
iron(III) nitrate		_____.___ g/mol	___.__ M	100 mL	
iron (II) sulfate		_____.___ g/mol	___.__ M	100 mL	
lead nitrate		_____.___ g/mol	___.__ M	100 mL	

Calculated Mass	Actual Mass	Actual M
25.22 g	25.34 g	1.005 M
_____.___ g	_____.___ g	__._____ M
_____.___ g	_____.___ g	__._____ M
_____.___ g	_____.___ g	__._____ M
_____.___ g	_____.___ g	__._____ M
_____.___ g	_____.___ g	__._____ M
_____.___ g	_____.___ g	__._____ M
_____.___ g	_____.___ g	__._____ M
_____.___ g	_____.___ g	__._____ M
_____.___ g	_____.___ g	__._____ M
_____.___ g	_____.___ g	__._____ M
_____.___ g	_____.___ g	__._____ M
_____.___ g	_____.___ g	__._____ M
_____.___ g	_____.___ g	__._____ M
_____.___ g	_____.___ g	__._____ M
_____.___ g	_____.___ g	__._____ M

FIGURE 7-1: *Weighing the sample of copper sulfate*

Compound	Formula	Formula Weight	Nominal M	Volume
magnesium sulfate		_____.____ g/mol	___.__ M	100 mL
manganese(II) sulfate		_____.____ g/mol	___.__ M	100 mL
nickel(II) nitrate		_____.____ g/mol	___.__ M	100 mL
potassium bromide		_____.____ g/mol	___.__ M	100 mL
potassium chromate		_____.____ g/mol	___.__ M	100 mL
potassium ferrocyanide		_____.____ g/mol	___.__ M	100 mL
potassium hydrogen tartrate		_____.____ g/mol	___.__ M	100 mL
potassium iodide		_____.____ g/mol	___.__ M	100 mL
potassium nitrate		_____.____ g/mol	___.__ M	100 mL
potassium permanganate		_____.____ g/mol	___.__ M	100 mL
potassium thiocyanate		_____.____ g/mol	___.__ M	100 mL
silver nitrate		_____.____ g/mol	___.__ M	100 mL
sodium acetate		_____.____ g/mol	___.__ M	100 mL
sodium bicarbonate		_____.____ g/mol	___.__ M	100 mL
sodium bisulfite		_____.____ g/mol	___.__ M	100 mL
sodium borate		_____.____ g/mol	___.__ M	100 mL
sodium carbonate		_____.____ g/mol	___.__ M	100 mL
sodium hydroxide		_____.____ g/mol	___.__ M	100 mL
sodium nitrite		_____.____ g/mol	___.__ M	100 mL
sodium phosphate, tribasic		_____.____ g/mol	___.__ M	100 mL
sodium sulfate		_____.____ g/mol	___.__ M	100 mL
sodium sulfite		_____.____ g/mol	___.__ M	100 mL
sodium thiosulfate		_____.____ g/mol	___.__ M	100 mL
strontium nitrate		_____.____ g/mol	___.__ M	100 mL
zinc nitrate		_____.____ g/mol	___.__ M	100 mL

Calculated Mass	Actual Mass	Actual M
_____.___ g	_____.___ g	__._____ M
_____.___ g	_____.___ g	__._____ M
_____.___ g	_____.___ g	__._____ M
_____.___ g	_____.___ g	__._____ M
_____.___ g	_____.___ g	__._____ M
_____.___ g	_____.___ g	__._____ M
_____.___ g	_____.___ g	__._____ M
_____.___ g	_____.___ g	__._____ M
_____.___ g	_____.___ g	__._____ M
_____.___ g	_____.___ g	__._____ M
_____.___ g	_____.___ g	__._____ M
_____.___ g	_____.___ g	__._____ M
_____.___ g	_____.___ g	__._____ M
_____.___ g	_____.___ g	__._____ M
_____.___ g	_____.___ g	__._____ M
_____.___ g	_____.___ g	__._____ M
_____.___ g	_____.___ g	__._____ M
_____.___ g	_____.___ g	__._____ M
_____.___ g	_____.___ g	__._____ M
_____.___ g	_____.___ g	__._____ M
_____.___ g	_____.___ g	__._____ M
_____.___ g	_____.___ g	__._____ M
_____.___ g	_____.___ g	__._____ M
_____.___ g	_____.___ g	__._____ M

THESE CHEMICALS ARE PARTICULARLY USEFUL TO KEEP ON HAND:

- potassium bromide
- potassium iodide
- potassium permanganate
- silver nitrate
- sodium carbonate
- sodium hydroxide

LABELS THE EASY WAY

Precalculated and preformatted labels for each of the stock and working solutions I use in my lab can be downloaded from **www.homechemlab.com** in Microsoft Word and OpenOffice.org formats. You can save several hours of calculation and typing by downloading these labels.

Note, however, that the solute masses on the labels assume you are using 100% pure anhydrous salts, so if you use hydrated salts or salts of lesser purity, you'll need to adjust the actual masses of the solutes. For example, the label for copper (II) sulfate specifies 159.6 mg/mL of solute, which is correct for anhydrous copper (II) sulfate of 100% purity. I actually used 99% pure copper (II) sulfate pentahydrate, which meant that I needed 25.22 g of that salt to make up 100 mL of nominal 1.00 M copper (II) sulfate solution. When I weighed out the salt, I actually ended up with 25.34 g, which meant that my final solution was 1.005 M.

REVIEW QUESTIONS

Q1: The molarity of a solution can never be more than a close approximation. List at least five factors that may cause the actual molarity of a solution to differ from the calculated molarity of that solution.

Q2: Although Table 7-1 assumes 100 mL of each solution, there are some solutions (such as silver nitrate) that you may want to make up in smaller batches, because the chemical is expensive or because you expect to use smaller amounts of the solution. Similarly, there are some solutions that you may want to make up in larger batches (such as sodium chloride), either because the chemical is inexpensive or because you expect to use larger amounts of the solution. For what other reasons might you decide to make up a smaller or larger batch of a particular solution?

Q3: What characteristic or characteristics of a particular chemical make it easier to obtain a known mass of that chemical by measuring a known volume of a solution of known concentration of that chemical rather than by weighing the solid chemical?

LABORATORY 7.2:

MAKE UP A MOLAL SOLUTION OF A SOLID CHEMICAL

In this laboratory, we make up 100 mL of a 0.5 molal stock solution of potassium hexacyanoferrate(III) (potassium ferricyanide), which is used in several of the other labs in this book. Although we won't standardize the solution, we will make every effort to achieve an accurate concentration by weighing the masses of the solute and solvent carefully.

As always when making up a solution, the first thing to determine is the concentration of a saturated solution of the chemical to make sure the chemical is soluble enough to make up a solution of the desired concentration. From that information, we can calculate how much potassium ferricyanide is required. Proceed as follows:

1. Looking up potassium ferricyanide in a reference book, we find that its room-temperature solubility is about 330 g/L and its formula weight is 329.24 g/mol.
2. Dividing 330 g/L by 329.24 g/mol tells us that a saturated solution of potassium ferricyanide contains about 1.00 mol/L, which is to say 1.00 M. (In dilute solutions, molality and molarity are nearly equal, so this calculation is valid even though we're making up the solution to a specified molality.)
3. To give us a safety margin against the solution crystallizing at lower temperatures, we decide to make our stock solution 0.50 M, or 0.50 mol/L. We could make up the solution to 0.60, 0.75, or some other molality (0.90 m would be pushing it), but round numbers are easier to handle when you're calculating equivalents and measuring solutions for use in other tasks.
4. A one molal (1 m) solution is defined as one mole of solute dissolved in one kilogram of solvent. Because we're making up a 0.50 molal solution with 100 g of water as the solvent, we need 0.05 moles of potassium ferricyanide.
5. The formula weight of potassium ferricyanide is 329.24 g/mol, so we need 16.46 g of potassium ferricyanide for our solution.

REQUIRED EQUIPMENT AND SUPPLIES

☐ goggles, gloves, and protective clothing

☐ balance and weighing papers

☐ beaker, 150 mL

☐ funnel

☐ eyedropper

☐ wash bottle (distilled or deionized water)

☐ labeled storage bottle

☐ potassium ferricyanide (16.63 g)

SUBSTITUTIONS AND MODIFICATIONS

- You may substitute a foam cup or similar container for the 150 mL beaker. In fact, because the combined mass of the potassium ferricyanide and 100 g of water in a one molal solution is 116.63 g, using a glass beaker may exceed the maximum capacity of your balance. In that case, a lightweight foam cup is a better choice.

6. The assay on our bottle of reagent grade potassium ferricyanide lists the contents as 99% potassium ferricyanide, meaning that this substance may contain up to 1% of something other than potassium ferricyanide. Dividing 16.46 g by 0.99 (99%) tells us that we need to weigh out 16.63 g of the substance in the bottle to get 16.46 g (0.0500 mol) of potassium ferricyanide.

With the calculations complete, we're ready to prepare the solution.

PROCEDURE

1. If you have not already done so, put on your splash goggles, gloves, and protective clothing.
2. Place a weighing paper or boat on the balance and tare the balance to read 0.00 g.
3. Transfer potassium ferricyanide crystals to the weighing paper until the balance reads as closely as possible to 16.63 g. Read the mass to 0.01 g and record it. (It's not important to have exactly 16.63 g of potassium ferricyanide on the weighing paper, but it is important to know the mass as exactly as possible.)
4. Transfer the potassium ferricyanide to the beaker, place the beaker on the balance pan, and tare the balance to read 0.00 g.
5. Use the wash bottle to transfer distilled water to the beaker until the balance reads close to 100.00 g. Use the dropper to add distilled water dropwise until the balance reads as closely as possible to 100.00 g (Figure 7-2). Depending on the dropper, one drop of water may weigh from about 0.05 g to about 0.02 g. When you have 100.00 g of water in the beaker, or as close to that as you can get, read the mass to 0.01 g and record it.
6. Swirl the beaker gently until all of the potassium ferricyanide has dissolved.
7. Place the funnel in the mouth of the labeled storage bottle, and carefully pour the potassium ferricyanide solution into the bottle.
8. Using the exact masses you recorded in steps 3 and 5, calculate the molality to the correct number of significant figures and record that molality on the label. Also record the date on the label.
9. Rinse the beaker, funnel, and volumetric flask and stopper with tap water and then with distilled water, and set them aside to dry.

FIGURE 7-2:
Using a Beral pipette to bring the water mass up to 100.00 g

 CAUTIONS

Potassium ferricyanide is irritating to skin and eyes. Wear splash goggles and gloves.

DISPOSAL: The weighing paper can be disposed of with household waste.

REVIEW QUESTIONS

Q1: In this lab, we made up a 0.5 molal solution of potassium ferricyanide. What additional information would you need to calculate the molarity of that solution? Describe at least two procedures that you could use to determine the molarity.

Q2: After we dissolved the potassium ferricyanide, we did not rinse the beaker to transfer all of the remaining potassium ferricyanide solution quantitatively into the storage bottle. Why?

Q3: In what circumstances would it be useful to know the molality of a solution rather than its molarity?

LABORATORY 7.3:

MAKE UP A MOLAR SOLUTION OF A LIQUID CHEMICAL

Some common laboratory chemicals, such as sulfuric acid, are liquid at standard temperature and pressure. Many other chemicals, such as ammonia and hydrochloric acid, are gases, but are ordinarily stored and handled in the form of concentrated aqueous solutions. Whether the chemical is actually a liquid or is supplied in the form of a concentrated aqueous solution, the same principles apply to diluting the chemical to make up working solutions of known molarities.

In this laboratory, we make up 100 mL of 1.00 M hydrochloric acid, which is a common bench solution. (I actually make this stuff up 500 mL at a time, but not everyone has a 500 mL volumetric flask.) We'll make up this solution by diluting concentrated hydrochloric acid. The label on the bottle of reagent-grade hydrochloric acid gives us several pieces of information about its concentration:

- The mass percentage is specified as 36% to 38%, which means that 100 g of the concentrated acid contains between 36g and 38 g of dissolved HCl gas.

- The density is specified as ranging between 1.179 g/mL and 1.189 g/mL, which tells us the mass of a specific volume of the concentrated acid. (The density is also specified as between 22° and 23° Baumé, which is an older and obsolescent method for specifying density or specific gravity.)

- The molarity is specified as between 11.64 M and 12.39 M.

If they were mixing up a 1.0 M bench solution, most chemists would assume that the concentrated hydrochloric acid was 12.0 M, and add one part of concentrated acid to 11 parts water to make up the 1 M solution. To make up 100 mL of 1 M acid, we could measure 8.33 mL of the 12 M acid, add it to 70 mL or so of distilled water in a 100 mL volumetric flask, and make up the solution to 100 mL. In practice, that's how it's usually done.

But there are advantages to using mass percentage and density when making up dilute bench solutions, including higher accuracy and precision. Also, it's important to know how to make up solutions based on mass percentage and density, because not all concentrated aqueous solutions list a molarity.

REQUIRED EQUIPMENT AND SUPPLIES

☐ goggles, gloves, and protective clothing

☐ balance and weighing papers

☐ volumetric flask, 100 mL

☐ graduated cylinder, 10 mL

☐ beaker, 150 mL

☐ funnel

☐ eye dropper

☐ wash bottle (distilled or deionized water)

☐ labeled storage bottle

☐ hydrochloric acid, 37% (8.33 mL)

SUBSTITUTIONS AND MODIFICATIONS

- You may substitute a 100 mL graduated cylinder for the 100 mL volumetric flask, with some loss in accuracy.

- You may substitute any convenient mixing container of the appropriate size for the 150 mL beaker.

To make up solutions based on the mass percentage of a stock solution, begin by calculating the required number of moles of solute. For 100 mL (0.1 L) of 1 M hydrochloric acid, we need 0.1 moles of HCl. The formula weight of HCl is 36.46 g/mol, so we need 3.646 g of HCl. If the mass percentage of the concentrated acid is 37% (0.37), that means we need 3.646/0.37 = 9.854 g of the concentrated acid. To get that, we tare a small beaker or other container on our balance, and add concentrated acid to the beaker until the balance reads 9.854 g.

If we prefer to measure the concentrated acid volumetrically rather than gravimetrically, we also need to know the density of the solution to determine how many mL of the concentrated acid has a mass of 9.854 g. To calculate that value, we divide the required mass by the density of the solution. If the concentrated acid has a density of 1.183 g/mL, that means we need 9.854/1.183 = 8.330 mL.

Because the density and mass percentage of a solution are related, it's possible to determine either of them, if the value of the other is known or can be determined. For example, the label on the bottle of reagent-grade concentrated hydrochloric acid may specify a fairly wide range for mass percentage (such as 36% to 38%) and density (such as 1.179 to 1.189 g/mL). To get a better value for actual concentration, I tared a 100 mL volumetric flask, filled it to the reference line with my concentrated hydrochloric acid, and determined the mass of the concentrated acid. That turned out to be 118.31 g, or 1.183 g/mL.

In this lab, we'll make up 100 mL of 1.00 M hydrochloric acid, using our mass percentage calculations to determine how much concentrated acid to use. (If the values listed on your own bottle of concentrated hydrochloric acid differ significantly from these values, recalculate amounts based on the actual values of your concentrated hydrochloric acid.)

 CAUTIONS

Concentrated hydrochloric acid is corrosive and emits toxic and irritating fumes. Wear splash goggles, gloves, and protective clothing at all times, and work in a well-ventilated area or under a fume hood.

DR. MARY CHERVENAK COMMENTS:

Always add acid to water. Because most concentrated acids are denser than water, they will mix more readily when dropped into water. Water added on top of concentrated acid may sit on top of the acid, resulting in an interface that can generate a lot of heat and actually cause the water to boil.

PROCEDURE

1. If you have not already done so, put on your splash goggles, gloves, and protective clothing.
2. Place a clean, dry 10 mL graduated cylinder on the balance and tare the balance to read 0.00 g.
3. Using the dropper, transfer concentrated hydrochloric acid into the graduated cylinder until the balance reads as closely as possible to 9.854 g, which it should when you have transferred just over 8.3 mL of acid to the graduated cylinder. Note the mass and volume of the concentrated HCl as accurately as possible and record them for future use.
4. Fill the 100 mL volumetric flask about two-thirds full with distilled water.
5. Place the funnel in the mouth of the volumetric flask, and carefully pour the concentrated acid into the volumetric flask.
6. Use the wash bottle to rinse the inside of the graduated cylinder with several mL of water, and add the rinse water to the volumetric flask, rinsing the funnel as you do so.
7. Repeat the rinse with another several mL of water to make sure that the transfer is quantitative.
8. Use the water bottle to rinse the funnel again, and bring the solution level in the volumetric flask up to within 1 cm of the reference line.
9. Remove the funnel from the mouth of the volumetric flask, and use the dropper to add water dropwise until the level of solution in the flask reaches the reference line.
10. Insert the stopper in the volumetric flask and invert the flask several times to mix the solution thoroughly.
11. Using the funnel, transfer the solution from the volumetric flask to the labeled storage bottle.
12. Calculate the molarity to the correct number of significant figures and record that molarity on the label. Also record the date on the label.
13. Rinse the beaker, funnel, and volumetric flask and stopper with tap water and then with distilled water, and set them aside to dry.

DISPOSAL: There are no items that require disposal.

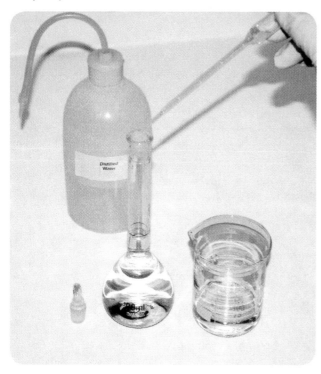

DR. MARY CHERVENAK COMMENTS:

Sulfuric and nitric acid solutions may require the donning of chemically impervious aprons. They eat through clothing, including lab coats. Even dilute solutions can ruin clothes (lots of tiny holes). I would recommend washing anything worn during this experiment separately from all other clothes. (Personal experience: in grad school, everything I owned was full of tiny holes.)

OPTIONAL ACTIVITIES

If you have time and the required materials, consider performing these optional activities:

- Make up 100 mL each of 1 M bench solutions of sulfuric acid, nitric acid, and aqueous ammonia.

- Look up values for the mass percentage and density of aqueous solutions of hydrochloric acid at various concentrations. Graph those values to determine whether the relationship is linear. Do the same for sulfuric acid.

REVIEW QUESTIONS

Q1: You need to make up 100 mL of 3 M nitric acid. The concentrated nitric acid you have on hand lists its mass percentage as 68%. Is this sufficient information for you to make up the 3 M bench solution? If not, why? If so, how much concentrated nitric acid would you dilute to make 100 mL of 3 M nitric acid?

Q2: You need to make up 100 mL of 0.5 M sulfuric acid. The concentrated sulfuric acid you have on hand lists its density as 1.84 g/mL. Is this sufficient information for you to make up the 0.5 M bench solution? If not, why? If so, how much concentrated sulfuric acid would you dilute to make 100 mL of 0.5 M sulfuric acid?

_____

_____

_____

_____

_____

_____

_____

Q3: You have on hand 100 mL of concentrated phosphoric acid. The label lists the mass percentage as 85.0% and the density as 1.68 g/mL. How much 2.00 M phosphoric acid solution could you make up from this amount of the concentrated acid?

_____

_____

_____

_____

Q4: You have on hand some concentrated muriatic acid from the hardware store that lists its contents as 31.45% HCl. You want to make up 1 L of 1.00 M HCl, but you are concerned that the mass percentage listed on the bottle may be inaccurate. You tare a 100 mL volumetric flask and use it to determine that 100 mL of your acid has a mass of 116.03 g. Based on that information, is the mass percentage listed on the bottle accurate? If not, by how much does it differ from the actual value?

_____

_____

_____

_____

_____

_____

_____

_____

_____

LABORATORY 7.4:
MAKE UP A MASS-TO-VOLUME PERCENTAGE SOLUTION

Mass-to-volume percentage solutions are more commonly used by pharmacists than by chemists, but these solutions still have a place in chemistry labs. The advantage of a mass-to-volume percentage solution is that it makes it easy to transfer a precise amount of solute by measuring the necessary volume of the solution. For example, a pharmacist may make up a solution that contains 3.00 g of a drug per 100 mL of solution. If the proper dosage of the drug is 150 mg, the pharmacist can direct the patient to take one teaspoon (5.0 mL). Similarly, if the proper dosage is 1.5 mg, the pharmacist can direct the patient to take one drop (0.05 mL).

The most common use of mass-to-volume solutions in chemistry labs is for indicators and similar solutions. For example, a chemist might keep on hand a 1% ethanol solution of phenolphthalein, a 0.2% aqueous solution of phenol red, and 0.4% aqueous solutions of bromocresol green and bromothymol blue. For most laboratory uses, the exact concentrations of these solutions is unimportant. For example, rather than weighing 1.000 g of phenolphthalein powder, dissolving it in ethanol, and making it up to 100 mL in a volumetric flask, most chemists would make up the solution by dissolving about a gram of the indicator powder in about 100 mL of ethanol, measured roughly in a beaker. We'll take a bit more care in this lab, but not much more.

REQUIRED EQUIPMENT AND SUPPLIES

☐ goggles, gloves, and protective clothing

☐ balance and weighing papers

☐ graduated cylinder, 100 mL

☐ beaker, 150 mL

☐ funnel

☐ labeled storage bottle

☐ labeled small dropper (Barnes) bottle

☐ ethanol (100 mL)

☐ phenolphthalein powder (1 g)

SUBSTITUTIONS AND MODIFICATIONS

• You may substitute drug-store rubbing alcohol (ethanol or isopropanol) for the ethanol.

CAUTIONS

Although none of the chemicals used in this lab are hazardous, it's good practice to wear splash goggles, gloves, and protective clothing at all times. (Phenolphthalein, formerly widely used as a laxative, was withdrawn from the market because of concerns about possible links with cancer, but the small amounts used as an indicator solution are not ingested and are no cause for concern.)

DISPOSAL: The weighing paper can be disposed of with household waste.

PROCEDURE

1. If you have not already done so, put on your splash goggles, gloves, and protective clothing.
2. Place a weighing paper on the balance and tare the balance to read 0.00 g.
3. Weigh out about 1.0 g of phenolphthalein powder, and transfer it to the beaker.
4. Fill the 100 mL graduated cylinder to 100.0 mL with ethanol, and transfer the ethanol to the beaker.
5. Swirl or stir the contents of the beaker until the phenolphthalein powder is dissolved and the solution is mixed thoroughly.
6. Using the funnel, transfer the solution from the beaker to fill the small dropper bottle. Transfer the remaining solution into the larger storage bottle.
7. Record the date on the labels of the bottles.
8. Rinse the beaker and funnel with tap water and then with distilled water, and set them aside to dry.

FIGURE 7-4:
Making up a bench solution of phenolphthalein indicator solution

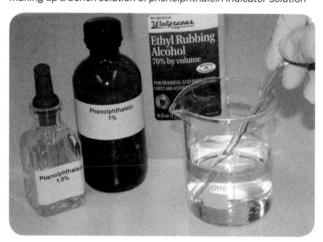

REVIEW QUESTIONS

Q1: What is the molarity of the 1% m/v solution of phenolphthalein solution? Why do you suppose that molarity is never used to specify the concentrations of solutions of phenolphthalein and similar indicators?

Q2: For which types of solutions is mass-to-volume percentage a more useful method of specifying concentration than molarity or other methods?

LABORATORY 7.5:

DETERMINE CONCENTRATION OF A SOLUTION BY VISUAL COLORIMETRY

Colorimetry is the study of color. As it applies to chemistry, colorimetry is used as a practical application of the *Beer-Lambert Law* (describing the relationship between chemical concentration and light transmitted; see your chemistry textbook) to determine the percentage of light that is transmitted by a solution that contains a colored solute, and from that to determine the concentration of the solute.

Nowadays, most colorimetry is done instrumentally, using one of the following instruments:

Colorimeter

A relatively inexpensive instrument—entry-level colorimeters are available for less than $1,000—that allows the light transmission of a solution to be tested with white light or with a limited number of discrete wavelengths. Older colorimeters use an incandescent white light source, which may be used as-is or may be filtered to allow testing transmission with colored light. Typically, such colorimeters are supplied with three filters: violet-blue, green-yellow, and orange-red. Newer calorimeters often substitute colored LEDs for the incandescent light source. Before use, a colorimeter is calibrated by filling the sample tube, called a *cuvette*, with distilled water and setting the colorimeter to read 100% transmission with that cuvette in place. The cuvette is then filled with the unknown solution and used to determine the percentage of light transmitted at one or several wavelengths. Colorimeters are widely used in industrial processes, such as wine making, where high accuracy and selectivity is not required.

Spectrophotometer

A *spectrophotometer* is a more sophisticated (and much more expensive) version of a colorimeter. Instead of using only white light or light at three or four discrete wavelengths, a spectrophotometer simultaneously tests a sample at many wavelengths. Inexpensive spectrophotometers may sample every 10 nanometers (nm) or 20 nm over the visible spectrum from 400 nm to 700 nm. Spectrophotometers used in university and industrial chemistry labs typically sample every nm (or better) over a range from 200 nm in the ultraviolet to 800 to 1,100 nm in the infrared. If a colorimeter is a meat-ax, a spectrophotometer is a scalpel. The many

REQUIRED EQUIPMENT AND SUPPLIES

☐ goggles, gloves, and protective clothing

☐ balance and weighing papers

☐ spatula or scoop

☐ pipette, 1.00 mL

☐ pipette, 5.00 mL

☐ test tubes (6) and rack

☐ copper sulfate pentahydrate (~5 g)

☐ distilled or deionized water

SUBSTITUTIONS AND MODIFICATIONS

- If you do not have a 5.00 mL pipette, you may use a 10.00 mL pipette.

- If you do not have a 5.00 mL pipette, you may use the 1.00 mL pipette repeatedly to transfer 5.00 mL of liquids, with some loss of accuracy.

- If you have no pipettes, you may substitute a 10 mL graduated cylinder for the pipettes with considerable loss in accuracy.

data points determined by a spectrophotometer can be plotted on a graph to yield a "fingerprint" of the transmission characteristics of a particular solution over a wide range of wavelengths. By comparing this fingerprint to the fingerprints of known compounds, spectrophotometry can be used to identify unknown compounds in solution, rather than just the concentration of a known compound.

Despite the ubiquity of colorimeters and spectrophotometers in modern laboratories, such instruments are not required to obtain useful data by colorimetry. Colorimetry was used long before such instruments became available, by employing one of the most sensitive instruments available for determining color: the Mark I human eyeball.

In this lab, we'll use our eyes to determine the concentration of a solution of copper sulfate by colorimetry.

PROCEDURE

This laboratory has two parts. In Part I, we'll make up solutions that contain known concentrations of copper sulfate. In Part II, we'll make up a solution with an unknown concentration of copper sulfate and estimate the concentration of that unknown solution using visual colorimetry.

PART I: MAKE UP REFERENCE SOLUTIONS

1. If you have not already done so, put on your splash goggles, gloves, and protective clothing.
2. Label six test tubes #1 through #6.
3. Use the 5.00 mL pipette to transfer 10.00 mL of distilled water into test tube #1, and 5.00 mL of distilled water into each of the other five test tubes.
4. Place a weighing paper on the balance and tare the balance to read 0.00 g.
5. Weigh out about 2.40 g of copper sulfate, record the mass to 0.01 g on line A of Table 7-2 under "Test tube #1," and transfer the copper sulfate to test tube #1.
6. Swirl test tube #1 until all of the copper sulfate has dissolved.
7. Use the 5.00 mL pipette to transfer 5.00 mL of the copper sulfate solution from test tube #1 to test tube #2.
8. Swirl test tube #2 until the distilled water and copper sulfate solution is thoroughly mixed.
9. Repeat steps 7 and 8 for test tubes #3, #4, and #5. When you finish, each of the five test tubes should contain a solution of copper sulfate at half the concentration of the preceding tube. Test tube #1 should have an intense blue color, with each succeeding test tube a paler blue.
10. Calculate the mass of copper sulfate in each of the tubes #2 through #5 and the mass-to-volume percentage of tubes #1 through #5, and enter those values in lines A and B of Table 7-2.

CAUTIONS:

Copper sulfate is moderately toxic by ingestion or inhalation and is irritating to eyes and skin. Wear splash goggles, gloves, and protective clothing at all times.

DISPOSAL: Copper sulfate solutions may be neutralized with sodium carbonate and the precipitate can then be disposed of with household waste. Alternatively, small amounts of copper sulfate solution may be flushed down the drain. (Copper sulfate is, after all, sold as root killer, for which purpose it is—you guessed it—flushed down the drain.)

TABLE 7-2:

Determine concentration of a solution by visual colorimetry (standard solutions): observed and calculated data

Item	Test tube #1	Test tube #2	Test tube #3	Test tube #4	Test tube #5
A. mass of copper sulfate	__.____ g	__.____ g	__.____ g	__.____ g	__.____ g
B. mass-to-volume percentage	__.____ %	__.____ %	__.____ %	__.____ %	__.____ %

PART II: DETERMINE THE CONCENTRATION OF AN UNKNOWN SOLUTION

1. If you have not already done so, put on your splash goggles, gloves, and protective clothing.
2. Use the 5.00 mL pipette to transfer 5.00 mL of distilled water into test tube #6.
3. Place a weighing paper on the balance and tare the balance to read 0.00 g.
4. Weigh out 2 to 3 g of copper sulfate and record the mass to 0.01 g on line A of Table 7-3.
5. Using the spatula, transfer about one quarter of the copper sulfate to test tube #6, and swirl the test tube until all of the copper sulfate has dissolved.
6. Compare the color of the solution in test tube #6 against the colors of the solutions in test tubes #1 through #5. You'll find it easiest to get an accurate comparison if you place a well-lit sheet of white paper behind the test tubes.
7. Estimate the concentration of the unknown solution in test tube #6 by judging how its color compares to the known sample solutions in test tubes #1 through #5. For example, you might judge that the color of your unknown solution is a bit more than halfway between the colors of test tubes # 3 and #4. By interpolation, estimate the mass-volume percentage of copper sulfate in test tube #6 as closely as possible, and enter that value on line B of Table 7-3.
8. Weigh the remaining copper sulfate and record the mass to 0.01 g on line C of Table 7-3.
9. Determine the mass of the copper sulfate sample that you dissolved in test tube #6 and enter the value to 0.01 g on line D of Table 7-3.
10. Calculate the actual mass-to-volume percentage for test tube #6, and enter that value on line E of Table 7-3.
11. Calculate the percentage error and enter that value on line F of Table 7-3.

TABLE 7-3:

Determine concentration of a solution by visual colorimetry (unknown solution): observed and calculated data

Item	Data
A. initial mass of copper sulfate	_____.___ g
B. interpolated mass-to-volume percentage estimate	_____%
C. mass of remaining copper sulfate	_____.___ g
D. mass of copper sulfate sample (A–C)	_____.___ g
E. actual mass-to-volume percentage	_____.___%
F. percentage error	ffl___.___%

OPTIONAL ACTIVITIES

If you have the time and the required materials, consider performing these optional activities:

- Refine your estimated mass-volume percentage for test tube #6 by making up additional known concentrations of copper sulfate that bracket your estimated value. For example, if your unknown falls about halfway between two knowns with concentrations of 6.0% and 12.0%, you might estimate its concentration at about 9%. Use the most concentrated solution (test tube #1) to make up reference solutions at, for example, 8.0%, 8.5%, 9.0%, 9.5%, and 10.0%, and use those solutions to determine how accurately you can estimate the mass-volume percentage of an unknown solution by using only your eyes.

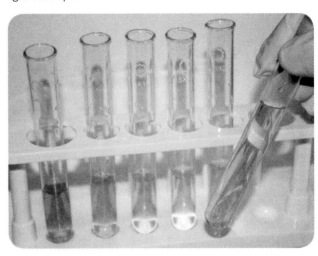

FIGURE 7-5: *Comparing the solution of unknown concentration against samples of known concentration*

REVIEW QUESTIONS

Q1: We used copper sulfate in this lab because it is strongly colored, inexpensive, and readily available. But colorimetry is not the best method for determining the concentration of a solution of copper sulfate. (We could instead simply evaporate the solution in test tube #6 and determine accurately the mass of the copper sulfate it contains.) Under what circumstances would colorimetry be a good choice for determining the concentration of a solution?

Q2: List at least five possible sources of error that may occur with this visual colorimetry procedure.

Q3: You are presented with an unknown solution that contains different concentrations of four food coloring dyes. Could you use visual colorimetry to determine the concentrations of those dyes? Why or why not?

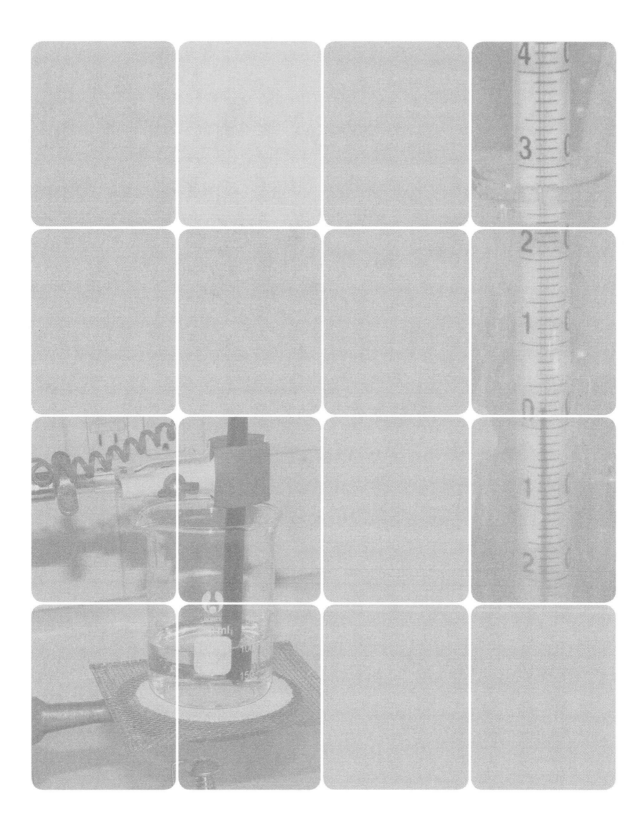

Laboratory:
Colligative Properties of Solutions

8

> The colligative properties of solutions are those affected by the number of solute particles (ions, atoms, molecules, or accretions of multiple molecules) in a solution rather than by the type of solute particles present. There are three colligative properties, as described here.

Vapor pressure reduction

Vapor pressure reduction occurs as the number of solute particles in a solution increases. In effect, some solute particles displace solvent particles at the surface of the solution (the liquid-gas boundary), making fewer solvent particles available for vaporization and thereby reducing the vapor pressure.

Boiling point elevation is sometimes considered a separate colligative property, but is in fact a special case of vapor pressure reduction. The boiling point is reached at the temperature where the vapor pressure of the liquid matches the pressure of the gas at the liquid-gas boundary. Because the presence of solute particles reduces the vapor pressure of the solution, it also increases the boiling point of the solution. Boiling point elevation (ΔT_b or ΔT_{bp}) is the product of the van't Hoff factor (i) of the solute, the *boiling point elevation constant* (also called the *ebullioscopic constant*) of the solvent (K_b or K_{bp}), and the molality (m) of the solution, and can be expressed as the formula $\Delta T = iK_b m$.

One familiar application of vapor pressure reduction is the use of antifreeze (which could just as easily be called anti-boil) in automobile radiators. The addition of high-boiling ethylene glycol and similar chemicals to the water in the radiator raises the boiling point of the solution above that of pure water, reducing the likelihood of the radiator boiling over.

Freezing point depression

Freezing point depression occurs as the number of solute particles in a solution increases. One familiar application of freezing point depression is the application of road salt (usually sodium chloride or calcium chloride) to melt ice on streets and sidewalks. As the ice dissolves the salt (yes, one solid can dissolve another solid) the solid ice is converted to a liquid solution of the salt, because the solution has a lower freezing point than the essentially pure water that makes up the ice. Freezing point depression is calculated in the same way as boiling point elevation, but substituting the *freezing point depression constant* (K_f or K_{fp}) for the boiling point elevation constant. Because the freezing point depression constant is conventionally expressed as an unsigned value, the formula for calculating freezing point depression, $\Delta T = -(iK_f m)$, adds a negative sign to indicate the reduction in freezing point.

Osmotic pressure

Osmotic pressure occurs when a differential concentration of solute particles causes pressure to be exerted across a semipermeable membrane. The phenomenon of osmotic pressure is exploited in many chemical and industrial processes as well as in medical procedures such as kidney dialysis, not to mention such routine bodily functions as your kidneys extracting waste products from your bloodstream. The behavior of solutions under osmotic pressure resembles the behavior of gases. In fact, the formula for calculating osmotic pressure uses the ideal gas constant: $\pi = (niRT)/V$, where π is the osmotic pressure, n the moles of solute, i the van't Hoff constant, R the ideal gas constant, T the absolute temperate in kelvins, and V the volume.

VAN'T WHAT?

The van't Hoff factor depends on the nature of the solute, and takes into account the fact that different solutes behave differently in solution. Ionic solutes dissociate in solution, and therefore have van't Hoff factors greater than 1. (For example, one molecule of sodium chloride in aqueous solution dissociates fully into two particles, one Na^+ ion and one Cl^- ion, so sodium chloride has a van't Hoff factor of 2.) Some solutes, such as sucrose in water, do not dissociate, and therefore have van't Hoff factors of exactly 1. A few solutes associate in solution, yielding fewer individual particles than the molality of the solution suggests. These solutes have van't Hoff factors of less than 1. Some solutes, such as weak acids and bases in aqueous solution, dissociate partially, giving fractional van't Hoff factors, typically greater than 1 but less than 2.

The van't Hoff factor depends on both solvent and solute. For example, when hydrogen chloride (HCl) gas dissolves in water, it dissociates completely into H^+ and Cl^- ions, giving a van't Hoff factor of 2. Conversely, when HCl gas dissolves in benzene, it remains in molecular form, giving a van't Hoff factor of 1 for that combination of solvent and solute.

Colligative properties change with changing concentration of a solution. For most procedures, chemists use molarity (moles per liter of *solution* or mol/L) to specify concentration, but when working with colligative properties, molality (moles per kilogram of *solvent* or mol/kg) is a more useful metric, because molality can be determined with extreme accuracy using only an analytical balance and because molality does not change with increasing or decreasing volume as the solution is heated or cooled.

In this laboratory, we'll examine all three colligative properties of solutions.

EVERYDAY COLLIGATIVE PROPERTIES

Although colligative properties may seem to be an abstract concept with applications only in chemistry labs, in fact colligative properties of solutions have everyday real-world applications.

- The blood of mammals, birds, fish, and insects native to arctic, antarctic, and other cold regions has been found to contain special proteins that act colligatively to lower the freezing point of their blood. Some studies suggest that humans who live in cold environments have different blood chemistry from people who live in warm climates. What's really interesting is that if someone from, say Nome, Alaska, moves to Honolulu, Hawaii, after some years he loses his "cold resistance." Conversely, someone from Honolulu who moves to Nome over some years develops this "cold resistance."

- The antifreeze in your car's radiator depresses the freezing point of the water, preventing the radiator from freezing up in cold weather. Conversely, it also raises the boiling point, preventing the radiator from boiling over during hot weather.

- In winter, highway departments scatter road salt on icy roads to melt the ice, a practical application of freezing point depression.

- Tears are salty for a reason. They keep your eyes sterile by killing microbes by osmotic dehydration.

- Millions of people have drinking water thanks to a practical implementation of the colligative properties of solutions. Many seawater desalinization plants use reverse osmosis to remove salt from seawater, producing potable water. In ordinary (forward) osmosis, water molecules migrate from the freshwater side of the semipermeable membrane to the seawater side, which is exactly the opposite of what's needed to desalinize water. In a reverse osmosis plant, the seawater side is placed under pressure, which forces water molecules through the semipermeable membrane to the freshwater side.

LABORATORY 8.1:

DETERMINE MOLAR MASS BY BOILING POINT ELEVATION

Dissolving a nonvolatile solute in a solvent increases the boiling point of that solvent by an amount proportional to the quantity of the solute. Although this phenomenon occurs with any solvent and solute, for reasons of safety and economy we'll use water as the solvent for this session, and sodium chloride (table salt) and sucrose (table sugar) as the solutes.

PROCEDURE

This lab has three parts. In Part I, we'll determine the boiling point of water under ambient pressure. In Part II, we'll prepare solutions of sodium chloride and sucrose of various molalities. In Part III, we'll determine the boiling points of those solutions.

PART I: DETERMINE THE BOILING POINT OF WATER UNDER AMBIENT PRESSURE

The boiling point of water under standard conditions is 100°C, but, as is true of any liquid, the actual boiling point depends on the ambient barometric pressure, which varies with weather and altitude. If the barometric pressure is below the standard pressure of 1000 millibars (mbar), the boiling point of water is less than 100°C; if the pressure is greater than 1000 mbar, the boiling point of water is greater than 100°C. Because barometric pressure is not constant, the actual boiling point of water may differ significantly from place to place and from day to day at a particular place.

CAUTIONS

Although the chemicals used in this experiment are not hazardous, you will be using open flame, so use caution and have a fire extinguisher handy. Handle the hot liquids used in this experiment with extreme care. If you use a microwave oven to prewarm the solutions so that they'll come to a boil faster over the alcohol lamp, be careful not to superheat the solutions. A superheated solution may unpredictably boil violently, ejecting hot liquid from the container. Always use a boiling chip to avoid superheating. Wear splash goggles, gloves, and protective clothing.

REQUIRED EQUIPMENT AND SUPPLIES

- ☐ goggles, gloves, and protective clothing
- ☐ balance and weighing paper
- ☐ beaker, 150 mL (6)
- ☐ beaker, 250 mL (1)
- ☐ graduated cylinder, 100 mL
- ☐ alcohol lamp or other heat source
- ☐ tripod stand and wire gauze
- ☐ beaker tongs
- ☐ stirring rod
- ☐ thermometer
- ☐ boiling chips
- ☐ sodium chloride (70.13 g)
- ☐ sucrose (410.76 g)
- ☐ tap water

SUBSTITUTIONS AND MODIFICATIONS

- You may use a kitchen stove burner for the heat source. Use a large tin can lid or similar item between the burner and glassware rather than putting the glassware in direct contact with the burner.
- You may substitute ordinary table salt for the sodium chloride.
- You may substitute ordinary table sugar for the sucrose.

DR. MARY CHERVENAK COMMENTS:

This is also the reason for special high-altitude baking instructions. Water boils at a lower temperature, so cakes lose moisture more quickly, causing them to dry out and collapse. The Canadian Rockies were high enough to cause some baking disasters for me!

It's important to know the actual boiling point of water under ambient pressure, because we'll use that boiling point as a baseline reference datum to determine the boiling point elevation (ΔT_b) of the various molal solutions and from those data the molar masses of the solutes. To determine the boiling point of water, take the following steps:

1. If you have not already done so, put on your splash goggles, gloves, and protective clothing.
2. Fill the 250 mL beaker about halfway with water, place it on the heat source, and heat the water until it comes to a full boil.
3. Immerse the thermometer in the beaker, making sure that it does not contact the beaker itself (suspend it with a stand and clamp, or find some other way), allow the thermometer to stabilize, and record the temperature reading in Table 8-1. (With most thermometers, you can interpolate a reading to 0.5°C or closer.)
4. Empty and dry the beaker.

PART II: PREPARE MOLAL SOLUTIONS OF SODIUM CHLORIDE AND SUCROSE

To test the effect of molality and dissociation on boiling point, we need to prepare solutions of ionic and molecular (covalent) compounds of known molality. I chose sodium chloride and sucrose because both of these chemicals are inexpensive, readily available, and extremely soluble in water. Sodium chloride is ionic. In solution, sodium chloride dissociates into sodium ions (Na^+) and chloride ions (Cl^-), and should therefore have a van't Hoff factor of 2. Sucrose is molecular, and should therefore have a van't Hoff factor of 1.

Coincidentally, the solubility of both sodium chloride and sucrose in water at room temperature is just over 6 mol/L. We'll therefore prepare 6 molal samples of both of these compounds, and then by serial dilution prepare 3 molal and 1.5 molal samples of each.

1. If you have not already done so, put on your splash goggles, gloves, and protective clothing.
2. Use the 100 mL graduated cylinder to transfer 200 mL of water to the 250 mL beaker.
3. Weigh 70.13 g of sodium chloride, add it to the 250 mL beaker, and stir until the sodium chloride is completely dissolved. This solution contains 6 mol/kg of sodium chloride (6 molal).
4. Transfer 100 mL of the 6 molal sodium chloride solution to beaker A.
5. Add 100 mL of water to the 250 mL beaker to dilute the remaining 100 mL of 6 molal sodium chloride solution to 200 mL of 3 molal solution.
6. Transfer 100 mL of the 3 molal sodium chloride solution to beaker B.

7. Add 100 mL of water to the 250 mL beaker to dilute the remaining 100 mL of 3 molal sodium chloride solution to 200 mL of 1.5 molal solution.
8. Transfer 100 mL of the 1.5 molal sodium chloride solution to beaker C and discard the remainder. (Or transfer all 200 mL; the quantity of the solution has no effect on the boiling point elevation.)
9. Use the 100 mL graduated cylinder to transfer 200 mL of water to the 250 mL beaker.
10. Weigh 410.76 g of sucrose, add it to the 250 mL beaker, and stir until the sucrose is completely dissolved. This solution contains 6 mol/kg of sucrose (6 molal).
11. Transfer 100 mL of the 6 molal sucrose solution to beaker D.
12. Add 100 mL of water to the 250 mL beaker to dilute the remaining 100 mL of 6 molal sucrose solution to 200 mL of 3 molal solution.
13. Transfer 100 mL of the 3 molal sucrose solution to beaker E.
14. Add 100 mL of water to the 250 mL beaker to dilute the remaining 100 mL of 3 molal sucrose solution to 200 mL of 1.5 molal solution.
15. Transfer 100 mL of the 1.5 molal sucrose solution to beaker F and discard the remainder. (Or transfer all 200 mL; the quantity of the solution has no effect on the boiling point elevation.)

PART III: DETERMINE THE BOILING POINTS OF SODIUM CHLORIDE AND SUCROSE SOLUTIONS

In this part of the lab, we determine the boiling points of the sodium chloride and sucrose solutions.

1. If you have not already done so, put on your splash goggles, gloves, and protective clothing.
2. Place beaker A on the heat source. Add one boiling chip, and apply heat with constant stirring until the contents of beaker A come to a full boil.
3. Immerse the thermometer in the beaker, making sure that it does not contact the beaker itself, allow the thermometer to stabilize, and record the temperature reading in Table 8-1. Place the beaker aside to cool.
4. Repeat steps 2 and 3 for beakers B, C, D, E, and F.

FIGURE 8-1: *Determining the boiling point of a solution*

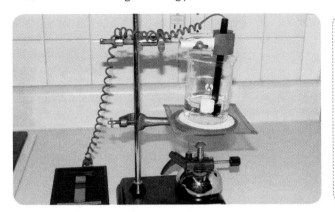

DISPOSAL: Retain all of the solutions for use in the following laboratory session. Allow all six beakers to cool to room temperature, and then place them in the refrigerator to chill them. A typical refrigerator maintains a temperature of about 5°C. Having the solutions at this temperature reduces the time required for the freezing point determinations in the following laboratory session. (I violate my general rule of avoiding mixing laboratory materials with kitchen materials because these beakers contain only sodium chloride and sucrose solutions, both of which are harmless.)

OPTIONAL ACTIVITIES

If you have time and the required materials, consider performing these optional activities:

- Repeat the experiment, using various concentrations of engine coolant (antifreeze). Most commercial antifreezes contain ethylene glycol (FW 62.07 g/mol), propylene glycol (FW 76.09 g/mol), or a mixture of the two. Assume that the solute is molecular rather than ionic, and use the boiling point elevation values you observe to calculate the formula weight of the commercial coolant solution. (Note that ethylene glycol is extremely toxic and must be handled with respect and disposed of properly. Propylene glycol is less toxic, but still must be handled properly.)

- Repeat the experiment, but dissolve both 70.13 g of sodium chloride and 410.76 g of sucrose in the initial 200 mL of water. (Warm the water if necessary to dissolve all of the solids.) Calculate the expected boiling point elevations and then determine the actual boiling point elevations for the original solution and for the 2X and 4X dilutions.

- Repeat the experiment, using an unknown solute provided by a parent, teacher, or lab partner. (In order to maximize the boiling point elevation, the unknown solute should be of as low a formula weight as possible, and as high a solubility as possible. Solutions of low molality are difficult to test with a standard thermometer, because the accordingly small value for boiling point elevation is subsumed by measurement error.) Record the value for boiling point elevation, and calculate the formula weight of the solute assuming a van't Hoff factor of 1.

TABLE 8-1: *Boiling point elevation—observed and calculated data*

Beaker/solution	Boiling point	Calculated formula weight
Water	_____._°C	
A. Sodium chloride, 6 mol/kg	_____._°C	_____.___ g/mol
B. Sodium chloride, 3 mol/kg	_____._°C	_____.___ g/mol
C. Sodium chloride, 1.5 mol/kg	_____._°C	_____.___ g/mol
D. Sucrose, 6 mol/kg	_____._°C	_____.___ g/mol
E. Sucrose, 3 mol/kg	_____._°C	_____.___ g/mol
F. Sucrose, 1.5 mol/kg	_____._°C	_____.___ g/mol

REVIEW QUESTIONS

Q1: Name and define the unit of concentration used in calculating values for colligative properties of solutions. Why is this unit of concentration used rather than molarity?

Q2: The K_{bp} for H_2O is 0.512°C/molal. Applying this value and the van't Hoff factors for sodium chloride (2) and sucrose (1) to the boiling point elevation values you recorded in Table 8-1, calculate the formula weights of sodium chloride and sucrose. Record these calculated values in Table 8-1.

Q3: What effect on boiling point would you expect if you dissolved sufficient ethanol in water to produce a 1 molal solution?

LABORATORY 8.2:
DETERMINE MOLAR MASS BY FREEZING POINT DEPRESSION

Just as a dissolved solute increases the boiling point of a solvent by an amount proportional to the quantity of the solute, it also decreases the freezing point of that solvent. The same formula is used to calculate either value, except that a negative sign is added for freezing point depression to account for the fact that the freezing point is lowered rather than raised.

Boiling point elevation $= \Delta T_b = iK_bm$
Freezing point depression $= \Delta T_f = -(iK_fm)$

In fact, some sources specify freezing point depression constants, K_f or K_{fp}, as negative signed values, eliminating the need to reverse the sign in the formula. In either case, the calculations are done the same way for boiling point elevation and freezing point depression. Only the values of the constants differ. For example, the K_b for water is 0.512°C/molal and the K_f for water is 1.858°C/molal. In other words, assuming a van't Hoff factor of 1 for the solute, a 1 molal solution has a boiling point 0.512°C higher than pure water, and a freezing point 1.858°C lower. A 2 molal solution has a boiling point 1.024°C higher than pure water, and a freezing point 3.716°C lower. And so on.

In this laboratory, we'll use the sodium chloride and sucrose solutions prepared in the preceding lab to test the effect of dissolved solutes on freezing points.

PROCEDURE
This lab has two parts. In Part I, we'll prepare an ice/salt bath and determine the freezing point of water. In Part II, we'll determine the freezing points of the solutions we prepared in the preceding laboratory.

PART I: PREPARE THE ICE BATH AND DETERMINE THE FREEZING POINT OF WATER
Unlike the boiling point of water, which varies significantly with ambient local pressure, the freezing point of water is almost unaffected by pressure. But thermometers vary in accuracy, so before we test freezing point depressions in the sodium chloride and sucrose solutions, we'll calibrate our thermometer by using it to determine the freezing point of water. (For example, if our thermometer indicates a freezing point for pure water of +0.5°C and we subsequently measure a depressed freezing point at −2.5°C, the actual freezing point depression is 3.0°C rather than 2.5°C.)

REQUIRED EQUIPMENT AND SUPPLIES

- [ ] goggles, gloves, and protective clothing
- [ ] beaker, 600 mL (1)
- [ ] test tubes (6)
- [ ] thermometer
- [ ] stirring rod
- [ ] chilled solutions A through F from preceding lab
- [ ] chipped ice sufficient to half fill beaker
- [ ] sodium chloride sufficient to make a 15 mm to 25 mm layer on top of crushed ice

SUBSTITUTIONS AND MODIFICATIONS

- You may substitute any glass or plastic wide-mouth container with a capacity of 1 pint (~500 mL) to 1 quart (~1 L) for the 400 mL or 600 mL beaker.
- You may substitute table salt or rock salt for the sodium chloride.

CAUTIONS

Wear splash goggles, gloves, and protective clothing.

Because the presence of a solute reduces the freezing point of an aqueous solution below 0°C, the freezing point of pure water, we need a way to cool the solutions below 0°C. In an elegant application of using freezing point depression to test freezing point depression, we'll use the phenomenon itself to provide the conditions necessary for the test. We'll make an ice/salt bath by mixing crushed ice with sodium chloride. The phenomenon of freezing point depression means that the temperature of this ice/salt bath will be lower than the temperature of a pure ice bath.

1. If you have not already done so, put on your splash goggles, gloves, and protective clothing.
2. Half fill the large beaker with crushed ice, pour a 15 mm to 25 mm layer of sodium chloride on top of the ice, and stir until the ice and salt are thoroughly mixed.
3. Immerse the thermometer in the ice/salt bath, allow a minute or so for it to stabilize, and record the temperature of the ice/salt bath in Table 8-2.
4. Half fill a test tube with water and carefully press it down into the ice bath until the water is completely below the surface of the ice bath. See Figure 8-2.
5. Gently stir the water with the thermometer continuously until ice crystals begin to form on the side of the test tube. You will probably have to remove the test tube from the ice bath periodically to check the status. Ice crystals tend to form first near the bottom of the tube.
6. When ice crystals begin to form, record the temperature of the water in Table 8-2.

> DISPOSAL: Dispose of the solutions in the test tubes by flushing them down the drain. Allow the beakers with the remainder of the six solutions to come to room temperature. These solutions will be used in the following laboratory session.

PART II: DETERMINE THE FREEZING POINTS OF SODIUM CHLORIDE AND SUCROSE SOLUTIONS

In this part of the lab, we determine the freezing points of the sodium chloride and sucrose solutions.

1. If you have not already done so, put on your splash goggles, gloves, and protective clothing.
2. Label six test tubes A through F.
3. Transfer approximately 10 mL (about half a test tube) of solution from beaker A to test tube A.
4. Repeat step 3 for the solutions in beakers B through F.
5. Carefully press test tube A down into the ice bath until the solution is completely below the surface of the ice bath.
6. Gently and continuously stir the solution with the thermometer until ice crystals begin to form on the side of the test tube. You will probably have to remove the test tube from the ice bath periodically to check the status. Ice crystals tend to form first near the bottom of the tube.
7. When ice crystals begin to form, record the temperature of the solution in Table 8-2. If no ice crystals appear within a few minutes, note that fact in Table 8-2, record the lowest temperature reached for that solution, and go on to the next solution.
8. Repeat steps 5 through 7 for test tubes B through F.

TABLE 8-2: Freezing point depression—observed and calculated data

Test tube/solution	Freezing point	Calculated formula weight
Ice/salt bath	_____.__ °C	
Water	_____.__ °C	
A. Sodium chloride, 6 mol/kg	_____.__ °C	_____.___ g/mol
B. Sodium chloride, 3 mol/kg	_____.__ °C	_____.___ g/mol
C. Sodium chloride, 1.5 mol/kg	_____.__ °C	_____.___ g/mol
D. Sucrose, 6 mol/kg	_____.__ °C	_____.___ g/mol
E. Sucrose, 3 mol/kg	_____.__ °C	_____.___ g/mol
F. Sucrose, 1.5 mol/kg	_____.__ °C	_____.___ g/mol

FIGURE 8-2: Determining the freezing point of a solution

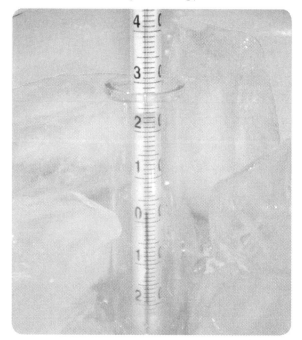

OPTIONAL ACTIVITIES

If you have time and the required materials, consider performing these optional activities:

- Repeat the experiment, using solutions of ethanol or isopropanol in water. (Try solutions of 1 mL alcohol in 9 mL water, 2 mL alcohol in 8 mL water, and 3 mL alcohol in 7 mL water.) Assume a van't Hoff factor for either alcohol of 1, and use the freezing point depression values that you observe to calculate the formula weight of the alcohol. (Note that ethanol is typically available as a 70% to 95% aqueous solution and isopropanol as a 70% to 91% aqueous solution, and take that into account when you make your calculations.)

- Repeat the experiment, using an unknown solute provided by a parent, teacher, or lab partner. (In order to maximize the freezing point depression, the unknown solute should be of as low a formula weight as possible, and as high a solubility as possible. Solutions of low molality are difficult to test with a standard thermometer, because the accordingly small value for freezing point depression is subsumed by measurement error.) Record the value for freezing point depression, and calculate the formula weight of the solute assuming a van't Hoff factor of 1.

REVIEW QUESTIONS

Q1: Did all of the solutions that you tested freeze? If not, propose and quantify an explanation for this observation.

_____

_____

_____

_____

Q2: In the early days of automobiles, methanol was used universally as antifreeze. In 1937, antifreeze solutions based on ethylene glycol were introduced, and quickly gained almost 100% market share, despite the fact that they cost much more than methanol. Propose an explanation for this rapid adoption of ethylene glycol antifreeze solutions. (Hint: look up the physical properties of methanol and ethylene glycol.)

_____

_____

_____

_____

Q3: The K_{fp} for H_2O is 1.858°C/molal. Applying this value and the van't Hoff factors for sodium chloride (2) and sucrose (1) to the freezing point depression values that you recorded in Table 8-2, calculate the formula weights of sodium chloride and sucrose. Record these calculated values in Table 8-2.

Q4: What effect on freezing point would you expect if you dissolved sufficient ethanol in water to produce a 1 molal solution?

_____

_____

_____

LABORATORY 8.3:

OBSERVE THE EFFECTS OF OSMOTIC PRESSURE

Osmotic pressure is the hydrostatic pressure that exists when two solutions of differing concentrations are in contact with the two sides of a semipermeable membrane. The semipermeable membrane allows small molecules (such as water) to pass through it freely, but physically prevents larger molecules (such as a solute) from passing through the membrane. In such a system, water passes from the side that contains the less-concentrated solution to the side with the more-concentrated solution until the concentrations of the solutions are equal, at which point the system is in equilibrium.

If a biological cell is placed in a *hypotonic* environment (one in which the solution inside the cell is more concentrated than the solution surrounding the cell), water flows into the cell from the outside solution through the semipermeable membrane that surrounds the cell, causing the cell to expand and gain mass. Conversely, in a *hypertonic* environment (one in which the outside solution is more concentrated), water flows outward from the cell, shrinking it and reducing its mass.

Children sometimes unwittingly illustrate the principle of osmotic pressure by pouring table salt on a garden slug. This creates an extreme hypertonic environment, sucks water out of the slug, and thereby kills it by quick and extreme dehydration. My editor told me to make sure that no slugs were harmed in running this experiment, so I decided to use celery instead. If that offends People for the Ethical Treatment of Vegetables (*http://www.petv.org*), I'm sorry, but I had to use something.

In this laboratory, we'll use the sodium chloride and sucrose solutions left over from the first laboratory session in this chapter to observe the effects of osmotic pressure on the semipermeable membrane that surrounds celery cells.

> DISPOSAL: Dispose of the solutions by flushing them down the drain.

REQUIRED EQUIPMENT AND SUPPLIES

- ☐ goggles, gloves, and protective clothing
- ☐ balance
- ☐ beaker, 150 mL (1)
- ☐ marking pen
- ☐ sharp knife
- ☐ paper towels
- ☐ solutions A through F from Laboratory 8.1
- ☐ celery stalk

SUBSTITUTIONS AND MODIFICATIONS

- You may substitute any container of similar size or larger for the 150 mL beaker.

CAUTIONS

Wear splash goggles, gloves, and protective clothing. Be careful with the knife.

PROCEDURE

1. If you have not already done so, put on your splash goggles, gloves, and protective clothing.
2. Cut the celery stalk into seven pieces of approximately equal mass.
3. Label the celery pieces A through G.
4. Weigh each piece of celery and record the mass of each in Table 8-3.
5. Drop each piece of celery, A through F, into the corresponding beaker, making sure that the celery is completely submerged.
6. Drop celery piece G into a container of pure water.
7. Allow the celery pieces to soak for at least one hour.
8. Remove each piece of celery from its beaker, rinse it briefly in running water, pat it dry with a paper towel, and reweigh it (Figure 8-3). Record the mass of each piece in Table 8-3.

TABLE 8-3: *Osmotic pressure—observed and calculated data*

Celery piece/solution % mass gain (loss)	Initial mass	Final mass	Mass gain (loss)
A. Sodium chloride, 6 mol/kg	_____.___ g	_____.___ g	_____.___ g
B. Sodium chloride, 3 mol/kg	_____.___ g	_____.___ g	_____.___ g
C. Sodium chloride, 1.5 mol/kg	_____.___ g	_____.___ g	_____.___ g
D. Sucrose, 6 mol/kg	_____.___ g	_____.___ g	_____.___ g
E. Sucrose, 3 mol/kg	_____.___ g	_____.___ g	_____.___ g
F. Sucrose, 1.5 mol/kg	_____.___ g	_____.___ g	_____.___ g
G. Water	_____.___ g	_____.___ g	_____.___ g

OPTIONAL ACTIVITIES

If you have time and the required materials, consider performing these optional activities:

- Repeat the experiment, using prunes rather than celery. Record the differences, if any, in mass loss/gain percentages.

- It's difficult to get quantitative results using vegetables. Assume that you have access to a commercial semipermeable membrane that passes water freely but blocks sucrose as well as an appropriate container to provide two chambers separated by the membrane. Design an experiment to provide more accurate data about osmotic pressure effects, including intermediate values for molality as the system comes to equilibrium.

FIGURE 8-3: *Determining the mass of a celery sample*

REVIEW QUESTIONS

Q1: Calculate the mass gain (or loss) for each of the samples, in grams and percentage, and enter the results in Table 8-3. Note which samples gained mass and which samples lost mass. Propose an explanation.

_____

_____

_____

_____

Q2: If you submerged a celery sample in a 70% ethanol solution, would you expect that sample to gain or lose mass? Why?

Q3: Using only the materials and equipment listed in this chapter, propose an experimental method to accurately determine the concentration of the solution in the celery cells.

Laboratory:
Introduction to Chemical Reactions and Stoichiometry

9

> A *chemical reaction* is a process during which one or more original substances (called *reactants*) are converted to one or more different substances (called *products*). A chemical reaction may occur quickly (such as a firecracker exploding) or slowly (such as iron rusting). Some chemical reactions occur spontaneously at normal temperatures and pressures, as soon as the reactants are brought into contact with each other. Other chemical reactions occur only if external energy is supplied in the form of heat, light, or an electrical spark.

There are many different types of chemical reactions. In this chapter, we examine the following reaction types:

Composition reaction

A *composition reaction*, also called a *combination reaction*, occurs when two or more reactants combine to yield one or more products. Composition reactions take the general form A + B → AB, where the reactants A and B may be elements or compounds and the compound AB is the resulting product. For example, if you heat a mixture of iron filings (A) and sulfur (B), those two elements react to form iron sulfide (AB).

Decomposition reaction

A *decomposition reaction* occurs when one reactant is broken down into two or more products, usually by the application of heat. Decomposition reactions take the general form AB → A + B, where the compound AB is the reactant and the products A and B may be elements or compounds. For example, if you heat sodium hydrogen carbonate (sodium bicarbonate or baking soda, $NaHCO_3$), it breaks down into sodium carbonate (Na_2CO_3), carbon dioxide (CO_2), and water (H_2O). In this

COMBUSTION REACTIONS

Although it is sometimes considered to be a separate type of reaction, a *combustion reaction* is really just a type of composition reaction. In a combustion reaction, a substance combines vigorously with oxygen, releasing energy in the process. For example, when you light a candle, you initiate a combustion reaction, and the energy manifests as heat and light, making the flame you see. A combustion reaction can be thought of as a fast *oxidation reaction*, in the sense that a substance is combining rapidly with oxygen. (Iron rusting or bread growing stale are examples of slower oxidation reactions.)

However, chemists no longer use the term "oxidation" to refer exclusively to reactions in which a substance combines with oxygen. Instead, oxidation is used in a more general sense to refer to a class of reactions that may not involve oxygen at all. See Chapter 10 for more details.

example, the general form AB → A + B is expanded to ABC → A + B + C or, more precisely, 2 ABC → A + B + C, where ABC is sodium bicarbonate, A is sodium carbonate, B is carbon dioxide, and C is water. The balanced equation for this reaction is:

$$2\ NaHCO_3 \rightarrow Na_2CO_3 + CO_2 + H_2O$$

Displacement reaction

A *displacement reaction*, also called a *replacement reaction*, occurs when two or more reactant substances recombine to form two or more different products.

Single displacement reaction

In a *single displacement reaction*, a more active element displaces a less active element in a compound, forming a new compound that incorporates the more active element and releasing the less active element in elemental form. Single displacement reactions take the general form A + BX → AX + B, where A is the more active element, B is the less active element, and X is an anion (such as chloride, sulfate, or nitrate). For example, if you react sodium metal (Na) with hydrochloric acid (HCl), the more active sodium displaces the less active hydrogen, according to the following balanced equation:

$$2\ Na + 2\ HCl \rightarrow 2\ NaCl + H_2$$

Double displacement reaction

In a *double displacement reaction*, also called a *metathesis reaction*, two compounds exchange elements or ionic species with each other. Double displacement reactions take the general form AX + BY → AY + BX, where A and B are cations and X and Y are anions. If AX and BY are both freely soluble and (for example) AY is insoluble, combining solutions of AX and BY results in a precipitate.

Any of these reaction types can be represented by *stoichiometrically* balanced equations. For example, the decomposition reaction of sodium hydrogen carbonate into sodium carbonate, carbon dioxide, and water can be represented as:

$$NaHCO_3(s) \rightarrow Na_2CO_3(s) + CO_2(g) + H_2O(g)$$

or, in balanced form:

$$2\ NaHCO_3(s) \rightarrow Na_2CO_3(s) + CO_2(g) + H_2O(g)$$

The laboratory sessions in this chapter explore various aspects of these types of chemical reactions.

DECOMPOSITION BY STAGES

In practice, the actual products from the decomposition of sodium hydrogen carbonate depend on the temperature and duration of heating. At higher temperatures and longer durations, the products are indeed anhydrous sodium carbonate, carbon dioxide gas and water vapor. At lower temperatures and durations, some or all of the water may be retained by the sodium carbonate as water of hydration.

This variation in products is common to decomposition reactions, particularly if the reactant is a complex organic compound. For example, the products of the decomposition of sucrose (table sugar) by heating depend on the temperature and duration. Heat sucrose gently and you get caramel. Heat sucrose intensely (or decompose it with concentrated sulfuric acid), and you end up with nothing more than carbon, carbon dioxide, and water.

EVERYDAY CHEMICAL REACTIONS

Chemical reactions are so much a part of everyday life that it's impossible to provide even a representative list in this space. Nearly everything with which we have daily contact is the product of a chemical reaction or reactions, and we're constantly surrounded by chemical reactions in progress.

If you make bread, for example, it rises because of chemical reactions, bakes because of other chemical reactions, and grows stale because of still other chemical reactions. When you eat the bread, chemical reactions in your body convert the starches it contains to glucose, which your cells then metabolize via other chemical reactions to provide the energy you use to complete the labs in this book. The carbon dioxide waste from those metabolic chemical reactions enters your bloodstream via other chemical reactions, and is eventually converted back into carbon dioxide gas by other chemical reactions and disposed of when you exhale. Our bodies are, in effect, chemical reaction vessels.

Civilization itself is fundamentally based on chemical reactions under human control. The two pillars of civilization are human manipulation of metals and of energy, both of which ultimately devolve to controlled chemical reactions.

LABORATORY 9.1:
OBSERVE A COMPOSITION REACTION

The reaction of iron and sulfur to form iron(II) sulfide is an example of the simplest type of composition reaction, one in which two elements react to form a compound. In its simplest form, the balanced equation can be represented as:

$$Fe(s) + S(s) \rightarrow FeS(s)$$

In fact, it's a bit more complicated. Although one iron atom can react with one sulfur atom to yield one molecule of iron(II) sulfide, solid sulfur exists as a molecule that comprises eight bound sulfur atoms. The balanced equation therefore becomes:

$$8\ Fe(s) + S_8(s) \rightarrow 8\ FeS(s)$$

But we're not finished yet. The reaction of iron and sulfur is an example of a class of reactions called *nonstoichiometric reactions*. In a stoichiometric reaction, the proportions of the reactants are fixed. For example, exactly one mole of sodium (Na) reacts stoichiometrically with exactly one mole of chlorine (Cl) to yield exactly one mole of sodium chloride (NaCl), or two moles of aluminum (Al) react with six moles of hydrogen chloride (HCl, hydrochloric acid) to yield two moles of aluminum chloride ($AlCl_3$) and three moles of molecular hydrogen gas (H_2). In a nonstoichiometric reaction, the proportions of the reactants are not fixed relative to those of the product. In other words, different amounts of iron can react with different amounts of sulfur to yield a product in which the proportions of iron and sulfur may vary according to the available quantities of the reactants and the conditions under which the reaction takes place.

Three products are possible, which may occur in relatively pure form or as a mixture. All three of these products occur naturally as minerals, and can be produced in the lab by adjusting the proportions of the reactants and the reaction conditions.

Troilite
Troilite is the most common natural form of iron(II) sulfide. It has the formula FeS, and contains stoichiometrically equivalent amounts of iron and sulfur.

Pyrrhotite
Pyrrhotite has the general formula $Fe_{(1-x)}S$, which indicates that it contains a stoichiometric deficiency of iron and therefore a stoichiometric excess of sulfur. In other words, even though there is insufficient iron present to react

REQUIRED EQUIPMENT AND SUPPLIES

- ☐ fume hood or exhaust fan (or perform this laboratory session outdoors)
- ☐ goggles, gloves, and protective clothing
- ☐ balance and weighing paper
- ☐ ring stand with ring
- ☐ large tin can lid
- ☐ beaker tongs
- ☐ test tube
- ☐ sand bed (large cake pan half filled with sand)
- ☐ large pail at least half full of water
- ☐ gas burner
- ☐ mortar and pestle
- ☐ spatula
- ☐ magnet
- ☐ iron filings (~3 g)
- ☐ sulfur (~2 g)
- ☐ concentrated hydrochloric acid (~1 mL)

SUBSTITUTIONS AND MODIFICATIONS

- You may substitute a large bolt and pipe cap (from the hardware store) for the mortar and pestle.
- You may substitute an equal mass of steel wool for the iron filings. If necessary, wash the steel wool first to remove soap or other contaminants. Fine steel wool may react very vigorously.
- You may substitute a propane torch for the gas burner.
- You may substitute a pair of pliers or similar gripping tool for the beaker tongs.
- You may substitute an old kitchen knife or putty knife for the spatula.

CAUTIONS

This experiment uses the hot flame provided by a gas burner. Be careful with the flame, and have a fire extinguisher readily available. The reaction may be vigorous and throw sparks that may ignite flammable materials. The sulfur may ignite and emit toxic and choking sulfur dioxide fumes, so perform this experiment under a fume hood or strong exhaust fan or do it outdoors. (Even without ignition, this reaction smells really, really bad.) The product, iron sulfide, is *pyrophoric* when in powdered form, which means that it may ignite spontaneously. Wear splash goggles, gloves, and protective clothing.

stoichiometrically with all of the sulfur present, the iron that is present reacts with all of the sulfur to form a nonstoichiometric form of iron sulfide in which the proportions differ from the stoichiometric form.

Mackinawite

Mackinawite has the general formula $Fe_{(1+x)}S$, which indicates that it contains a stoichiometric excess of iron and therefore a stoichiometric deficiency of sulfur.

In this lab, we'll prepare iron(II) sulfide using quantities of reactants that are stoichiometrically balanced. Our product should therefore be primarily the FeS (troilite) form of the compound, with perhaps minor amounts of the other two forms as contaminants. We'll use a slight excess of sulfur, because unless the reaction occurs under an inert atmosphere, it's common for the heat of the reaction to ignite the sulfur, converting some of it to sulfur dioxide gas.

PROCEDURE

1. If you have not already done so, put on your splash goggles, gloves, and protective clothing.
2. Set up your tripod stand in the sand bed with your gas burner set up to direct the hottest part of its flame at the center of the ring.
3. Weigh the tin can lid to 0.01 g and record its mass on line A of Table 9-1.
4. Center the tin can lid flat on the tripod ring.
5. Weigh out 3.0 g of iron filings and 2.0 g of sulfur to within 0.01 g and record the masses on lines B and C of Table 9-1.
6. Calculate the number of moles of iron filings and sulfur, and enter those values on lines D and E of Table 9-1.
7. Determine the limiting reagent from the values you calculated in the preceding step. Assume that one mole of iron reacts with one mole of sulfur to form one mole of

FeS, calculate the expected mass of FeS and enter that value on line F of Table 9-1.
8. Combine the iron filings and sulfur in the center of a folded sheet of ordinary paper and flex the paper back and forth until the two chemicals are well mixed.
9. Pour the mixed iron and sulfur into a small pile in the center of the can lid. Try to make the pile as compact as possible. If the material is spread out, it's likely that the sulfur will catch fire and burn off to sulfur dioxide gas before the reaction between the iron and sulfur begins.
10. If you are not performing this experiment outdoors, turn on the fume hood or exhaust fan. (And you will probably regret doing this indoors, even with an exhaust fan.)
11. Light the gas burner, and make sure that it's centered under the pile of iron and sulfur. At first, nothing appears to happen. As you continue heating the mixture, some of the sulfur begins to melt. Soon after, you should notice a glow begin to develop in the mixture as the iron and sulfur begin to react. Once that glow becomes obvious, turn off the gas burner and allow the reaction to proceed to completion on its own. See Figure 9-1.

CAUTIONS

Sulfur dioxide is a toxic gas with a sharp, choking odor. You'll certainly get a whiff of it when you do this experiment. It will probably remind you of the smell produced by firecrackers. If too much sulfur dioxide is produced or the reaction becomes too vigorous, stop the reaction by gripping the tin can lid with the tongs and dropping the lid and reaction mixture into the pail of water. Evacuate the area until the exhaust fan clears the air.

12. Once the reaction has completed, allow several minutes for the iron(II) sulfide and the equipment to cool.
13. Once the lid and iron(II) sulfide have cooled, weigh them to 0.01 g and record the combined mass on line G of Table 9-1. Calculate the actual mass of the product by subtracting line A from line G and record this mass to 0.01 g on line H of Table 9-1.
14. Assuming that the product is pure FeS, calculate the percent yield by dividing the actual yield (line H) by the theoretical yield (line F) and multiplying by 100. Enter this value on line I of Table 9-1.
15. Use the spatula to scrape the iron(II) sulfide into your mortar. The product is a fused mass of lumps. Use the pestle to crush the lumps into a coarse powder.
16. Use the magnet to test the iron(II) sulfide for ferromagnetism (attraction to a magnet). If the reaction

CAUTIONS

Iron(II) sulfide powder is pyrophoric, which means that it can catch fire spontaneously. Crush it gently, and don't attempt to powder it too finely. If it does catch fire, dump the whole mass into the pail of water.

has proceeded to completion and there is not an excess of elemental iron present, the product should not be attracted to the magnet.

17. Sulfur does not react with hydrochloric acid under normal conditions. Iron reacts with hydrochloric acid to form hydrogen gas. Iron(II) sulfide reacts with hydrochloric acid to form hydrogen sulfide, which has an intense odor of rotten eggs. Test for the presence of iron(II) sulfide by adding a few sand-size grains of your product to a test tube that contains about 1 mL of concentrated hydrochloric acid and checking for the odor of hydrogen sulfide. (Note that concentrated hydrochloric acid itself has a strong, distinctive odor, but the odor of hydrogen sulfide is quite different.) Don't put the test tube under your nose to check for hydrogen sulfide. Instead, hold the test tube well away from your face and use your hand to waft the odor from the test tube toward your nose.

CAUTIONS

In addition to smelling awful, hydrogen sulfide gas is extremely toxic—more so than hydrogen cyanide. Hydrogen sulfide is particularly pernicious, because the human nose rapidly becomes desensitized to its odor. That means it's possible to be exposed to a lethal concentration of hydrogen sulfide without noticing a particularly strong odor. Use only a few tiny grains of iron(II) sulfide, and limit your exposure to hydrogen sulfide to a short, wafted sniff. *Never* put the test tube under your nose.

DISPOSAL: The waste material from this lab can be ground to powder and flushed down the drain with plenty of water.

FIGURE 9-1: *Iron and sulfur reacting to form iron sulfide*

TABLE 9-1:
Observe a composition reaction—observed and calculated data.

Item	Data
A. Mass of tin can lid	____.____ g
B. Mass of iron filings	____.____ g
C. Mass of sulfur	____.____ g
D. Moles of iron filings	___.____ moles
E. Moles of sulfur	___.____ moles
F. Expected mass of FeS	____.____ g
G. Mass of tin can lid + product	____.____ g
H. Actual mass of product (G – A)	____.____ g
I. Percent yield ($100 \cdot H/F$)	____.____ %

REVIEW QUESTIONS

Q1: List at least five sources of error in this experiment.

_____

_____

_____

_____

_____

_____

Q2: If you repeated this experiment using 2.0 g of iron and 1.5 g of sulfur as the reactants, would you expect the product to be attracted to a magnet? Why?

_____

_____

_____

Q3: Upon running this experiment, a student obtains an actual yield of 113.4%. Ruling out weighing errors, what is the most likely source of this error?

_____

_____

_____

Q4: How might you adjust the conditions of this experiment to maximize the amount of FeS produced as product while minimizing the production of pyrrhotite ($Fe_{(1-x)}S$) and mackinawite ($Fe_{(1+x)}S$)?

_____

_____

_____

_____

_____

_____

_____

LABORATORY 9.2:
OBSERVE A DECOMPOSITION REACTION

The *refractory* (using heat) decomposition of sodium hydrogen carbonate to sodium carbonate, carbon dioxide, and water is a typical decomposition reaction, in which a compound reacts to form two or more other elements and/or compounds. The balanced equation for this reaction is:

$$2\ NaHCO_3(s) \rightarrow Na_2CO_3(s) + CO_2(g) + H_2O(g)$$

I didn't choose this example decomposition reaction arbitrarily. The refractory decomposition of sodium bicarbonate to sodium carbonate is the second most important decomposition reaction used in industrial processes, and is crucial to the world's economy. This reaction is the final step in the Solvay process, which is used to produce about 75% of the world's supply of sodium carbonate, also known as washing soda or soda ash. (The remaining 25% is mined or extracted from brine.) Worldwide production of sodium carbonate is more than 40 billion kilograms annually, about 7 kilograms for every person on the planet. Sodium carbonate is an essential component of glass, and is also used for hundreds of other purposes, including processing wood pulp to make paper, as an ingredient in soaps and detergents, as a buffer, and as a neutralizer.

Unlike many decomposition reactions, which are essentially irreversible, the refractory decomposition of sodium hydrogen carbonate is easily reversible, and in fact a considerable amount of sodium hydrogen carbonate is produced commercially by this reverse reaction.

$$Na_2CO_3(s) + CO_2(g) + H_2O(g) \rightarrow 2\ NaHCO_3(s)$$

Carbon dioxide gas is bubbled through a saturated solution of sodium carbonate and reacts to form sodium hydrogen carbonate, which is less soluble than sodium carbonate. The sodium hydrogen carbonate precipitates out, and is isolated by filtration.

In this lab, we'll produce sodium carbonate from sodium hydrogen carbonate via a decomposition reaction.

REQUIRED EQUIPMENT AND SUPPLIES

☐ goggles, gloves, and protective clothing

☐ balance and weighing paper

☐ ring stand

☐ support ring

☐ clay triangle

☐ gas burner

☐ crucible and lid

☐ test tube

☐ crucible tongs

☐ test tube holder or clamp

☐ sodium bicarbonate (~10 g)

☐ cobalt chloride test strip (optional)

☐ toothpicks or wood splints

SUBSTITUTIONS AND MODIFICATIONS

- If you do not have a crucible and lid, you can substitute a Pyrex dessert cup with a Pyrex saucer, watch glass, or similar heat-resistant item as a lid.

- If you do not have the recommended equipment needed to heat the samples, you can substitute a large tin can lid placed flat on a stove burner or hotplate.

- You can make your own cobalt chloride test strips by soaking ordinary paper in a saturated solution of cobalt chloride and then drying it thoroughly. If any moisture is present, the test strip is pink. In the absence of moisture, the test strip is blue. If you don't have any cobalt chloride, look for one of those silica gel drying packets that changes color from blue when dry to pink when it needs to be recharged by heating it in the oven. You can use a few of the blue crystals to test for the presence of water, which turns them pink.

- You may substitute a propane torch for the gas burner.

- You may substitute a pair of pliers or similar gripping tool for the crucible tongs. (Be careful; crucibles are fragile.)

CAUTIONS

This experiment uses heat. Be careful with the heat source and when handling hot objects, and have a fire extinguisher readily available if your heat source uses flame. Wear splash goggles, gloves, and protective clothing.

THE MOST IMPORTANT DECOMPOSITION REACTION

If you're wondering what decomposition reaction could possibly be run on a larger scale than the industrial production of sodium carbonate, think about it the next time you're in an automobile. The most important decomposition reaction is the refining of crude petroleum to gasoline and other fuels, which dwarfs the production of sodium carbonate both in mass and in economic value. Most refining is done by a decomposition process called *catalytic cracking*, which converts the long-chain hydrocarbons present in crude petroleum into the shorter-chain hydrocarbons used in gasoline and other fuels and lubricants.

FIGURE 9-2: *A burning splint is immediately extinguished by the carbon dioxide gas evolved during the reaction.*

PROCEDURE

This laboratory has two parts. In Part I, we'll do qualitative testing to determine the products of the reaction. In Part II, we'll do some semi-quantitative testing to verify that the mass of the solid product (sodium carbonate) closely matches the mass we would expect from examining the balanced equation for the reaction.

PART I:

1. If you have not already done so, put on your splash goggles, gloves, and protective clothing.
2. Add sodium hydrogen carbonate to the test tube until it is about one quarter full.
3. Light your gas burner (or alcohol lamp).
4. Using the test tube holder or clamp, hold the test tube so that the flame heats that part of the test tube that contains the sodium hydrogen carbonate. Try to play the flame evenly over the entire sample. (Using a clamp mounted on a ring stand rather than a test tube holder makes the following steps easier, because it frees both hands.)
5. After you have heated the sample for at least 15 to 30 seconds, ignite a toothpick or wood splint and insert the burning end into the top of the test tube. If the flame or ember is snuffed out (Figure 9-2), that's a good indication that the reaction is producing carbon dioxide as expected. If the splint continues to burn, continue heating the sample until enough carbon dioxide is produced to displace the air inside the test tube and try using the splint again.
6. As you heat the sample, you should notice a liquid condensing on the cooler inside upper surface of the test tube, at first as fogging and later as actual droplets. Touch a blue cobalt chloride test strip to this liquid. Does it turn pink? If so, that confirms that the liquid is water.

PART II:

In Part II, we'll determine the mass loss that occurs when sodium hydrogen carbonate is heated to decompose it to sodium carbonate. In theory, all of the mass loss is attributable to the outgassing of carbon dioxide and water during the heating process, and the remaining mass should represent only sodium carbonate. In practice, things can be a bit more complicated. Sodium carbonate exists in four hydration states: anhydrous (Na_2CO_3), the monohydrate ($Na_2CO_3 \cdot 1H_2O$), the heptahydrate ($Na_2CO_3 \cdot 7H_2O$), and the decahydrate ($Na_2CO_3 \cdot 10H_2O$). If we don't heat the sample strongly enough, it's possible that some of the water produced by the decomposition will be retained by the sodium carbonate as water of hydration. Accordingly, we'll heat our sample at a high enough temperature and for long enough to ensure that it is completely converted to anhydrous sodium carbonate.

1. If you have not already done so, put on your splash goggles, gloves, and protective clothing.
2. Place the covered crucible in the clay triangle, and adjust the height of the supporting ring to put the bottom of the crucible in the hottest part of the flame. Make sure that the crucible lid is positioned to allow gases to be vented. Heat the empty crucible and lid strongly for at least a minute to drive off any water and volatile contaminants present.
3. After allowing them to cool completely, weigh the crucible and lid and record the mass to 0.01 g on line A of Table 9-2.
4. Add approximately 5.0 g of sodium hydrogen carbonate to the crucible, reweigh the crucible and lid, and record the mass to 0.01 g on line B of Table 9-2.
5. Subtract the empty mass of the crucible and lid from the combined mass of the crucible, lid, and contents, and record the mass of the sodium hydrogen carbonate to 0.01 g on line C of Table 9-2.
6. Place the covered crucible in the clay triangle, make sure that the crucible lid is positioned to allow gases to be vented, and heat the crucible strongly for 15 minutes.
7. After allowing the crucible, lid, and contents to cool completely, reweigh the crucible, lid, and contents and record the mass to 0.01 g on line D of Table 9-2.
8. Determine the mass of the product remaining in the crucible by subtracting the mass of the empty crucible and lid from the combined mass of the crucible, lid, and contents after heating. Record the mass of the sodium carbonate to 0.01 g on line E of Table 9-2.
9. Calculate the mass loss attributable to outgassing of carbon dioxide and water vapor by subtracting the mass of the sodium carbonate product (line E) from the mass of the sodium hydrogen carbonate reactant (line C). Enter the mass loss to 0.01 g on line F of Table 9-2. Calculate the mass loss percentage by multiplying F by 100, then dividing by C. Enter the result on line G.

TABLE 9-2:

Observe a decomposition reaction—observed and calculated data.

Item	Data
A. Mass of crucible and lid	_____._____ g
B. Mass of crucible, lid, and sample (before heating)	_____._____ g
C. Mass of sodium hydrogen carbonate (B − A)	_____._____ g
D. Mass of crucible, lid, and sample (after heating)	_____._____ g
E. Mass of sodium carbonate (D − A)	_____._____ g
F. Mass loss (C − E)	_____._____ g
G. Mass loss percentage (100 · F/C)	_____._____ %

OPTIONAL ACTIVITIES

If you have extra time, consider performing these optional activities:

- Determine the effect of time and temperature on the decomposition reaction. Use four crucibles (or Pyrex custard cups). Weigh each crucible to 0.01g and record the mass. Add 5.0 g of sodium hydrogen carbonate, replace the lid, reweigh each crucible, and record the combined mass of crucible, lid, and sample. Preheat the kitchen oven to its lowest setting (anything over about 50°C or 120°F will do) and place all four crucibles on a baking tin into the oven. Start your timer. Using crucible tongs, remove one crucible after 15 minutes, and the remaining crucibles at 30 minutes, one hour, and two hours. After allowing the crucibles to cool, determine how much mass has been lost from each crucible and calculate the percentage of mass loss. Compare the mass loss percentage to the value you determined in Part II. Propose an explanation for your findings.

- Repeat the preceding experiment, but with the oven preheated to 450°F or higher. Do the results differ? Propose an explanation.

DISPOSAL: Sodium hydrogen carbonate and sodium carbonate are not hazardous. Flush them down the drain with plenty of water.

REVIEW QUESTIONS

Q1: Based on the original mass of sodium hydrogen carbonate used in Part II and the balanced equation, calculate the theoretical yield of sodium carbonate.

Q2: Assuming that the product in Part II is pure anhydrous sodium carbonate, calculate the percent yield. Would you expect the actual yield of sodium carbonate to closely approach the theoretical yield? Why or why not?

LABORATORY 9.3:

OBSERVE A SINGLE DISPLACEMENT REACTION

In a single displacement reaction, a more active element displaces a less active element in a compound, forming a new compound that incorporates the more active element and releasing the less active element in elemental form. The reaction of an active metal with an acid is a common type of single displacement reaction in which the metal is oxidized to a cationic oxidation state and the hydrogen present in the acid is reduced to elemental hydrogen. The fizzing and bubbling that occurs when an active metal comes into contact with a strong acid is caused by the emission of gaseous hydrogen.

A nonchemist might believe that the metal "dissolves" in the acid. That's not true, of course. When we dissolve table salt or sugar in water, for example, no chemical reaction occurs. We can recover the solute unchanged, simply by evaporating the solvent. But when we "dissolve" a metal in an acid, evaporating the liquid does not allow us to recover the metal unchanged. Instead, the metal has been converted to a salt by a single displacement reaction.

Different metals have different reactivity, as shown in Table 9-3. The most active metals, such as lithium and potassium, react (sometimes explosively) even with cold water to release hydrogen gas. Less active metals, such as magnesium and aluminum, do not react with cold water, but do react with steam, particularly if the metal is finely divided and the steam is superheated. Still less active metals, such as cobalt and nickel, do not react with water in any form, but do react with acids to release hydrogen gas. Finally, the least active metals, from antimony and bismuth to platinum and gold, do not react even with acids under normal circumstances. (Even platinum and gold do react with *aqua regia*, a mixture of concentrated hydrochloric and nitric acids.)

Another common type of single displacement reaction is the reaction of a solid metal immersed in a solution of a salt of another metal. For example, an iron (or steel) nail immersed in a solution of copper sulfate becomes plated, because the copper ions are reduced to metallic copper while the iron metal is oxidized to iron ions. If sufficient metallic iron is available, eventually all of the copper ions are reduced, leaving metallic copper in a solution of iron sulfate.

REQUIRED EQUIPMENT AND SUPPLIES

- ☐ goggles, gloves, and protective clothing
- ☐ balance and weighing paper
- ☐ test tube (2)
- ☐ test tube rack
- ☐ graduated cylinder, 10 mL
- ☐ sandpaper or emery board
- ☐ toothpicks or wood splints
- ☐ thread or string (~15 cm)
- ☐ aluminum granules, turnings, or shot (~1 g)
- ☐ iron or steel nail
- ☐ hydrochloric acid, concentrated (~5 mL)
- ☐ copper sulfate pentahydrate (~2.5 g)
- ☐ water

SUBSTITUTIONS AND MODIFICATIONS

- You may substitute any type of scrap aluminum (such as aluminum foil or aluminum beverage cans) for the aluminum granules or shot, although the reaction rate may be lower. If you substitute scrap aluminum, use sandpaper or a similar abrasive to remove ink, paint, plastic coating, the oxidation layer, and other contaminants from its surface. Alternatively, you can produce aluminum filings by using a file on a piece of scrap aluminum stock.

- You may substitute any form of metallic iron or steel for the nail, including iron filings or steel wool (be sure to wash steel wool to remove soap and other contaminants).

- You may substitute muriatic acid from the hardware store for the hydrochloric acid. If you use muriatic acid, make sure the concentration is 30% or higher.

In this laboratory, we react aluminum metal with hydrochloric acid to form aluminum(III) chloride and hydrogen gas. The balanced equation for this reaction is:

$$2\,Al(s) + 6\,HCl(aq) \rightarrow 2\,AlCl_3(aq) + 3\,H_2(g)$$

We'll observe the reaction and test for the presence of hydrogen gas by using a burning splint. We also react iron metal with a solution of copper sulfate to form iron(II) sulfate and copper metal. The balanced equation for this reaction is:

$$Fe(s) + CuSO_4(aq) \rightarrow Cu(s) + FeSO_4(aq)$$

TABLE 9-3: *Reactivity series of common metals, from most to least reactive*

Element	Displace H_2 from acids?	Displace H_2 from steam?	Displace H_2 from water?
Lithium	●	●	●
Potassium	●	●	●
Barium	●	●	●
Calcium	●	●	●
Sodium	●	●	●
Strontium	●	●	●
Magnesium	●	●	○
Aluminum	●	●	○
Manganese	●	●	○
Zinc	●	●	○
Chromium	●	●	○
Iron	●	●	○
Cadmium	●	●	○
Cobalt	●	○	○
Nickel	●	○	○
Tin	●	○	○
Lead	●	○	○
Hydrogen	–	–	–
Antimony	○	○	○
Bismuth	○	○	○
Copper	○	○	○
Mercury	○	○	○
Silver	○	○	○
Platinum	○	○	○
Gold	○	○	○

FIGURE 9-3: *Aluminum metal reacting with hydrochloric acid*

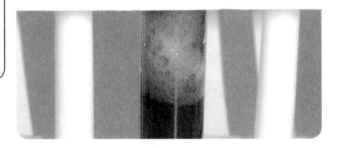

CAUTIONS

Concentrated hydrochloric acid is corrosive and emits hazardous fumes. Keep the test tube pointed away from you and in a safe direction in case the reaction proceeds too vigorously. Spilled acid can be neutralized with an excess of solid sodium bicarbonate (baking soda). Hydrogen gas is extremely flammable. Although only a small amount of hydrogen is generated in this experiment, it may "pop" when ignited, startling you and causing you to drop the test tube. Do not sniff the test tube during or after the reaction, because the heat of the reaction liberates irritating hydrogen chloride gas. Perform this experiment outdoors, or under an efficient fume hood or exhaust fan. Wear splash goggles, gloves, and protective clothing.

PROCEDURE

This laboratory has two parts. In Part I, we'll observe a single-displacement reaction in which aluminum metal displaces hydrogen from hydrochloric acid. In Part II, we'll observe another single-displacement reaction in which iron metal displaces copper ions from a solution of copper sulfate.

PART I:

1. If you have not already done so, put on your splash goggles, gloves, and protective clothing.
2. Transfer about 5 mL of concentrated hydrochloric acid to one of the test tubes in your test tube rack.
3. Ignite a toothpick or wood splint, and hold it just above the mouth of the test tube to verify that no flammable gas is present.
4. Weigh about 1 g of aluminum granules, and transfer them to the test tube with the hydrochloric acid. (Depending on the fineness of the aluminum metal, the reaction can be quite vigorous, so make sure that the mouth of the test tube is pointing in a safe direction in case liquid is ejected.) It takes a moment for the hydrochloric acid to remove the aluminum oxide that coats the aluminum metal. After this oxide coating is removed, the reaction proceeds vigorously (Figure 9-3).
5. Ignite another toothpick or wood splint and hold the burning end just above the top of the test tube. The hydrogen gas produced by the reaction ignites. (Don't be startled if the gas ignites with a noticeable pop; the small amount of hydrogen gas produced by this reaction is harmless.)

DR. PAUL JONES COMMENTS:

One can also do this on a slightly larger scale and catch the hydrogen in a balloon, which can later be ignited, to amusing effect, in a suite-mate's shower. Or so I'm told.

PART II:

1. If you have not already done so, put on your splash goggles, gloves, and protective clothing.
2. Weigh about 2.5 g of copper sulfate pentahydrate and add it to a test tube in the rack.
3. Transfer about 10 mL of water to the test tube, and swirl or shake the test tube until the copper sulfate dissolves completely.
4. Use the sandpaper or emery board to clean the surface of the nail thoroughly. Remove all oxidation and leave a shiny, bright surface.
5. Tie the thread around the head of the nail, and carefully submerge the nail in the solution of copper sulfate, keeping the thread outside the tube.
6. After 15 minutes and then again every 15 minutes, withdraw the nail, and observe and note any visible changes, and then return it to the copper sulfate solution. Continue this until no further change is evident (Figure 9-4).

FIGURE 9-4:

Metallic copper plated on an iron nail by displacement

DISPOSAL: All of the waste solutions can be flushed down the drain with plenty of water. Dispose of the solid waste with household waste.

REVIEW QUESTIONS

Q1: Would you expect the reaction rate of iron with hydrochloric acid to be faster or slower than the rate you observed for aluminum with hydrochloric acid? Why?

Q2: U.S. pennies made after 1982 consist of a zinc core with thin copper plating. What reaction would you expect if you dropped an uncirculated U.S. penny into concentrated hydrochloric acid? Why? If testing a badly worn and scratched coin had different results, how would you explain that difference?

Q3: In Part II, if you had submerged a silver teaspoon in a copper sulfate solution, what reaction would you expect? Why?

Q4: Brass is an alloy (mixture) of copper metal and zinc metal. If you added brass filings to a solution of concentrated hydrochloric acid, what reaction(s) would you expect to occur and what products would you expect the reaction(s) to produce?

LABORATORY 9.4:
STOICHIOMETRY OF A DOUBLE DISPLACEMENT REACTION

In a double displacement reaction, two compounds exchange elements or ionic species with each other. Double displacement reactions take the general form AX + BY → AY + BX, where A and B are cations (positively charged ions) and X and Y are anions (negatively charged ions). Note that the reactants combine in a fixed proportion—1:1 in this case—and that the products are also in a fixed proportion, again 1:1 in this case.

The proportions are not always 1:1. For example, copper(II) sulfate reacts with sodium hydroxide to form copper(II) hydroxide and sodium sulfate, according to the following balanced equation:

$$CuSO_4(aq) + 2NaOH(aq) \rightarrow Cu(OH)_2(s) + Na_2SO_4(aq)$$

or, looking at the ionic species:

$$Cu^{2+}(aq) + SO_4^{2-}(aq) + 2Na^+(aq) + 2OH^-(aq)$$
$$\rightarrow Cu(OH)_2(s) + 2Na^+(aq) + SO_4^{2-}(aq)$$

Note that on both sides of both equations the number of molecules (or moles) of the cationic species (copper and sodium) and the anionic species (sulfate and hydroxide) are the same and that the total net charge is zero.

In this lab, we'll react solutions of copper(II) sulfate and sodium hydroxide to produce copper(II) hydroxide and sodium sulfate. Knowing the mass of the copper(II) sulfate we used and after determining the masses of the copper(II) hydroxide and sodium sulfate products, we'll examine the stoichiometric relationship of the reactants and products and determine whether our balanced equation accurately represents the actual reaction.

To illustrate that reactions are useful for practical purposes as well as learning purposes, we'll use the copper(II) hydroxide produced by this reaction as one of the reactants needed to synthesize a new compound that we'll use in a later laboratory session. This compound, tetraamminecopper dihydroxide , $[Cu(NH_3)_4](OH)_2$, or *Schweizer's Reagent* for short, has the interesting property of being able to dissolve cellulose (for example, wood pulp, cotton, paper), and is used to produce rayon and other semisynthetic fibers.

REQUIRED EQUIPMENT AND SUPPLIES

- ☐ goggles, gloves, and protective clothing
- ☐ balance and weighing paper
- ☐ beaker, 150 mL (2)
- ☐ graduated cylinder, 10 mL
- ☐ graduated cylinder, 100 mL
- ☐ stirring rod
- ☐ wash bottle (water)
- ☐ funnel and filter paper
- ☐ ring stand
- ☐ support ring
- ☐ evaporating dish
- ☐ hotplate
- ☐ storage bottle labeled Schweizer's Reagent
- ☐ copper(II) sulfate pentahydrate (24.97 g)
- ☐ sodium hydroxide (8.00 g)
- ☐ aqueous household ammonia (~70 mL)
- ☐ water

CAUTIONS

Copper sulfate is moderately toxic. Sodium hydroxide is corrosive and caustic. Aqueous ammonia is corrosive and irritating. Wear splash goggles, gloves, and protective clothing.

STOICHIOMETRY

Stoichiometry is used to calculate quantitative relationships between the reactants and products in a chemical reaction. Stoichiometry is based upon the *Law of Conservation of Mass*, the *Law of Constant Proportions* (also called the *Law of Definite Proportions*), and the *Law of Multiple Proportions*.

In most chemical reactions, the reactants and products combine in fixed proportions. For example, one molecule of reactant A may combine with two molecules of reactant B to yield one molecule of product C and two molecules of product D. Because a chemical reaction can neither create nor destroy matter, nor transmute one element into another, the total number of atoms of each element present in the reactants must equal the total number of atoms of each element present in the products.

Chemists use stoichiometry to balance equations—as we did for the reaction of copper(II) sulfate and sodium hydroxide—and to specify the *mole ratio* (or *molar proportions*) of elements in a compound. For example, the mole ratio of copper:sulfur:oxygen in copper(II) sulfate is 1:1:4.

SUBSTITUTIONS AND MODIFICATIONS

- You may substitute a large Pyrex saucer or similar heat-resistant container for the evaporating dish.

- You may substitute for the hotplate another heat source, such as a kitchen stove burner or alcohol lamp with wire gauze and stand.

- For this laboratory session, we've specified quantities of chemicals sufficient to make a useful amount of Schweizer's Reagent. If you do not intend to make or use the Schweizer's Reagent, you may reduce the quantities of chemicals to one-tenth (2.50 g of copper(II) sulfate pentahydrate and 0.80 g of sodium hydroxide) and substitute test tubes or similar containers for the beakers.

- You may substitute 15.96 g of anhydrous copper(II) sulfate for the 24.97 g of copper(II) sulfate pentahydrate. You may also substitute raw copper sulfate from the hardware store (sold as root killer) or the copper sulfate you purified by recrystallization in an earlier laboratory session **Lab 6.3, "Recrystallization: Purify Copper Sulfate"**.

PROCEDURE

This laboratory has two parts. In Part I, we'll react solutions of copper(II) sulfate and sodium hydroxide to produce copper(II) hydroxide and sodium sulfate, determine the masses of the products, and calculate the stoichiometric relationships for the reaction. In Part II, we'll use the copper(II) hydroxide we produced in Part I to synthesize Schweizer's Reagent.

PART I:

1. If you have not already done so, put on your splash goggles, gloves, and protective clothing.
2. Weigh about 24.97 g of copper(II) sulfate pentahydrate, and record the mass to 0.01 g on line A of Table 9-4.
3. Calculate the number of moles of copper(II) sulfate represented by this mass, and record the number of moles on line B of Table 9-4.
4. Transfer the copper sulfate to a 150 mL beaker, add about 100 mL of water, and stir until the copper sulfate dissolves completely.
5. Weigh about 8.00 g of sodium hydroxide, and record the mass to 0.01 g on line C of Table 9-4.
6. Calculate the number of moles of sodium hydroxide represented by this mass, and record the number of moles on line D of Table 9-4.
7. Transfer about 10 mL of water to the second 150 mL beaker, and add the sodium hydroxide to the water with constant stirring or swirling. (Caution: this reaction is extremely exothermic.)
8. With constant stirring, pour the sodium hydroxide solution into the beaker of copper(II) sulfate solution. Rinse the sodium hydroxide beaker two or three times with a few mL of water and transfer the rinse water to the beaker of copper sulfate solution to make sure that you've transferred all of the sodium hydroxide (Figure 9-5).
9. Rinse the empty beaker thoroughly and set up your funnel holder and funnel with the empty beaker as the receiving container.
10. Fanfold a piece of filter paper, weigh it, record its mass to 0.01 g on line E of Table 9-4, and place the filter paper in the funnel.
11. Stir or swirl the beaker that contains the copper(II) hydroxide to make sure as much as possible is suspended in the solution, and then pour the solution into the filter funnel, catching the filtrate in the empty 150 mL beaker. Rinse the reaction beaker two or three times with a few mL of water and transfer the rinse water into the filter funnel to make sure all of the copper(II) hydroxide is retained by the filter paper.
12. Squirt a few mL of water onto the retained copper(II) hydroxide in the filter paper to rinse as much of the soluble sodium sulfate as possible into the receiving container.

13. Carefully remove the filter paper from the funnel and set it aside to dry. You can speed drying by placing the filter paper and product under an incandescent lamp or by drying it gently in an oven at the lowest heat setting.

14. Weigh the evaporating dish and record its mass to 0.01 g on line F of Table 9-4.

15. Place the evaporating dish on the hotplate, transfer as much of the filtrate as the dish can comfortably contain, and turn the hotplate on, set to low heat. As the solution in the dish evaporates, transfer more of the filtrate until all of it has been transferred to the dish. Finally, rinse the receiving beaker with a few mL of water, and transfer the rinse water to the evaporating dish. When the liquid in the evaporating dish is nearly gone, watch the dish carefully and continue heating until the solution has been evaporated to dryness.

16. While you are waiting for the copper(II) hydroxide and sodium sulfate to dry, calculate the expected masses for both of them based on the initial masses of the copper(II) sulfate and sodium hydroxide. (Remember that one mole of copper(II) sulfate reacts with two moles of sodium hydroxide, and remember to determine the limiting reagent before you calculate the expected masses.) Enter the expected masses of copper(II) hydroxide and sodium sulfate on lines G and H of Table 9-4, respectively.

17. When the copper(II) hydroxide is dry, weigh the filter paper and product, and record their mass to 0.01 g on line I of Table 9-4. Subtract the mass of the filter paper (line E) from the combined mass of the filter paper and product (line I) to determine the mass of copper(II) hydroxide, and record that value to 0.01 g on line J of Table 9-4.

18. When the sodium sulfate is dry, weigh the evaporating dish and product, and record their mass to 0.01 g on line K of Table 9-4. Subtract the mass of the evaporating dish (line F) from the combined mass of the evaporating dish and product (line K) to determine the mass of sodium sulfate, and record that value to 0.01 g on line L of Table 9-4.

19. Calculate the actual percent yields of copper(II) hydroxide and sodium sulfate, and enter those values on lines M and N of Table 9-4, respectively.

FIGURE 9-5: *The precipitate of copper(II) hydroxide*

TABLE 9-4: *Stoichiometry of a double displacement reaction—observed and calculated data*

Item	Data
A. Mass of copper sulfate pentahydrate	____.____ g
B. Moles of copper sulfate pentahydrate	__.____ moles
C. Mass of sodium hydroxide	____.__ g
D. Moles of sodium hydroxide	__.____ moles
E. Mass of filter paper	____.__ g
F. Mass of evaporating dish	____.__ g
G. Expected mass of copper(II) hydroxide	____.__ g
H. Expected mass of sodium sulfate	____.__ g
I. Mass of filter paper + copper(II) hydroxide	____.__ g
J. Actual mass of copper(II) hydroxide (I − E)	____.__ g
K. Mass of evaporating dish + sodium sulfate	____.__ g
L. Actual mass of sodium sulfate (K − F)	____.__ g
M. Percent yield of copper(II) hydroxide $(100 \cdot J/G)$	____.__ %
N. Percent yield of sodium sulfate $(100 \cdot L/H)$	____.__ %

PART II:

1. If you have not already done so, put on your splash goggles, gloves, and protective clothing.
2. Transfer as much as possible of the dried copper(II) hydroxide from the filter paper to a clean 150 mL beaker.
3. Transfer approximately 70 mL of 6 M aqueous ammonia to the beaker, and stir until all of the copper(II) hydroxide has reacted with the ammonia to form tetraamminecopper dihydroxide (Schweizer's Reagent). If all of the copper(II) hydroxide does not react, add a bit more aqueous ammonia and stir until no copper(II) hydroxide remains on the bottom of the beaker (Schweizer's Reagent, Figure 9-6).
4. Transfer the Schweizer's Reagent to a storage bottle. Label the bottle with its contents and date it. (If you prefer, you can evaporate the solution to dryness and store the tetraamminecopper dihydroxide as crystals.)

DISPOSAL: The sodium sulfate produced in Part I can be stored for later use, or can be disposed of with household waste. If you did not complete Part II, you can store the copper hydroxide in an airtight container for later use. (If exposed to air, the blue copper hydroxide is gradually oxidized to black copper oxide.)

FIGURE 9-6: *Schweizer's Reagent, formed by reacting copper(II) hydroxide with aqueous ammonia*

REVIEW QUESTIONS

Q1: What are the theoretical mole ratios of copper(II) hydroxide and sodium sulfate?

Q2: If you react one mole of sodium hydroxide with excess copper(II) sulfate, how many moles of copper(II) hydroxide are produced?

Q3: Write a balanced equation for the reaction of copper(II) sulfate with sodium carbonate (Na_2CO_3) to produce copper(II) carbonate ($CuCO_3$) and sodium sulfate.

Q4: Write a balanced equation for the reaction of copper(II) hydroxide with aqueous ammonia (NH_3) to produce tetraamminecopper dihydroxide.

10

Laboratory:
Reduction-Oxidation (Redox) Reactions

A *reduction-oxidation reaction* or *redox reaction* is a chemical reaction during which the *oxidation state* of two or more of the reactants change. Note that the word "oxidation" in this context has nothing to do with oxidation in the sense of combining with oxygen. Most redox reactions do not involve oxygen at all.

The oxidation state of an atom or ion can be thought of as its electrical charge. In simple terms, a neutral atom is in oxidation state 0. An ion with a positive charge has a positive oxidation state overall (although one or more of the component atoms in that ion may have nonpositive oxidation states. For example, if a neutral sodium atom (oxidation state 0, Na^0) loses an electron, it has been oxidized to oxidation state 1+, and becomes a Na^+ ion. Conversely, a carbonate ion, CO_3^{2-}, is made up of one carbon atom in oxidation state 4+ and three oxygen atoms, each in oxidation state 2−, for a net charge (oxidation state) of 2−.

Oxidation in the context of a redox reaction refers to the oxidation state of an atom being increased. Conversely, *reduction* refers to the oxidation state of an atom being reduced. We saw an example of a redox reaction in the preceding chapter, when we reacted aluminum metal with hydrochloric acid to produce aluminum chloride and hydrogen gas. The balanced equation for that reaction is:

$$2\ Al(s) + 6\ HCl(aq) \rightarrow 2\ AlCl_3(aq) + 3\ H_2(g)$$

or, with the oxidation states of each species as a superscript:

$$2\ Al^0(s) + 6\ H^+(aq) + 6\ Cl^-(aq) \rightarrow 2\ Al^{3+}(aq) + 6\ Cl^-(aq) + 3\ H_2^0(g)$$

In this reaction, two atoms of aluminum metal (oxidation state 0) each lose three electrons and are oxidized to aluminum ions (oxidation state +3). The six electrons lost by the two aluminum atoms combine with six hydrogen ions (oxidation state +1), reducing the hydrogen ions to form six atoms of hydrogen gas (oxidation state 0), which then combine to form three molecules of diatomic hydrogen gas (still in oxidation state 0). The six chloride ions (oxidation state −1) do not change oxidation states, and so are the same on both sides of the equation.

Many chemical reactions do not involve any change in oxidation state. For example, when we reacted aqueous solutions of copper(II) sulfate and sodium hydroxide to form a precipitate of copper(II) hydroxide and a solution of sodium sulfate, no changes in oxidation state occurred:

$$CuSO_4(aq) + 2\ NaOH(aq) \rightarrow Cu(OH)_2(s) + Na_2SO_4(aq)$$

or, showing the oxidation states:

$$Cu^{2+}(aq) + SO_4{}^{2-}(aq) + 2\ Na^+(aq) + 2\ OH^-(aq)$$

$$\rightarrow [Cu^{2+}(OH^-)_2](s) + 2\ Na^+(aq) + SO_4{}^{2-}(aq)$$

On both sides of the equation, copper is in the +2 oxidation state, sulfate −2, sodium +1, and hydroxide −1.

Oxidation state 0 is the neutral form of any element, such as (at standard conditions) aluminum in its metallic form or chlorine in its gaseous form. An element in oxidation state 0 possesses the same number of negatively charged electrons in its shell as the number of positively charged protons in its nucleus, leaving a net charge of 0.

If a chlorine atom at oxidation state 0 gains an electron, that chlorine atom becomes a chlorine ion with a charge of −1, and an oxidation state of −1. Conversely, if a chlorine atom at oxidation state 0 loses an electron, that chlorine atom becomes a chlorine ion with a charge of +1, and an oxidation state of +1. With the exception of the noble gases—helium, neon, argon, krypton, xenon, and radon, all in column 18 of the periodic table—all elements are commonly found in at least one oxidation state other than 0. Some elements have only one or two known nonzero oxidation states. Other elements are commonly found in many oxidation states. For example, in its natural form and in its many compounds, carbon can be found in oxidation states −4, −3, −2, −1, 0, +1, +2, +3, and +4.

Oxidation and reduction also apply to atoms that are in oxidation states other than 0. For example, iron is found in three oxidation states. Metallic iron is in oxidation state 0. Metallic iron can be oxidized to the Fe^{2+} ion, also called the ferrous ion or the iron(II) ion. The Fe^{2+} ion can be further oxidized to the Fe^{3+} ion—also called the ferric ion or the iron(III) ion—or it can be reduced to metallic iron. The Fe^{3+} ion can be reduced to the Fe^{2+} ion by gaining one electron, or all the way to Fe^0 (iron metal) by gaining three electrons.

Ions of a particular element in different oxidation states can have very different chemical and physical properties. For example, solutions that contain vanadium ions in different oxidation states show a rainbow of colors. Solutions of V^{5+} ions are yellow; those of V^{4+} are blue; V^{3+}, green; and V^{2+}, violet. Solutions of ions in different oxidations states of chromium, manganese, and other elements show similar color differences.

It's important to understand that in a redox reaction oxidation and reduction are two sides of the same coin. If one atom is oxidized during the reaction, another must be reduced, and vice versa. Furthermore, the total number of electrons lost by the atom or atoms being oxidized must equal the total number of electrons gained by the atom or atoms that are being reduced. Charge must be conserved. For example, when aluminum reacts with the H^+ ions from hydrochloric acid to form Al^{3+} ions, the three electrons lost when one aluminum atom is oxidized reduce three H^+ ions to H^0 atoms.

The laboratory sessions in this chapter explore various aspects of redox reactions.

EVERYDAY REDOX REACTIONS

Redox reactions, including some that we literally couldn't live without, are common in everyday life. Here are just a few examples:

- Metals are usually mined in the form of an oxide, carbonate, or other salt of the metal. Refining metals—whether by smelting, electrolysis, or some other method—uses redox reactions; specifically, reduction of the metal in a positive oxidation state to the neutral metallic form.

- Water purification uses redox reactions to oxidize colored, bad-tasting, or otherwise objectionable impurities to forms that are safe and acceptable.

- Bleaches used in laundry, papermaking, and other processes all depend upon redox reactions to oxidize colored stains and impurities to colorless compounds.

- Photographic film works by a redox reaction initiated by light. When photons strike the tiny grains of silver bromide and other silver halides present in the emulsion, they cause silver ions to be reduced to microscopic specks of metallic silver. During development, those tiny specks of silver catalyze reduction of the silver halide grains surrounding them. After washing away the unreduced silver, what remains are the tiny black grains that make the picture.

- Corrosion and rusting are examples of redox reactions that are familiar to everyone. In particular, unprotected iron and steel are prone to react with atmospheric oxygen in the presence of moisture to form iron oxides, familiarly called *rust*. Many forms of protection against corrosion take advantage of redox reactions as well. For example, plating iron with zinc (called *galvanizing*) protects the underlying iron, because it will not oxidize while in contact with the more active metal.

WHAT ABOUT FE⁴⁺ AND FE⁺?

It might seem that an Fe^{3+} ion could lose one more electron to produce an Fe^{4+} ion. That doesn't happen, because of the configuration of iron's electron shell. Only tightly bound electrons remain when iron is in the +3 state. Similarly, it might seem that Fe^0 could lose one electron (or Fe^{2+} gain one electron) to form Fe^+. That doesn't happen, because the electron shell configuration of Fe^+ is unstable. Any Fe^+ ions that fleetingly form during an oxidation-reduction reaction immediately shed an electron to form Fe^{2+} ions or grab an electron to form Fe^0.

KEEPING OXIDATION AND REDUCTION STRAIGHT

Many beginning chemists have trouble remembering whether oxidation means gaining or losing an electron. Generations of chemists have kept this straight using the mnemonic **LEO the lion says GER**, which you can expand to **L**ose **E**lectron = **O**xidation and **G**ain **E**lectron = **R**eduction.

LABORATORY 10.1:

REDUCTION OF COPPER ORE TO COPPER METAL

It may surprise you to learn that at least one redox reaction was discovered and used thousands of years ago. Like many discoveries in chemistry, this one was purely accidental. We can speculate about what happened. One evening long ago, some of our remote ancestors gathered around a fire built in a circle of stones. Those stones happened to be chunks of a copper-bearing ore, probably copper carbonate.

All of the required elements were present. The copper ore provided the copper ions. The charred wood provided the carbon in the form of charcoal. The fire provided the heat needed to initiate and sustain the reaction. Here's what happened:

$$2 CuCO_3(s) + C(s) \rightarrow 2 Cu(s) + 3 CO_2(g)$$

or, showing the oxidation states:

$$2 [Cu^{2+}[C^{4+}(O^{2-}_3)]^{2-}](s) + C^0(s) \rightarrow 2 Cu^0(s) + 3 [C^{4+}(O^{2-}_2)](g)$$

As the fire burned during the night, some of the copper ore and charcoal disappeared and, as if by magic, were replaced by a shiny puddle of metallic copper. Chances are that this had happened many times before and no one thought much about it, if they even noticed. But this time someone did notice and started wondering. What was this shiny new substance? Could it be used for anything? Can we get more of it just by heating rocks?

REQUIRED EQUIPMENT AND SUPPLIES

- ☐ goggles, gloves, and protective clothing
- ☐ balance and weighing paper
- ☐ ring stand
- ☐ support ring
- ☐ clay triangle
- ☐ crucible and lid
- ☐ crucible tongs
- ☐ gas burner
- ☐ copper carbonate (~5 g)
- ☐ activated charcoal (~3 g)

At the time, metallic copper was known, but the only source was small, rare deposits of natural metallic copper. The discovery that copper metal could be obtained by heating copper ore in conjunction with charcoal, a process called **smelting**, marked the end of the Stone Age.

In this lab, we'll recreate that first accidental smelting of copper ore to copper metal.

CAUTIONS

This experiment uses the hot flame provided by a gas burner. Be careful with the flame, have a fire extinguisher readily available, and use care in handling hot objects (a hot crucible looks exactly like a cold crucible). Wear splash goggles, gloves, and protective clothing.

PROCEDURE

1. If you have not already done so, put on your splash goggles, gloves, and protective clothing.
2. Weigh a clean, dry crucible and lid and record the mass to 0.01 g on line A of Table 10-1.
3. Weigh about 5.0 g of copper carbonate and record the mass to 0.01 g on line B of Table 10-1.
4. Weigh about 3.0 g of activated charcoal and record the mass to 0.01 g on line C of Table 10-1.
5. Mix the copper carbonate and activated charcoal thoroughly, and transfer them to the crucible.
6. Weigh the crucible and contents and record the mass to 0.01 g on line D of Table 10-1.
7. Set up your tripod stand or ring stand with the clay triangle, put the covered crucible in place, and place the gas burner to direct the hottest part of its flame at the bottom of the crucible.
8. Heat the crucible strongly for at least 15 minutes.
9. Allow the crucible to cool to room temperature. You can check the temperature of the crucible by placing your hand near, but not touching, its surface to check for radiated heat.
10. Reweigh the crucible and contents and record the mass to 0.01 g on line E of Table 10-1.
11. Calculate the mass loss (line D minus line E) and enter the value to 0.01 g on line F of Table 10-1.

FIGURE 10-1: *Mixed copper carbonate and charcoal before smelting*

SUBSTITUTIONS AND MODIFICATIONS

- You may substitute a test tube and clamp or test tube holder for the crucible, support stand, and clay triangle. If you do use a test tube, choose one that you don't mind sacrificing, as it will probably not be reusable.
- You may substitute a pair of pliers or similar gripping tool for the crucible tongs. (Be careful. Crucibles are physically fragile.)
- You may substitute a propane torch for the gas burner.
- If you do not have copper carbonate, you can make it by dissolving about 11 g of copper sulfate pentahydrate in about 50 mL of water and adding a solution of about 5 g of anhydrous sodium carbonate in about 20 mL of water. Filter the precipitate of copper carbonate, rinse it two or three times with water, and dry it before use.
- You may substitute an equivalent weight of any form of carbon for the activated charcoal. If you decide to use 3 g of diamonds, send them to me instead, and I'll send you a whole kilogram of activated charcoal in exchange.

DISPOSAL: The waste material from this lab can be ground to powder and flushed down the drain with plenty of water.

TABLE 10-1: *Reduction of copper ore to copper metal—observed and calculated data*

Item	Data
A. Mass of crucible + lid	_____._____ g
B. Mass of copper carbonate	_____._____ g
C. Mass of carbon (charcoal)	_____._____ g
D. Mass of crucible, lid, and reactants (A + B + C)	_____._____ g
E. Mass of crucible, lid, and products	_____._____ g
F. Mass loss (D – E)	_____._____ g

REVIEW QUESTIONS

Q1: Observe and describe the appearance of the products after heating the crucible. (If necessary, use a mortar and pestle to crush the product.) Which products can you identify visually?

_____

_____

_____

_____

_____

Q2: From the balanced equation for this reaction and knowing that the molar masses of carbon and copper(II) carbonate are 12.01 g/mol and 123.56 g/mol respectively, which reactant was in excess, and by how much? Why might we have chosen to use this reactant in excess?

_____

_____

_____

_____

_____

Q3: Assuming that the carbon reacted stoichiometrically with the copper(II) carbonate and that none of the carbon reacted with atmospheric oxygen to form carbon dioxide, calculate the expected mass loss. How does the calculated mass loss compare with the mass loss you observed?

_____

_____

_____

_____

Q4: More than 2,000 years passed from when copper ore was first smelted in about 3500 BC until iron ore was first smelted in about 1200 BC, and another 3,000 years passed before aluminum metal was successfully reduced from aluminum ore in the nineteenth century? Why did it take so long?

_____

_____

_____

_____

_____

LABORATORY 10.2:

OBSERVE THE OXIDATION STATES OF MANGANESE

Many elements can exist in several oxidation states, which may differ noticeably in color. In the introduction, we mentioned vanadium as one such element. Manganese is another element whose various oxidation states have different colors, including MnO_4^- (+7, violet), MnO_4^{2-} (+6, green), MnO_2 (+4, orange), Mn_2O_3 (+3, violet), and Mn^{2+} (+2, pale pink).

In this laboratory, we'll begin with a solution of potassium permanganate, in which manganese is in the +7 oxidation state, and reduce the oxidation state to observe the color changes associated with various oxidation states of manganese. To achieve these changes in oxidation state, we'll react potassium permanganate with sodium bisulfite in neutral, basic (sodium hydroxide), and acid (sulfuric acid) solutions, which are represented by the equations shown in Table 10-2.

REQUIRED EQUIPMENT AND SUPPLIES

- ☐ goggles, gloves, and protective clothing
- ☐ test tubes (4)
- ☐ test tube rack
- ☐ graduated cylinder, 10 mL
- ☐ dropper or beral pipette
- ☐ stirring rod
- ☐ potassium permanganate ($KMnO_4$) solution, 0.01 M (50 mL)
- ☐ sodium hydroxide solution, 1.0 M (8 mL)
- ☐ sulfuric acid solution, 1.0 M (6 mL)
- ☐ sodium bisulfite ($NaHSO_3$) solution, 0.01 M (as required)

CAUTIONS

Solutions of sodium hydroxide and sulfuric acid are corrosive. Wear splash goggles, gloves, and protective clothing.

TABLE 10-2: *Oxidation states of manganese*

Solution	Oxidation state	Ion	Color	Equation
Neutral	+4	MnO_2	orange	$2\,MnO_4^-(aq) + 3\,HSO_3^- + OH^- \rightarrow 2\,MnO_2(s) + 3\,SO_4^{2-}(aq) + 2\,H_2O(l)$
Basic (sodium hydroxide)	+6	MnO_4^{2-}	green	$2\,MnO_4^-(aq) + HSO_3^- + 3\,OH^- \rightarrow 2\,MnO_4^{2-}(aq) + SO_4^{2-}(aq) + 2\,H_2O(l)$
Acid (sulfuric acid)	+2	Mn^{2+}	pink	$2\,MnO_4^-(aq) + 5\,HSO_3^- + H^+ \rightarrow 2\,Mn^{2+}(aq) + 5\,SO_4^{2-}(aq) + 3\,H_2O(l)$

SUBSTITUTIONS AND MODIFICATIONS

- If you do not have a 0.01 M solution of potassium permanganate, make one up by diluting your potassium permanganate stock solution as required. Alternatively, dissolve 0.08 g of potassium permanganate in about 40 mL of water and make it up to 50 mL. (The exact concentration is not critical for this lab.)

- If you do not have a bench solution of 1.0 M sodium hydroxide, dissolve 4.00 g of sodium hydroxide in 90 mL of water and make it up to 100 mL. (Be careful; this reaction is extremely exothermic.)

- If you do not have a bench solution of 1.0 M sulfuric acid, carefully add 5.6 mL of concentrated (18 M) sulfuric acid to 90 mL of water and make it up to 100 mL. (Be careful; this reaction is extremely exothermic.).

- If you do not have a 0.01 M solution of sodium bisulfite, dissolve 0.10 g of sodium bisulfite in 90 mL of water and make up the solution to 100 mL.

FIGURE 10-2: *Oxidation states of manganese*

DISPOSAL: You can flush the dilute solutions used in this lab down the drain with plenty of water.

PROCEDURE

1. If you have not already done so, put on your splash goggles, gloves, and protective clothing.
2. Label four test tubes A through D, place them in the test tube rack, and transfer about 10 mL of 0.01 M potassium permanganate solution to each of the tubes.
3. Test tube A will contain only the potassium permanganate solution, and will serve as the control and example of manganese in oxidation state +7. Record the color of this solution on line A of Table 10-3.
4. With stirring, slowly add 0.01 M sodium bisulfite solution to test tube B until no further color change occurs. Record the color and the presence or absence of any precipitate on line B of Table 10-3.
5. Add 8 mL of 1.0 M sodium hydroxide solution to test tube C.
6. With stirring, slowly add 0.01 M sodium bisulfite solution to test tube C until no further color change occurs. Record the color and the presence or absence of any precipitate on line C of Table 10-3.
7. Add 6 mL of 1.0 M sulfuric acid solution to test tube D.
8. With stirring, slowly add 0.01 M sodium bisulfite solution to test tube D until no further color change occurs. Record the color and the presence or absence of any precipitate on line D of Table 10-3.
9. Comparing the appearances of the contents of the four test tubes, decide which oxidation state of manganese is represented by the contents of each test tube and enter that value on the appropriate line of Table 10-3.

TABLE 10-3. *Observe the oxidation states of manganese—observed and calculated data*

Test tube	Color	Precipitate?	Oxidation state
A. Potassium permanganate only			
B. Potassium permanganate + sodium bisulfite			
C. Potassium permanganate + sodium hydroxide + sodium bisulfite			
D. Potassium permanganate + sulfuric acid + sodium bisulfite			

OPTIONAL ACTIVITIES

If you have extra time, consider performing this optional activity:

- Potassium permanganate has many uses, nearly all of which are based on the fact that it is an extremely strong oxidizer. You can demonstrate this by placing several layers of filter paper or paper towel in the bottom of a Petri dish or large beaker and pouring the excess 0.01 M potassium permanganate solution you prepared for this lab onto the paper. Initially, the paper will have the strong violet color of the permanganate ion (+7). As the permanganate ion oxidizes the cellulose in the paper, it is reduced from the +7 oxidation state to the green +6 oxidation state. If you allow the reaction to continue long enough, the +7 permanganate ion is eventually reduced to the +4 state as orange/brown manganese dioxide (MnO_2).

REVIEW QUESTIONS

Q1: What are the final oxidation states of manganese in test tubes A, B, C, and D?

Q2: What are the oxidation states of each atom in the following species: NH_3, H_2O, H_2O_2, $K_2Cr_2O_7$, $NaNO_3$, $FeCl_2$, and $AgNO_3$?

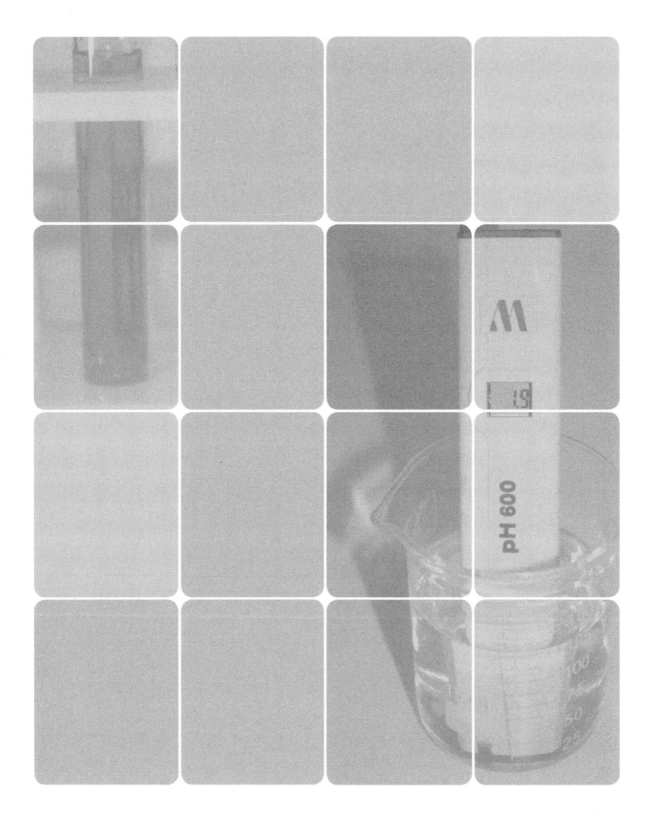

Laboratory:
Acid-Base Chemistry

11

Acids and bases are two of the most important classes of compounds in chemistry. Surprisingly, it took chemists many years to agree on just what made an acid an acid and a base a base.

From experience, we understand that acids and bases have certain characteristics. For example, acid solutions taste sour and base solutions bitter. (Obviously, you should never taste any laboratory chemical, but the sour taste of edible acids such as lemon juice and vinegar is familiar to most people, as is the bitter taste of edible bases such as baking soda and quinine water). Acid solutions generally have an astringent feel on the skin, and base solutions feel slimy. Acid solutions turn blue litmus paper red, and base solutions turn red litmus paper blue. Acids and bases combine to form salts. And so on.

But these characteristics are insufficient to define acids and bases. Around the turn of the nineteenth century, French and European chemists believed that all acids must contain oxygen. (They were wrong, but echoes of that error persist; for example, the German word for oxygen is Sauerstoff, which translates literally as "sour material.") Davy and other English chemists were much closer to the mark. They believed that all acids must contain hydrogen. And, although that statement is not absolutely true for all substances considered to be acids by modern definitions, it is true that all acids known to be acids at that time do contain hydrogen.

The first good working definition of acids and bases was proposed in the late nineteenth century by the Swedish chemist Svante August Arrhenius, who defined an acid as a substance that when dissolved in water increases the concentration of the hydronium (H_3O^+) ion and a base as a substance that when dissolved in water increases the concentration of the hydroxide (OH^-) ion. Although that definition limits acids and bases to compounds that are water-soluble, it was a pretty good definition for the time and remains a useful definition even today.

But not all compounds that behave chemically as acids can dissociate to produce hydrogen ions, and not all compounds that behave chemically as bases can dissociate to produce hydroxide ions. In 1923, the Danish chemist Johannes Nicolaus Brønsted and the English chemist Thomas Martin Lowry redefined an acid as a proton (hydrogen nucleus) donor, and a base as a proton acceptor. Under the *Brønsted-Lowry definition of acids and bases*, an acid and its corresponding base are referred to as a *conjugate acid-base pair*. The conjugate acid is the member of the pair that donates a proton, and the conjugate base the member of the pair that accepts a proton. For example, hydrochloric acid dissociates in water to form chloride ions and hydronium ions:

$$HCl + H_2O \Leftrightarrow H_3O^+ + Cl^-$$

In the forward reaction, the acid reactant (HCl) and the base reactant (H_2O) form the acid product (H_3O^+) and the base product (Cl^-). In the reverse reaction, the acid reactant (H_3O^+) and the base reactant (Cl^-) form the acid product (HCl) and the base product (H_2O). If the conjugate acid (HCl on the left side of the equation and H_3O^+ on the right side) is strong, the conjugate base (H_2O on the left side of the equation and Cl^- on the right side) is weak, and vice versa. At equilibrium, the weaker acid is favored.

For aqueous solutions, the Arrhenius definition and the Brønsted-Lowry definition are essentially the same. The value of the Brønsted-Lowry definition is that it extends the concept of acids and bases to compounds that are not soluble in water.

The same year that Brønsted and Lowry defined acids and bases as proton donors and proton acceptors, respectively, the American chemist Gilbert N. Lewis extended the definition of acids and bases to include compounds that behaved chemically as acids and bases without donating or accepting a proton. Under the *Lewis definition of acids and bases*, a *Lewis acid* (also called an *electrophile*) is a substance that accepts an electron pair, and a *Lewis base* (also called a *nucleophile*) is a substance that donates an electron pair. (For example, the compound ferric chloride, $FeCl_3$, behaves as an acid, although it has no proton to donate, and so is classified as a Lewis acid.) Lewis acids and bases are particularly important to organic chemists, who make use of them in many syntheses.

In this chapter, we'll examine the properties of acids and bases.

EVERYDAY ACID-BASE CHEMISTRY

Acid-base chemistry is inextricably tied with our everyday lives. Here are just a few examples:

- Without the hydrochloric acid in our stomachs, we could not digest our food. Most people are surprised to learn that the pH of gastric juice (see Lab 11.1) can be as low as 1, a very strong acid capable of "dissolving" some metals.

- Mineral (inorganic) acids, including hydrochloric, nitric, phosphoric, and sulfuric acids, are among the most important industrial chemicals. Millions of tons of these acids are produced worldwide, and they are critical to the manufacture of thousands of important compounds, including foods and food supplements, drugs, fertilizers, soaps and detergents, dyes and pigments, explosives, metals, and fuels.

- Bases react with fats by a process called *saponification* to produce soaps, a reaction that is used to produce commercial soap products. But we also use saponification directly every time we use a window cleaner or other cleaning product that contains ammonia. The basic ammonia reacts with greasy dirt to produce soluble compounds (soaps) that can be rinsed away.

ACTIVITY VERSUS CONCENTRATION

Technically, the pH of a solution depends on the activity of the hydronium ions rather than their concentration, but it's easy to measure concentration accurately and very difficult to measure activity accurately. (Activity can be thought of as the effective concentration of hydronium ions, which is lower than the actual concentration, because in solution some of the hydronium ions are "screened" by other ions present in the solution and therefore unable to participate in reactions.) In very concentrated solutions, activity may be considerably lower than the actual concentration. In dilute solutions, concentration and activity correlate very closely, so the concentration is typically used for calculations.

DR. PAUL JONES COMMENTS

Most beginning students are aware of acids and their dangers but are more or less ignorant of the dangers of alkali (base). For instance, aqueous sodium hydroxide can blind you in a matter of minutes if not cleansed thoroughly and I've seen lots of kids who are quick to put on goggles to work with 0.01 M HCl but throw 6 M NaOH around like it's candy. Aqueous bases are every bit as dangerous as aqueous acids, if not more so, but many students treat aqueous bases as though they were innocuous.

LABORATORY 11.1:
DETERMINE THE EFFECT OF CONCENTRATION ON PH

pH is a metric used to specify the acidity (or basicity, also called alkalinity) of an aqueous solution. The pH of a solution is determined by the relative activities of the hydronium (H_3O^+) ions and the hydroxide (OH^-) ions in the solution. The pH of a solution in which the activities of these two ions are equal—such as pure water at 25°C—is 7.00. A solution in which the activity of the hydronium ions is higher than the activity of the hydroxide ions has a pH lower than 7 and is acidic. A solution in which the activity of the hydroxide ions is higher than the activity of the hydronium ions has a pH greater than 7, and is basic.

pH is specified on a log_{10} scale, which allows a very wide range of activities (concentrations) to be specified using a small range of numbers. A difference of one pH number corresponds to a difference of ten times in acidity or basicity. For example, a solution with a pH of 5 is ten times (10^1) more acidic than a solution with a pH of 6, and a solution with a pH of 9 is ten times (10^1) more basic than a solution with a pH of 8. Similarly, a solution with a pH of 2 is 10,000 (10^4) times more acidic than a solution with a pH of 6, and a solution with a pH of 12 is 10,000 (10^4) times more basic than a solution with a pH of 8. Although the range of pH values is usually considered to be 0 through 14, an extremely acidic solution (such as a concentrated solution of hydrochloric acid) can have a pH lower than 0, and an extremely basic solution (such as a concentrated solution of sodium hydroxide) can have a pH greater than 14.

For relatively dilute solutions of strong acids and bases, you can estimate pH using the formula:

$$pH = -log_{10}[H_3O^+]$$

where $[H_3O^+]$ is the concentration of the hydronium ion in mol/L. For example, hydrochloric acid dissolves in water according to the following equation:

$$HCl + H_2O \rightarrow H_3O^+ + Cl^-$$

Because HCl is a strong acid, the reaction proceeds to completion, which is to say that essentially all of the HCl reacts to

REQUIRED EQUIPMENT AND SUPPLIES

☐ goggles, gloves, and protective clothing

☐ balance and weighing papers

☐ beaker, 150 mL (6)

☐ volumetric flask, 100 mL

☐ pipette, 10 mL

☐ pH meter

☐ hydrochloric acid, 1 M (100 mL)

☐ sulfuric acid, 1 M (100 mL)

☐ acetic acid, 1 M (100 mL)

☐ sodium hydroxide, 1 M (100 mL)

☐ sodium carbonate, 1 M (100 mL)

☐ distilled or deionized water (boil and cool before use)

form hydronium ions and chloride ions. The approximate pH of a 0.01 M solution of hydrochloric acid is:

$$pH = -log_{10}[0.01] = 2$$

In calculating that approximate pH, we assume that the hydrochloric acid fully dissociates in solution into H_3O^+ ions and Cl^- ions. For strong acids like hydrochloric acid, that's a reasonable assumption. For weak acids, such as acetic acid, that assumption is not valid, because weak acids dissociate only partially in solution. The concentration of hydronium ions in a solution of a weak acid is lower (perhaps much lower) than the concentration of the acid itself.

When acetic acid dissolves in water, the dissociation reaction looks like this:

$$CH_3COOH + H_2O \Leftrightarrow H_3O^+ + CH_3COO^-$$

This reaction reaches an equilibrium, with reactants being converted to products and vice versa at the same rate. Therefore, in a 1.0 M solution of acetic acid (about the concentration of household vinegar), the actual concentration of the hydronium ion is something less than 1.0 M, because some of the acetic acid

SUBSTITUTIONS AND MODIFICATIONS

- You may substitute foam cups or similar containers for the beakers.
- You may substitute a 100 mL graduated cylinder for the 100 mL volumetric flask and a 10 mL graduated cylinder for the 10 mL pipette, with some loss of accuracy. (Unless your pH meter is very accurate, instrumental error will be greater than volumetric error anyway.)
- You may substitute narrow-range pH testing paper for the pH meter, with some loss of accuracy. (Typical inexpensive pH meters are accurate to 0.1 or 0.2 pH, while typical narrow-range pH testing paper is accurate to 0.4 or 0.5 pH.)
- You may reduce the quantities of the acid and base solutions, depending on the size and type of probe used by your pH meter, and use smaller containers. If your meter has a very thin probe (or if you are using pH paper), you can reduce the quantities from 100 mL to 10 mL and use test tubes.
- If you don't have a 1 M bench solution of hydrochloric acid available, you can make it up by diluting 8.33 mL of 12 M (37%) concentrated hydrochloric acid to 100 mL. If you have hardware-store muriatic acid that lists the contents as 31.45% HCl, you can make up a 1 M solution by diluting 9.71 mL of acid to 100 mL.
- If you don't have a 1 M bench solution of sulfuric acid available, you can make it up by diluting 5.56 mL of 18 M (98%) concentrated sulfuric acid to 100 mL. If you have hardware-store battery acid that lists the contents as 35% sulfuric acid, you can make up a 1 M solution by diluting 15.57 mL of acid to 100 mL.
- If you don't have a 1 M bench solution of acetic acid available, you can make it up by diluting 5.75 mL of 17.4 M (99.8%) concentrated (glacial) acetic acid to 100 mL. Alternatively, most distilled white vinegar contains 5% to 6% acetic acid, which is close enough to 1 M to use for this experiment.
- If you don't have a 1 M bench solution of sodium hydroxide available, you can make it up by dissolving 4.00 g of sodium hydroxide in water and making up the solution to 100 mL.
- If you don't have a 1 M bench solution of sodium carbonate available, you can make it up by dissolving 10.60 g of anhydrous sodium carbonate in water and making up the solution to 100 mL.
- I suggest boiling and cooling the distilled or deionized water before use to eliminate any dissolved carbon dioxide gas, which forms carbonic acid and lowers the pH below the 7.00 pH of pure water. Alternatively, you can use distilled or deionized water from a freshly opened container or one that has been kept tightly capped.

remains undissociated. Based on the previous calculation, we know that the pH of a 1.0 M solution would be about 0 if the acid had fully dissociated, so we know that the actual pH of the 1.0 M acetic acid will have some value higher than 0.

If we assume for a moment that at equilibrium only 10% of the acetic acid has dissociated, we can calculate the approximate pH of the solution. If only 10% of the acid has dissociated in a 1.0 M solution of acetic acid, the hydronium ion concentration is 0.1 M. Filling in the formula, we get:

$$pH = -\log_{10}[0.1] = 1$$

But to estimate the pH of this solution accurately, we need a better value than our guesstimate of 10% dissociation. We get that value by looking up the equilibrium constant for the dissociation reaction shown above. In the context of pH, the equilibrium constant is referred to as the *acidity constant*, *acid dissociation constant*, or *acid ionization constant*, and is abbreviated K_a.

With the acid dissociation constant for acetic acid known to be $1.74 \cdot 10^{-5}$, and ignoring the tiny contribution to $[H_3O^+]$ made by the water, we can calculate the pH of a 1.0 M solution of acetic acid as follows:

$$K_a = 1.74 \cdot 10^{-5} = ([H_3O^+] \cdot [CH_3COO^-]/[CH_3COOH])$$

An unknown amount of the acetic acid has dissociated, which we'll call x. That means that the concentration of the acetic acid, or $[CH_3COOH]$ is $(1.0 - x)$, while the concentrations of the dissociated ions, $[H_3O^+]$ and $[CH_3COO^-]$, are both x. Filling in the formula gives us:

$$1.74 \cdot 10^{-5} = ([x] \cdot [x]/[1.0 - x])$$
or
$$1.74 \cdot 10^{-5} = x^2/(1.0 - x)$$
or
$$(1.74 \cdot 10^{-5}) \cdot (1.0 - x) = x^2$$
or
$$1.74 \cdot 10^{-5} - (1.74 \cdot 10^{-5} \cdot x) = x^2$$

K_A VERSUS PK_A

The acidity constant may be specified directly as K_a or as the negative logarithm of K_a, which is abbreviated as pK_a. For example, at 25°C, the K_a of acetic acid is $1.74 \cdot 10^{-5}$ and the pK_a is 4.76. Either value can be used by adjusting your calculations accordingly.

DR. PAUL JONES COMMENTS

I'd say that pK_a is used far more today than K_a. Also, you might mention that any compound with hydrogen can, plausibly, be an acid. However, if the pK_a of the compound is above 14, it essentially won't dissociate at all in water. Also, when $pH = pK_a$, the acid is 50% dissociated; another good reason to use pK_a.

Solving for x gives us a value of about $4.17 \cdot 10^{-3}$ or about 0.00417. Our original estimate that 10% of the acetic acid would dissociate was wildly high. In fact, only about 0.417% of the acetic acid dissociates. Knowing that value tells us that the concentration of the hydronium ion in a 1 M acetic acid solution is 0.00417 M. Plugging that value into the formula allows us to calculate the approximate pH of the 1.0 M acetic acid solution:

$$pH = -\log_{10}[0.00417] = 2.38$$

Although we've focused until now on acids rather than bases, remember that the concentrations of the hydronium ion and the hydroxide ion are related. We know that pure water at 25°C has a pH of 7.00 and that the concentrations of hydronium ions and hydroxide ions are equal. A pH of 7.00 tells us that the concentration of hydronium ions, $[H_3O^+]$, is $1.00 \cdot 10^{-7}$, which means that the concentration of hydroxide ions, $[OH^-]$, must also be $1.00 \cdot 10^{-7}$. We know that hydronium ions and hydroxide ions react to form water:

$$H_3O^+ + OH^- \Leftrightarrow H_2O$$

According to Le Chatelier's Principle (see Chapter 13), in a system at equilibrium, increasing the concentration of one reactant forces the reaction to the right, producing more product. Increasing $[H_3O^+]$ reduces $[OH^-]$ proportionately, and vice versa. Expressed as a formula, the equilibrium constant is:

$$K = [H_3O^+] \cdot [OH^-]$$
or
$$K = (1.00 \cdot 10^{-7}) \cdot (1.00 \cdot 10^{-7}) = 1.00 \cdot 10^{-14}$$

In other words, the product of the concentrations of the hydronium ions and the hydroxide ions always equals $1.00 \cdot 10^{-14}$.

CAUTIONS

All of the chemicals used in this laboratory can be hazardous, particularly in concentrated solutions. Check the MSDS for each of these chemicals before you proceed. Wear splash goggles, gloves, and protective clothing at all times.

If you increase the concentration of hydronium ions by a factor of 10 (or 10,000), the concentration of hydroxide ions decreases by a factor of 10 (or 10,000), and vice versa.

In this laboratory, we'll determine the pH of solutions of two strong acids, a weak acid, a strong base, and a weak base at various concentrations.

PROCEDURE

1. If you have not already done so, put on your splash goggles, gloves, and protective clothing.
2. Label six beakers #1 through #6.
3. Pour about 100 mL of 1 M hydrochloric acid into beaker #1.
4. Using the 10 mL pipette, transfer 10.00 mL of the 1 M hydrochloric acid to the 100 mL volumetric flask.
5. Fill the volumetric flask to the reference line with distilled water, mix thoroughly, and transfer the 0.1 M solution of hydrochloric acid to beaker #2.
6. Proceeding from beaker to beaker, repeat steps 4 and 5 until you have 0.01 M, 0.001 M, 0.0001 M, and 0.00001 M solutions of hydrochloric acid in beakers #3 through #6, respectively.
7. Read and follow the directions for your pH meter with respect to calibrating it, rinsing the electrode between measurements and so on. Use the pH meter to measure the pH of the solutions in beakers #1 through #6, and record those values on line A of Table 11-1.
8. Repeat steps 3 through 7 for the solutions of sulfuric acid, acetic acid, sodium hydroxide, and sodium carbonate.
9. Based on your observations, calculate the pK_a values for the acids and the pK_b values for the bases, and enter your calculated values in Table 11-1.

DISPOSAL: Retain the 1.0 M acid and sodium hydroxide solutions for use in the following lab. Neutralize the other solutions, beginning with the most dilute samples, by pouring them into a bucket or similar container. The contents of the bucket can be flushed down the drain.

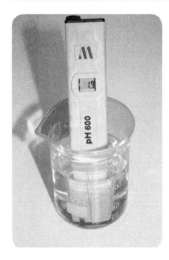

FIGURE 11-1:
Using a pH meter to determine the pH of a solution

TABLE 11-1: *Determine the effect of concentration on pH—observed and calculated data*

Solute	Beaker #1 (1.0 M)	Beaker #2 (0.1 M)	Beaker #3 (0.01 M)	Beaker #4 (0.001 M)	Beaker #5 (0.0001 M)	Beaker #6 (0.00001 M)	pK
A. Hydrochloric acid	___.___ pH	___.___ pH	___.___ pH	___.___ pH	___.___ pH	___.___ pH	___.___
B. Sulfuric acid	___.___ pH	___.___ pH	___.___ pH	___.___ pH	___.___ pH	___.___ pH	___.___
C. Acetic acid	___.___ pH	___.___ pH	___.___ pH	___.___ pH	___.___ pH	___.___ pH	___.___
D. Sodium hydroxide	___.___ pH	___.___ pH	___.___ pH	___.___ pH	___.___ pH	___.___ pH	___.___
E. Sodium carbonate	___.___ pH	___.___ pH	___.___ pH	___.___ pH	___.___ pH	___.___ pH	___.___

REVIEW QUESTIONS

Q1: Chart pH against concentration for all five compounds on one piece of graph paper. What do you conclude about the effect of concentration for a strong acid (hydrochloric acid or sulfuric acid) versus a weak acid (acetic acid) and for a strong base (sodium hydroxide) versus a weak base (sodium carbonate)?

_____

_____

_____

_____

_____

_____

Q2: What pH values would you expect for 10 M solutions of hydrochloric acid and acetic acid? Why?

Q3: If acids A and B have pK_a values of -9.32 and 2.74, respectively, which is the stronger acid? Why?

Q4: The $[H_3O^+]$ of a solution is known to be 0.000413 mol/L. What is the pH of the solution?

Q5: The $[OH^-]$ of a solution is known to be 0.000413 mol/L. What is the pH of the solution?

Q6: A 0.1 M solution of an acid is found to have a pH of 2.37. What is the pK_a of that acid?

Q7: Knowing only the molarity and pH of an unknown acid solution, how might you identify the acid?

LABORATORY 11.2:
DETERMINE THE PH OF AQUEOUS SALT SOLUTIONS

When an acid reacts with a base, it forms a salt and water. For example, reacting hydrochloric acid with sodium hydroxide produces the salt sodium chloride (common table salt) and water. Reacting nitric acid with potassium hydroxide produces potassium nitrate and water. Reacting acetic acid with aqueous ammonia produces ammonium acetate and water. And so on. Such reactions are referred to as *neutralizing* the acid with the base (or vice versa).

Unfortunately, the word "neutralize" is a common source of confusion among beginning chemists, many of whom assume that reacting equivalent amounts of an acid and base should yield a solution that contains only the "neutral" salt that is neither acidic nor basic, and therefore has a pH of 7.0. That's not how it works.

The pH of a neutralized solution depends on the particular acid and base that are reacted. Reacting equivalents of a strong acid with a strong base in fact does produce a salt solution that has a pH at or near 7.0, as does reacting a weak acid with a weak base. But if the strengths of the acid and base are very different—as occurs, for example, if you react a strong acid with a weak base or vice versa—the pH of the neutralized solution will not be 7.0. The greater the difference in the strengths of the acid and base, the greater the difference in the pH of the neutralized solution from 7.0.

For example, if you neutralize hydrochloric acid (a very strong acid) with aqueous ammonia (a relatively weak base), the resulting solution of ammonium chloride will have a pH less than 7.0. Conversely, if you neutralize acetic acid (a relatively weak acid) with sodium hydroxide (a very strong base), the resulting neutralized solution of sodium acetate will have a pH greater than 7.0.

In this lab, we'll determine the pH of aqueous solutions of various salts.

> ### SUBSTITUTIONS AND MODIFICATIONS
> • You may substitute a foam cup or similar container for the beaker.

> ### REQUIRED EQUIPMENT AND SUPPLIES
> ☐ goggles, gloves, and protective clothing
> ☐ beaker, 150 mL
> ☐ pH meter
> ☐ ammonium acetate, 0.1 M (100 mL)
> ☐ ammonium chloride, 0.1 M (100 mL)
> ☐ sodium acetate, 0.1 M (100 mL)
> ☐ sodium chloride, 0.1 M (100 mL)

> ### SUBSTITUTIONS AND MODIFICATIONS
> • You may substitute narrow-range pH testing paper for the pH meter, with some loss of accuracy. (Typical inexpensive pH meters are accurate to 0.1 or 0.2 pH, while typical narrow-range pH testing paper is accurate to 0.4 or 0.5 pH.)
> • You may reduce the quantities of the solutions, depending on the size and type of probe used by your pH meter, and use smaller containers. If your meter has a very thin probe (or if you are using pH paper), you can reduce the quantities from 100 mL to 10 mL and use test tubes.
> • If you have bench solutions of any of the required chemicals, you can make up the required solutions by diluting the bench solutions appropriately.
> • You can make 100 mL of 0.1 M ammonium acetate by mixing 10 mL of 1.0 M aqueous ammonia with 10 mL of 1.0 M acetic acid and 80 mL of water.
> • You can make 100 mL of 0.1 M ammonium chloride by mixing 10 mL of 1.0 M aqueous ammonia with 10 mL of 1.0 M hydrochloric acid and 80 mL of water.
> • You can make 100 mL of 0.1 M sodium acetate by mixing 10 mL of 1.0 M sodium hydroxide with 10 mL of 1.0 M acetic acid and 80 mL of water.
> • You can make 100 mL of 0.1 M sodium chloride by mixing 10 mL of 1.0 M sodium hydroxide with 10 mL of 1.0 M hydrochloric acid and 80 mL of water.

CAUTIONS

Although none of the salts used in this lab are particularly hazardous, it's good practice to wear splash goggles, gloves, and protective clothing at all times. If you make up the salt solutions from acid and base solutions as described in the Substitutions and Modifications section, use the normal precautions for working with acids and bases.

DISPOSAL: All of the solutions used in this laboratory can be flushed down the drain with plenty of water.

PROCEDURE

1. If you have not already done so, put on your splash goggles, gloves, and protective clothing.
2. Using what you know about strong and weak acids and bases, predict the approximate pH values for the solutions of the four chemicals and enter your predicted values in Table 11-2 by circling one of the pH numbers. If you are uncertain, circle a range of numbers.
3. Pour about 100 mL of 0.1 M ammonium acetate into the beaker.
4. Read and follow the directions for your pH meter with respect to calibrating it, rinsing the electrode between measurements and so on. Use the pH meter to measure the pH of the ammonium acetate solution, and record the observed value on the appropriate line of Table 11-2.
5. Repeat steps 3 and 4 for the solutions of ammonium chloride, sodium acetate, and sodium chloride.

TABLE 11-2: Determine the pH of aqueous salt solutions—observed and calculated data

Solute	Predicted pH	Observed pH
A. Ammonium acetate	0 1 2 3 4 5 6 7 8 9 10 11 12 13 14	____.__ pH
B. Ammonium chloride	0 1 2 3 4 5 6 7 8 9 10 11 12 13 14	____.__ pH
C. Sodium acetate	0 1 2 3 4 5 6 7 8 9 10 11 12 13 14	____.__ pH
D. Sodium chloride	0 1 2 3 4 5 6 7 8 9 10 11 12 13 14	____.__ pH

REVIEW QUESTIONS

Q1: What general conclusions can you draw from the pH values that you observed for the four solutions?

Q2: If you mixed equal amounts of a 0.1 M ammonium chloride solution and a 0.1 M sodium acetate solution, would you expect the pH of the resulting solution to be strongly acidic, weakly acidic, about neutral, weakly basic, or strongly basic? Why?

LABORATORY 11.3:
OBSERVE THE CHARACTERISTICS OF A BUFFER SOLUTION

A **buffer solution** is usually a solution of a weak acid and its conjugate base or, less commonly, a solution of a weak base and its conjugate acid. A buffer solution resists changes in the concentrations of the hydronium ion and hydroxide ion (and therefore pH) when the solution is diluted or when small amounts of an acid or base are added to it. The resistance of a buffer solution to pH change is based upon Le Chatelier's Principle and the common ion effect.

One common example of a buffer solution is a solution of acetic acid (the weak acid) and sodium acetate (its conjugate base). In solution, acetic acid reaches an equilibrium illustrated by the following equation.

$$CH_3COOH(aq) + H_2O(l) \Leftrightarrow CH_3COO^- + H_3O^+$$

As we learned earlier in this chapter, acetic acid does not dissociate completely in solution. For example, in a 1 M solution of acetic acid, only about 0.4% of the acetic acid molecules dissociate into hydronium and acetate ions, leaving most of the acetic acid in molecular form. Dissolving sodium acetate in the acetic acid solution forces the equilibrium to the left, reducing the hydronium ion concentration and therefore increasing the pH of the solution.

Consider what happens if you add a small amount of a strong acid or strong base to this buffer solution. Ordinarily, you would expect adding a small amount of a strong acid or base to cause a large change in the pH of a solution. But if you add hydrochloric acid (a strong acid) to the acetate/acetic acid buffer solution, the hydronium ions produced by the nearly complete dissociation of the hydrochloric acid react with the acetate ions to form molecular (non-dissociated) acetic acid. According to Le Chatelier's Principle, the equilibrium is forced to the left, reducing the concentration of hydronium and acetate ions, and increasing the concentration of the molecular acetic acid in the solution.

The acid dissociation constant for this buffer is:

$$K_a = [H_3O^+] \cdot [CH_3COO^-]/[CH_3COOH]$$

If the buffer solution contains equal amounts of acetic acid and sodium acetate, we can assume that the sodium acetate is fully dissociated and that dissociation of the acetic acid is negligible

REQUIRED EQUIPMENT AND SUPPLIES

☐ goggles, gloves, and protective clothing

☐ beaker, 150 mL (2)

☐ beaker, 250 mL

☐ graduated cylinder, 100 mL

☐ graduated cylinder, 10 mL

☐ pipette, 1.00 mL

☐ stirring rod

☐ pH meter

☐ acetic acid, 1.0 M (100 mL)

☐ sodium acetate, 1.0 M (100 mL)

☐ hydrochloric acid, 6.0 M (20 mL)

☐ sodium hydroxide, 6.0 M (20 mL)

☐ distilled or deionized water (boil and cool before use)

(because the high concentration of acetate ions from the sodium acetate drives the dissociation equilibrium for acetic acid far to the left). We can therefore assume that the concentrations of CH_3COO^- and CH_3COOH are essentially identical, and simplify the equilibrium equation to:

$$K_a = [H_3O^+]$$

which means that the pH of this buffer solution is equal to the pK_a.

If we add hydrochloric acid to the buffer solution, the HCl ionizes completely in solution, yielding hydronium ions and chloride ions. According to Le Chatelier's Principle, the increase in hydronium ions forces the acetic acid equilibrium to the left, decreasing the concentration of hydronium ions and increasing the concentration of molecular acetic acid. This equilibrium shift changes the effective number of moles of acetic acid and acetate ions, which can be calculated as follows:

final CH_3COO^- moles = initial CH_3COOH moles − initial HCl moles

final CH_3COOH moles = initial CH_3COO^- moles + initial HCl moles

SUBSTITUTIONS AND MODIFICATIONS

- You may substitute foam cups or similar containers for the beakers.
- You may substitute a calibrated beral pipette for the 1.00 mL transfer pipette. (To calibrate the beral pipette, count the number of drops required to reach 10.00 mL in a graduated cylinder, and then calculate the number of drops per mL.)
- You can make up 1.0 M acetic acid by diluting 5.7 mL of concentrated (glacial) acetic acid to 100 mL. Alternatively, you may use distilled white vinegar, which is close to 1 M straight out of the bottle.
- You can make 100 mL of 1.0 M sodium acetate by dissolving 8.20 g of anhydrous sodium acetate or 13.61 g of sodium acetate trihydrate in some distilled water and then making up the solution to 100 mL.
- You can make 20 mL of 6.0 M hydrochloric acid by mixing 10.00 mL of concentrated (37%, 12 M) hydrochloric acid with 10.00 mL of water. If you have hardware-store muriatic acid (31.45%, 10.3 M), mix 11.65 mL of the acid with 8.35 mL of water.
- You can make 20 mL of 6.0 M sodium hydroxide by dissolving 4.80 g of sodium hydroxide, stirring constantly, in 90 mL of water and making up the solution to 100 mL. (Caution: this reaction is extremely exothermic.)

CAUTIONS

Hydrochloric acid and sodium hydroxide are corrosive. Wear splash goggles, gloves, and protective clothing at all times.

In this lab, we'll make up a buffer solution of acetic acid and sodium acetate and examine the effects of adding hydrochloric acid and sodium hydroxide to this buffer solution.

PROCEDURE

1. If you have not already done so, put on your splash goggles, gloves, and protective clothing.
2. Make up the buffer solution by mixing 100 mL of 1.0 M acetic acid and 100 mL of 1.0 M sodium acetate in the 250 mL beaker.
3. Transfer 100 mL of the buffer solution to one of the 150 mL beakers, and 100 mL of distilled water to the second 150 mL beaker.
4. Read and follow the directions for your pH meter with respect to calibrating it, rinsing the electrode between measurements and so on.
5. Use the pH meter to measure the pH of the buffer solution, and record the observed value on line A in the second and fourth columns of Table 11-3.
6. Use the pH meter to measure the pH of the distilled water, and record the observed value on line A in the third and fifth columns of Table 11-3.
7. Use the 1.00 mL pipette to transfer 1.00 mL of 6.0 M hydrochloric acid to the beaker that contains the buffer solution and another 1.00 mL of hydrochloric acid to the beaker that contains the distilled water. Stir or swirl the beakers to mix the solutions thoroughly. (If you use only one stirring rod, rinse it thoroughly before using it in the other beaker.)
8. Use the pH meter to determine the pH value for the buffer solution, and record that value on line B in the second column of Table 11-3.
9. Use the pH meter to determine the pH value for the water solution, and record that value on line B in the third column of Table 11-3.
10. Repeat steps 7 through 9, adding 1.00 mL of hydrochloric acid each time until you have added a total of 10.00 mL of hydrochloric acid to the buffer solution and water solution.
11. Rinse the beakers and pipette thoroughly.
12. Transfer the remaining 100 mL of buffer solution to one of the 150 mL beakers, and 100 mL of distilled water to the second 150 mL beaker.
13. Repeat steps 7 through 11 using the 6.0 M sodium hydroxide solution.

Conversely, if we add sodium hydroxide to the buffer solution, the NaOH ionizes completely in solution, yielding hydroxide ions and sodium ions. According to Le Chatelier's Principle, the increase in hydroxide ions forces the acetic acid equilibrium to the right, decreasing the concentration of hydroxide ions and increasing the concentration of acetate ions. This equilibrium shift changes the effective number of moles of acetic acid and acetate ions, which can be calculated as follows:

final CH_3COO^- moles = initial CH_3COOH moles + initial NaOH moles

final CH_3COOH moles = initial CH_3COO^- moles − initial NaOH moles

In either case, after you calculate the number of moles of acetic acid and acetate ions, you can use the final volume of the solution to determine the concentration of the acetic acid and acetate ions and plug those values into the *Henderson-Hasselbalch equation* to determine the new pH.

$pH = pK_a + \log_{10}([CH_3COO^-]/[CH_3COOH])$

TABLE 11-3: *Observe the characteristics of a buffer solution—observed data*

Acid/base added	Buffer + HCl	Water + HCl	Buffer + NaOH	Water + NaOH
A. 0.00 mL	___.___ pH	___.___ pH	___.___ pH	___.___ pH
B. 1.00 mL	___.___ pH	___.___ pH	___.___ pH	___.___ pH
C. 2.00 mL	___.___ pH	___.___ pH	___.___ pH	___.___ pH
D. 3.00 mL	___.___ pH	___.___ pH	___.___ pH	___.___ pH
E. 4.00 mL	___.___ pH	___.___ pH	___.___ pH	___.___ pH
F. 5.00 mL	___.___ pH	___.___ pH	___.___ pH	___.___ pH
G. 6.00 mL	___.___ pH	___.___ pH	___.___ pH	___.___ pH
H. 7.00 mL	___.___ pH	___.___ pH	___.___ pH	___.___ pH
I. 8.00 mL	___.___ pH	___.___ pH	___.___ pH	___.___ pH
J. 9.00 mL	___.___ pH	___.___ pH	___.___ pH	___.___ pH
K. 10.00 mL	___.___ pH	___.___ pH	___.___ pH	___.___ pH

DISPOSAL:
All of the solutions used in this laboratory can be flushed down the drain with plenty of water.

REVIEW QUESTIONS

Q1: How did the pH change compare in the buffer solution versus water?

_____

_____

_____

Q2: Calculate the expected pH of the buffer solution. How does the calculated value correspond to the observed value?

_____

_____

_____

_____

Q3: Graph the amounts of added acid and base against the pH of the buffer solution. At what point does the buffer begin to lose effectiveness? Why?

_____

_____

_____

_____

Q4: You have on hand 0.5 M solutions of acetic acid and sodium acetate. You want to make up 100 mL of a pH 5.0 buffer solution to calibrate your pH meter. Calculate the amounts of acetic acid solution and sodium acetate solution required to make up this buffer solution.

_____

_____

_____

_____

_____

_____

_____

_____

_____

_____

_____

_____

_____

_____

_____

_____

_____

LABORATORY 11.4:

STANDARDIZE A HYDROCHLORIC ACID SOLUTION BY TITRATION

The process used to determine the concentration of a solution with very high accuracy is called *standardizing a solution*. To standardize an unknown solution, you react that solution with another solution whose concentration is already known very accurately.

For example, to standardize the hydrochloric acid solution that we made up in a preceding lab, we might very carefully measure a known quantity of that solution (called an *aliquot*) and neutralize that aliquot with a solution of sodium carbonate whose concentration is already known very accurately. Adding a few drops of an indicator, such as phenolphthalein or methyl orange, to the solution provides a visual indication (a color change) when an *equivalence point* is reached, when just enough of the standard solution has been added to the unknown solution to neutralize it exactly. By determining how much of the sodium carbonate solution is required to neutralize the hydrochloric acid, we can calculate a very accurate value for the concentration of the hydrochloric acid. This procedure is called *titration*.

Titration uses an apparatus called a *burette* (or *buret*), which is a very accurately graduated glass cylinder with a stopcock or pinchcock that allows the solution it contains to be delivered in anything from a rapid stream to drop-by-drop. Because titration is a volumetric procedure, the accuracy of the results depends on the concentration of the reagent used to do the titration. For example, if 5.00 mL of 1.0000 M sodium carbonate is required to neutralize a specific amount of the unknown acid, that same amount of acid would be neutralized by 50.00 mL of 0.10000 M sodium carbonate. If our titration apparatus is accurate to 0.1 mL, using the more dilute sodium carbonate reduces our level of error by a factor of 10, because 0.1 mL of 0.10000 M sodium carbonate contains only one tenth as much sodium carbonate as 0.1 mL of 1.0000 M sodium carbonate. For that reason, the most accurate titrations are those performed with a relatively large amount of a relatively dilute standard solution.

The obvious question is how to obtain an accurate reference solution. For work that requires extreme accuracy, the best answer is often to buy premade standard solutions, which are made to extremely high accuracy (and the more accurate, the more expensive). None of the work done in a home lab requires that level of accuracy, so the easiest and least expensive method is to make up your own standard solutions. (In fact, to illustrate the principles of standardization and titration, you don't even

REQUIRED EQUIPMENT AND SUPPLIES

☐ goggles, gloves, and protective clothing

☐ balance and weighing papers

☐ beaker, 150 mL (2)

☐ volumetric flask, 100 mL

☐ funnel

☐ pipette, 10 mL

☐ burette, 50 mL

☐ ring stand

☐ burette clamp

☐ storage bottle, 100 mL (labeled "sodium carbonate, 1.500 M")

☐ hydrochloric acid, 1 M bench solution (~10 mL)

☐ sodium carbonate, anhydrous (15.90 g)

☐ phenolphthalein indicator solution (a few drops)

☐ distilled or deionized water

need a truly accurate reference solution; you can simply pretend that a 1 M solution is actually 1.0000 M and proceed on that basis. Your results won't be accurate, but the principles and calculations are the same.)

When you make up a standard solution, take advantage of the difference between absolute errors and relative errors. For example, if your balance is accurate to 0.01 g, that means that any sample you weigh may have an absolute error of as much as 0.01 g. But that absolute error remains the same regardless of the mass of the sample. If you weigh a 1.00 g sample, the absolute error is 1% (0.01 g/1.00 g · 100). If you weigh a 100.00 g sample, the absolute error is 0.01% (0.01 g/100.00 g · 100). Similarly, volumetric errors are absolute, regardless of the volume you measure. For example, a 10.00 mL pipette may have an absolute error of 0.05 mL. If you use that pipette to measure 10.00 mL, the relative error is (0.05 mL/10.00 mL · 100) or 0.5%. If you measure only 1.00 mL, the relative error is ten times as large, (0.05 mL/1.00 mL · 100) or 5.0%.

SUBSTITUTIONS AND MODIFICATIONS

- You may substitute any suitable containers of similar size for the 150 mL beakers.
- You may substitute a 100 mL graduated cylinder for the 100 mL volumetric flask and/or a 10 mL graduated cylinder for the 10 mL pipette, with some loss in accuracy.
- If you do not have a burette, ring stand, and burette clamp, you may substitute a 100 mL graduated cylinder, with some loss in accuracy. To do so, fill the graduated cylinder with titrant and record the starting level. Transfer titrant to the reaction beaker by pouring carefully until the endpoint is near (evidenced by a nonpersistent color change in the indicator that disappears when you swirl the beaker). Then use a dropper or beral pipette to transfer titrant dropwise until the end point is reached. Transfer any remaining titrant from the dropper or beral pipette back into the graduated cylinder, record the ending level, and subtract to determine how much titrant was required to reach the endpoint.
- If you do not have anhydrous sodium carbonate, you may substitute equivalent weights of the monohydrate, heptahydrate, or decahydrate.
- You may substitute cresol red, thymol blue, or a similar indicator for the phenolphthalein indicator.

CAUTIONS

Hydrochloric acid is corrosive. Wear splash goggles, gloves, and protective clothing at all times. (Phenolphthalein, formerly widely used as a laxative, was withdrawn from the market because of concerns about possible links with cancer, but the small amount used in an indicator solution is not ingested and is no cause for concern.)

FIGURE 11-4:
Methyl orange (left) and phenolphthalein at their equivalence points

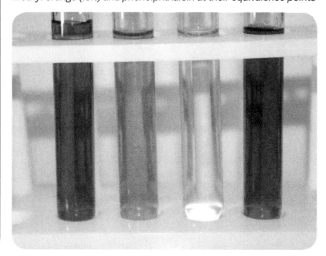

One way to minimize the scale of errors is to use a relatively large amount of solute to make a starting solution and then use *serial dilution* to make a dilute standard solution to use as the *titrant*. Serial dilution simply means repeatedly diluting small aliquots of known volume. For example, we might start with a 1.5 M solution of a chemical. We use a pipette to take a 10.00 mL aliquot of that solution and dilute it to 100.0 mL in a volumetric flask to make a 0.15 M solution. We then take a 10.00 mL aliquot of the 0.15 M solution and dilute it again to 100.0 mL, yielding a 0.015 M solution.

For example, a reference book tells us that the formula weight of anhydrous sodium carbonate is 105.99 g/mol and its solubility at 20°C is about 200 g/L. A saturated solution of sodium carbonate is therefore about 1.9 M, because (200 g/L)/(105.99 g/mol) = 1.88+ M. In this lab, we'll make up a 1.5 M solution of sodium carbonate and then use serial dilution to produce a 0.15 M solution to use as our titrant. We'll titrate our unknown HCl solution once using phenolphthalein as the indicator, and a second time using methyl orange as the indicator. Why two passes with two separate indicators?

As it happens, the neutralization of sodium carbonate by hydrochloric acid is a two-step process. In the first step, one mole of sodium carbonate reacts with one mole of hydrochloric acid to produce one mole of sodium hydrogen carbonate (sodium bicarbonate) and one mole of sodium chloride:

$$Na_2CO_3(aq) + HCl(aq) \rightarrow NaHCO_3(aq) + NaCl(aq)$$

In the second step, a second mole of hydrochloric acid reacts with the sodium hydrogen carbonate formed in the first step:

$$NaHCO_3(aq) + HCl(aq) \rightarrow NaCl(aq) + CO_2(g) + H_2O(l)$$

Each of these reactions has an equivalence point. The first occurs when the first mole of HCl has reacted with the sodium carbonate to form one mole each of sodium bicarbonate and sodium chloride. The pH at the equivalence point of this reaction happens to correspond very closely to the pH range where phenolphthalein changes color, but is well above the pH range of methyl orange. The second equivalence point occurs when the second mole of

HCl has reacted with the sodium bicarbonate to form one mole each of sodium chloride, carbon dioxide, and water. The pH at the equivalence point of this reaction happens to correspond very closely to the pH range where methyl orange changes color, but is well below the pH range of phenolphthalein.

Because these reactions occur with exactly a 2:1 proportion of hydrochloric acid, the amount of titrant needed to reach the indicated equivalence point with methyl orange is exactly twice as much as the amount of titrant needed to reach the indicated equivalence point with phenolphthalein. In other words, if you use methyl orange, you miss the first equivalence point completely. If you use phenolphthalein, you are misled into believing that the first equivalence point is the final equivalence point. For this reason, the neutralization of sodium carbonate with hydrochloric acid is often used as a (literal) textbook example of the importance of choosing the proper indicator.

Note that for titrations with multiple equivalence points, it makes a difference which solution is used for the titrant. For example, if we titrate a solution of sodium carbonate by adding hydrochloric acid, we could observe the first (higher pH) equivalence point by using phenolphthalein as an indicator. When sufficient HCl has been added to convert the sodium carbonate to sodium bicarbonate, the phenolphthalein changes from pink to colorless. We could then add some methyl orange indicator to observe the color change from yellow to red at the second (lower pH) equivalence point, when the sodium bicarbonate is converted to sodium chloride. Conversely, if we titrate a solution of hydrochloric acid with sodium carbonate as the titrant, the acid is always in excess until sufficient sodium carbonate has been added to completely neutralize the acid. In this case, we use phenolphthalein as the indicator, because only one equivalence point exists for this reaction, and it is at the higher pH when sodium carbonate is in slight excess.

In this lab, we'll standardize an approximately 1 M solution of hydrochloric acid by titrating a known volume of the acid with a sodium carbonate solution of known molarity.

PROCEDURE

This laboratory has three parts. In Part I, you'll make up a stock reference solution of 1.500 M sodium carbonate. In Part II, you'll use serial dilution to make up a working reference solution of 0.1500 M sodium carbonate solution to use as a titrant. In Part III, you'll use that tritrant to standardize an approximately 1 M bench solution of hydrochloric acid by titration.

PART I: MAKE UP A STOCK REFERENCE SOLUTION OF ~1.500 M SODIUM CARBONATE

1. If you have not already done so, put on your splash goggles, gloves, and protective clothing.
2. Place a weighing paper on the balance and tare the balance to read 0.00 g.
3. Weigh out about 15.90 g of anhydrous sodium carbonate powder, and record the mass to 0.01 g on line A of Table 11-4.
4. Using the funnel, transfer the sodium carbonate to the 100 mL volumetric flask.
5. Rinse the funnel with a few mL of distilled or deionized water to transfer any sodium carbonate that remains in the funnel into the volumetric flask.
6. Fill the volumetric flask with distilled or deionized water to a few cm below the reference line.
7. Stopper the flask and invert it repeatedly until all of the sodium carbonate dissolves.
8. Finish filling the volumetric flask with water until the bottom of the meniscus is just touching the reference line.
9. Stopper the flask and invert it several times to mix the solution thoroughly.
10. Transfer the sodium carbonate solution to the 100 mL storage bottle and cap the bottle.
11. Use the actual mass of sodium carbonate from step 3 to calculate the actual molarity of the sodium carbonate solution to the appropriate number of significant figures, and record that molarity on line B of Table 11-4.
12. Label the bottle with its contents, molarity, and the date.

PART II: USE SERIAL DILUTION TO MAKE UP A WORKING REFERENCE SOLUTION OF ~0.1500 M SODIUM CARBONATE

1. If you have not already done so, put on your splash goggles, gloves, and protective clothing.
2. Rinse the 100 mL volumetric flask, first with tap water and then with distilled or deionized water.
3. Use the 10 mL pipette to transfer about 10 mL of the ~1.500 M sodium carbonate solution you made up in Part I to the volumetric flask. Record the volume to 0.01 mL on line C of Table 14-4.
4. Fill the volumetric flask with water until the bottom of the meniscus just touches the reference line.
5. Stopper the flask and invert it several times to mix the solution thoroughly.
6. Use the actual volume of sodium carbonate solution from step 3 to calculate the actual molarity of the working sodium carbonate solution to the appropriate number of significant figures, and record that molarity on line D of Table 11-4.

PART III: STANDARDIZE A BENCH SOLUTION OF ~1 M HYDROCHLORIC ACID

1. If you have not already done so, put on your splash goggles, gloves, and protective clothing.
2. Use the 10 mL pipette to transfer about 10 mL of the ~1.0 M hydrochloric acid bench solution to a 150 mL beaker. Record the volume to 0.01 mL on line E of Table 11-4.
3. Add about 25 mL of water to the beaker and swirl the beaker to mix the solution.
4. Add a drop or two of phenolphthalein indicator to the beaker and swirl to mix the solution.
5. Calculate the approximate amount of titrant you expect to need to neutralize the 10 mL of ~1.0 M hydrochloric acid solution, and enter that value on line F of Table 11-4.
6. Rinse the burette thoroughly, first with water and then with a few mL of the titrant solution.
7. Install the burette in the burette clamp and fill it to or above the top (0.00 mL) line with titrant solution.
8. Run a few mL of titrant through the burette, making sure that no air bubbles remain and that the level of titrant is at or below the top index line at 0.00 mL. (It's not important that the initial reading be exactly 0.00 mL, but it is important to know the initial reading as closely as possible.) Record the initial reading as accurately as possible on line G of Table 11-4.
9. While swirling the beaker, use the burette to dispense into the beaker a few mL less of the titrant than you estimated in step 5 will be required to neutralize the hydrochloric acid.
10. Begin adding the titrant dropwise but quickly and with continuous swirling. As you approach the equivalence point, you'll see the solution turn pink where the titrant is being added, but the pink color will disappear with swirling. This indicates that you are rapidly approaching the equivalence point. Using a sheet of white paper or other white background under or behind the beaker makes it easier to detect the first hint of a color change.
11. Continue adding titrant slowly dropwise, with swirling, until the solution in the beaker shows an overall very slightly pink color that does not disappear when the solution is swirled. A permanent slight pink coloration indicates that you've reached the equivalence point. All of the hydrochloric acid is neutralized, and there is a tiny excess of sodium carbonate. A dark pink color (or any color other than pale pink) indicates that the equivalence point has been met and exceeded, which means you need to redo the titration. If you use a different indicator, such as universal indicator, the indicative color change may be slightly different.
12. Record the final burette reading on line H of Table 11-4. Subtract line G from line H and record the difference on line I as the volume of titrant required to neutralize the aliquot of hydrochloric acid.
13. Calculate the number of moles of sodium carbonate contained in the volume of titrant you used, and record that value on line J of Table 11-4.
14. Calculate the number of moles of hydrochloric acid present in the aliquot (remember that two moles of hydrochloric acid react with one mole of sodium carbonate) and enter that value on line K of Table 11-4.
15. Calculate the molarity of the HCl bench solution, and enter that value on line L of Table 11-4.

DISPOSAL: Retain the stock sodium carbonate solution and the standardized hydrochloric acid solution for later use. The other solutions can be flushed down the drain with plenty of water.

OPTIONAL ACTIVITIES

If you have time and the required materials, consider performing these optional activities:

- Repeat the titration once or twice and compare these results with your initial results. If the results are not in close agreement, run additional titrations until you have an accurate value for the molarity of the hydrochloric acid bench solution.

- Use your standard sodium carbonate solution and your newly standardized hydrochloric acid solution to determine accurate concentrations for various household acids and bases. For example, you might use the sodium carbonate solution (after serial dilution) to titrate aliquots of distilled vinegar (acetic acid) and lemon juice (citric acid), and the standardized HCl solution to titrate aliquots of household ammonia and liquid drain cleaner (sodium hydroxide).

TABLE 11-4: *Standardize a solution by titration—observed and calculated data*

Item	Data
A. Mass of sodium carbonate	_____.___ g
B. Titrant stock solution molarity	___._____ mol/L
C. Volume of titrant stock solution	____.___ mL
D. Titrant working solution molarity	___._____ mol/L
E. Volume of ~1.0 M HCl bench solution	____.___ mL
F. Estimated volume of titrant required	____.___ mL
G. Initial burette reading	____.___ mL
H. Final burette reading	____.___ mL
I. Actual volume of titrant used (H − G)	____.___ mL
J. Moles of sodium carbonate required	__._____ moles
K. Moles of HCl present in aliquot	__._____ moles
L. Molarity of HCl bench solution	___._____ mol/L

REVIEW QUESTIONS

Q1: Best practice when titrating an unknown is to do multiple titrations. Why?

_____

_____

_____

_____

_____

_____

_____

_____

Q2: List at least five possible sources of error in the procedure you followed in this laboratory.

Q3: You are titrating an unknown base solution and have standardized solutions of hydrochloric acid and acetic acid available. Is the acid you choose for the titration likely to affect the accuracy and/or precision of your results? If so, why?

Q4: Other than using an indicator solution, how might you determine the equivalence point(s) for a titration?

Laboratory:
Chemical Kinetics

12

Chemical kinetics, also called *reaction kinetics*, is the study of *reaction rates* in chemical reactions. In simple terms, if two reactant molecules are to interact with each other, they must collide with sufficient energy to initiate the reaction. Two molecules that do not collide cannot react, regardless of how high their energies are. Conversely, two molecules that do collide, but with insufficient energy, cannot react.

Intuitively, it's easy to understand that several factors affect reaction rates. Temperature affects reaction rates because molecules at a higher temperature have higher energies, and therefore any particular collision between reactant molecules is more likely to have the energy needed to initiate the reaction. Concentration affects reaction rates, because the number of reactant molecules is higher in more concentrated solutions, and therefore collisions between reactant molecules are more likely to occur. When one or more of the reactants is a gas, pressure affects reaction rates, because gases at higher pressures contain more molecules in a given volume (in effect, their concentration is higher). Finally, when one or more of the reactants is a solid, surface area affects reaction rates, because a larger surface area exposes more of the reactant molecules to collisions with the other reactant or reactants.

In this chapter, we'll examine the effects of temperature, concentration, and surface area on reaction rates.

EVERYDAY CHEMICAL KINETICS

We make use of the principles of chemical kinetics in many aspects of everyday life. Here are just a few examples:

- Cooking is actually an application of the effect of temperature on reaction rates. When we cook food, the higher temperature causes many chemical reactions to occur quickly within the food; primarily, the denaturation of proteins. For example, when you hard-boil an egg, the liquid albumin protein is denatured to a different chemical compound. This conversion of proteins during cooking is important not only for palatability, but because the denatured forms of proteins are often more easily metabolized than the natural forms.

- Sterilization and disinfection by heat is another application of chemical kinetics. When we heat an object to sterilize it, the heat increases the rate of various reactions among the chemical compounds present in microbes, converting those compounds to useless forms.

- Chemical kinetics processes in the form of catalysis are used to minimize the amount of air pollution produced by automobile exhausts (via catalytic converters) and factory smokestacks (via catalytic scrubbers).

- Catalysts are essential for the production of many important industrial chemicals, including mineral acids, fertilizers, and many other compounds.

- Catalytic cracking is used to produce gasoline and other short-chain hydrocarbon fuels from the heavy, long-chain hydrocarbons present in crude petroleum.

LABORATORY 12.1:

DETERMINE THE EFFECT OF TEMPERATURE ON REACTION RATE

Reactions proceed faster at higher temperatures because increasing the temperature also increases the average kinetic energy of the reactant molecules, making it more likely that a collision between two reactant molecules will have sufficient energy to initiate the reaction. An old rule of thumb states that increasing the temperature by 10°C doubles the rate of reaction. (So, of course, reducing the temperature by 10°C halves the reaction rate.)

We'll test that rule of thumb in this lab by reacting Alka-Seltzer tablets with water at different temperatures. When an Alka-Seltzer tablet is dropped into water, it emits carbon dioxide as it dissolves and its components react. We could get basic data about the reaction rate simply by determining how long it takes a tablet to dissolve completely in water at various temperatures. But using that method gives us data only about the starting and ending points. It doesn't tell us, other than in the most general terms, whether the reaction rate is linear or if it changes over time.

> **REQUIRED EQUIPMENT AND SUPPLIES**
>
> ☐ goggles, gloves, and protective clothing
>
> ☐ balance and weighing papers
>
> ☐ thermometer
>
> ☐ timer
>
> ☐ foam cup
>
> ☐ Alka-Seltzer tablets (3)
>
> ☐ water (hot and cold tap water)

We know that this reaction evolves gaseous carbon dioxide, which means that the mass of the reaction vessel and contents decreases as the reaction proceeds. By recording the mass of the reaction vessel and its contents periodically as the reaction progresses, we can gather the data needed to determine whether the reaction rate is linear and whether reaction temperature affects the linearity of the reaction rate.

PROCEDURE

1. If you have not already done so, put on your splash goggles, gloves, and protective clothing.
2. Weigh one of the tablets to 0.01 g and record its mass on line A of Table 12-1.
3. Add about 100 mL of cold tap water to the foam cup. (Use a foam cup rather than a beaker to keep the mass of the reaction vessel and its contents under the maximum capacity of your balance. If the combined mass of the foam cup, Alka-Seltzer tablet, and 100 mL of water is greater than the maximum capacity of your balance, reduce the quantity of water accordingly.)
4. Weigh the foam cup and water and record the mass on line B of Table 12-1.
5. Measure and record the temperature of the water on line C of Table 12-1.
6. With the cup and water still on the balance, drop the tablet into the cup.
7. Note the combined mass of the cup, water, and tablet every five seconds and record each mass in Table 12-1. (It may be helpful to have one person watching the clock while another calls out the mass reading at each 5-second milestone.)
8. Continue recording the changing mass until you reach one minute or until the reaction completes, as evidenced by the cessation of bubbling.
9. When the reaction completes, record the final mass of the cup, water, and tablet on line Q of Table 12-1.
10. Dispose of the spent solution and rinse out the cup.
11. Repeat steps 1 through 9, using hot tap water.
12. Repeat steps 1 through 9, using a mixture of half hot and half cold tap water.

FIGURE 12-1: *Observing the mass change as the Alka-Seltzer tablet reacts and carbon dioxide is evolved*

CAUTIONS

Although the chemicals used in this laboratory are not hazardous, it is good practice to wear splash goggles, gloves, and protective clothing at all times.

DISPOSAL: All of the solutions from this laboratory can be flushed down the drain with plenty of water.

SUBSTITUTIONS AND MODIFICATIONS

- You may substitute any brand of fizzy tablets for the Alka-Seltzer tablets.
- If you don't have Alka-Seltzer or similar tablets, you may substitute small sticks of chalk or calcium carbonate antacid tablets for the Alka-Seltzer tablets and cold and warm dilute (~3 M) solutions of hydrochloric acid for the water. If you do that, remember that hydrochloric acid is corrosive and hazardous to handle.

TABLE 12-1: *Effect of temperature on reaction rate—observed and calculated data*

Item	Trial A	Trial B	Trial C
A. Mass of tablet	_____.__ g	_____.__ g	_____.__ g
B. Mass of cup + water	_____.__ g	_____.__ g	_____.__ g
C. Temperature of water	_____._ °C	_____._ °C	_____._ °C
D. Mass at 0:00 (A + B)	_____.__ g	_____.__ g	_____.__ g
E. Mass at 0:05	_____.__ g	_____.__ g	_____.__ g
F. Mass at 0:10	_____.__ g	_____.__ g	_____.__ g
G. Mass at 0:15	_____.__ g	_____.__ g	_____.__ g
H. Mass at 0:20	_____.__ g	_____.__ g	_____.__ g
I. Mass at 0:25	_____.__ g	_____.__ g	_____.__ g
J. Mass at 0:30	_____.__ g	_____.__ g	_____.__ g
K. Mass at 0:35	_____.__ g	_____.__ g	_____.__ g
L. Mass at 0:40	_____.__ g	_____.__ g	_____.__ g
M. Mass at 0:45	_____.__ g	_____.__ g	_____.__ g
N. Mass at 0:50	_____.__ g	_____.__ g	_____.__ g
O. Mass at 0:55	_____.__ g	_____.__ g	_____.__ g
P. Mass at 1:00	_____.__ g	_____.__ g	_____.__ g
Q. Mass at completion of reaction	_____.__ g	_____.__ g	_____.__ g
R. Mass loss (D – Q)	_____.__ g	_____.__ g	_____.__ g
S. Mass loss percentage [(R/A) · 100]	_____.__ %	_____.__ %	_____.__ %

OPTIONAL ACTIVITIES

If you have time and the required materials, consider performing these optional activities:

- Graph your results for the trials using cold, hot, and warm water. Determine whether the reaction rate is linear over time and whether the reaction temperature affects linearity.

- The main component of fizzy tablets is sodium bicarbonate, usually mixed with citric acid, aspirin, binders, and other inactive components. When it comes into contact with water, the citric acid and sodium bicarbonate react to form (among other products) carbon dioxide. The carbon dioxide is evolved as a gas, which accounts for the mass loss. Write a balanced equation for this reaction, and use the data from Table 12-1 to calculate the approximate percentage of sodium bicarbonate by weight in the tablet.

REVIEW QUESTIONS

Q1: What effect did you observe temperature to have on reaction rate?

Q2: Based on the data you recorded in Table 12-1, does the 10°C rule of thumb provide a reasonably close approximation of the observed reaction rates in this laboratory? How far do your data depart from the expected values based on the rule of thumb?

Q3: Based on the data you recorded in Table 12-1, does reaction rate appear to be approximately linear over time? If you noticed an increase or decrease in reaction rate over time, propose at least one possible explanation.

LABORATORY 12.2:
DETERMINE THE EFFECT OF SURFACE AREA ON REACTION RATE

If at least one of the reactants is a solid, the reaction proceeds faster if the solid is finely divided, because the surface area is larger in a finely divided solid, exposing more of that reactant to the other reactant or reactants. For example, a 50-pound bag of flour is essentially inert, because the flour, although finely ground, exposes little of its surface area to the air. But that same amount of flour dispersed as airborne dust, if ignited by a spark, explodes with force sufficient to flatten a large building. (Military fuel-air explosives, or FAEs, use this principle by vaporizing a liquid fuel and detonating it. During Operation Desert Storm, British troops reportedly sent a flash-priority report of a nuclear detonation, mistaking the detonation of an FAE for a tactical nuke.)

In the last laboratory, we examined reaction rates of Alka-Seltzer tablets, keeping the initial surface area constant and varying the temperature. In this laboratory, we'll explore the effect on reaction rates of varying the surface area at constant temperature.

PROCEDURE

1. If you have not already done so, put on your splash goggles, gloves, and protective clothing.
2. Weigh one of the tablets to 0.01 g and record its mass on line A of Table 12-2.
3. Add about 100 mL of cold tap water to the foam cup. Again, make sure the combined mass of the foam cup, Alka-Seltzer tablet, and 100 mL of water is less than the maximum capacity of your balance. If not, reduce the quantity of water accordingly.
4. Weigh the foam cup + water and record the mass on line B of Table 12-2.
5. Measure and record the temperature of the water on line C of Table 12-2.
6. With the cup and water still on the balance, drop the tablet into the cup.
7. Note the combined mass of the cup, water, and tablet every five seconds and record each mass in Table 12-2. (It may be helpful to have one person watching the clock while another calls out the mass reading at each 5-second milestone.)

REQUIRED EQUIPMENT AND SUPPLIES

☐ goggles, gloves, and protective clothing

☐ balance and weighing papers

☐ thermometer

☐ timer

☐ foam cup

☐ ruler

☐ Alka-Seltzer tablets (3)

☐ water

SUBSTITUTIONS AND MODIFICATIONS

• You may substitute any brand of fizzy tablets for the Alka-Seltzer tablets.

• If you don't have Alka-Seltzer or similar tablets, you may substitute small sticks of chalk or calcium carbonate antacid tablets for the Alka-Seltzer tablets and a dilute (~3 M) solution of hydrochloric acid for the water. If you do that, remember that hydrochloric acid is corrosive and hazardous to handle.

8. Continue recording the changing mass until you reach one minute or until the reaction completes, as evidenced by the cessation of bubbling.
9. When the reaction completes, record the final mass of the cup, water, and tablet on line Q of Table 12-2.
10. Dispose of the spent solution and rinse out the cup.
11. Repeat steps 1 through 9, using a tablet that you have split into quarters.
12. Repeat steps 1 through 9, using a tablet that you have crushed into powder. (This reaction may be too fast to time accurately; just do your best.)

CAUTIONS

Although the chemicals used in this laboratory are not hazardous, it is good practice to wear splash goggles, gloves, and protective clothing at all times.

DISPOSAL: All of the solutions from this laboratory can be flushed down the drain with plenty of water.

TABLE 12-2: *Effect of surface area on reaction rate—observed and calculated data*

Item	Trial A (solid)	Trial B (chunks)	Trial C (powder)
A. Mass of tablet	_____.__ g	_____.__ g	_____.__ g
B. Mass of cup + water	_____.__ g	_____.__ g	_____.__ g
C. Temperature of water	_____.__ °C	_____.__ °C	_____.__ °C
D. Mass at 0:00 (A + B)	_____.__ g	_____.__ g	_____.__ g
E. Mass at 0:05	_____.__ g	_____.__ g	_____.__ g
F. Mass at 0:10	_____.__ g	_____.__ g	_____.__ g
G. Mass at 0:15	_____.__ g	_____.__ g	_____.__ g
H. Mass at 0:20	_____.__ g	_____.__ g	_____.__ g
I. Mass at 0:25	_____.__ g	_____.__ g	_____.__ g
J. Mass at 0:30	_____.__ g	_____.__ g	_____.__ g
K. Mass at 0:35	_____.__ g	_____.__ g	_____.__ g
L. Mass at 0:40	_____.__ g	_____.__ g	_____.__ g
M. Mass at 0:45	_____.__ g	_____.__ g	_____.__ g
N. Mass at 0:50	_____.__ g	_____.__ g	_____.__ g
O. Mass at 0:55	_____.__ g	_____.__ g	_____.__ g
P. Mass at 1:00	_____.__ g	_____.__ g	_____.__ g
Q. Mass at completion of reaction	_____.__ g	_____.__ g	_____.__ g
R. Mass loss (D − Q)	_____.__ g	_____.__ g	_____.__ g
S. Mass loss percentage [(R/A) · 100]	_____.__ %	_____.__ %	_____.__ %

REVIEW QUESTIONS

Q1: What effect did you observe surface area to have on reaction rate?

_____

_____

_____

_____

_____

_____

_____

Q2: Use the ruler to measure the diameter and thickness of a tablet. Calculate its approximate surface area. Make the same calculation for surface area of the quartered tablet. Does the increase in reaction rate you observed with the quartered tablet approximately correspond to the increase in surface area? If not, propose an explanation.

_____

_____

_____

_____

_____

_____

_____

Q3: Most fizzy tablet remedies direct the user to dissolve two tablets in a glass of water. Why don't they simply make the tablets larger and recommend dissolving only one of the larger tablets?

_____

_____

_____

_____

_____

_____

LABORATORY 12.3:
DETERMINE THE EFFECT OF CONCENTRATION ON REACTION RATE

Reactions proceed faster at higher concentrations because more reactant molecules are available and therefore collisions between reactant molecules are more likely. In the two preceding labs, we used Alka-Seltzer tablets and water to demonstrate the effects of temperature and surface area on reaction rates. Obviously, we'll have to use some other method to demonstrate the effect of concentration on reaction rates, because the amounts of the citric acid and sodium bicarbonate reactants in a fizzy tablet are fixed and cannot be changed.

In this lab, we'll use another OTC medicine with varying concentrations of hydrochloric acid to demonstrate the effect of concentration on reaction rates. Some (but not all) antacid tablets contain primarily calcium carbonate, along with flavoring, binders, and other inactive ingredients. These tablets neutralize excess stomach acid, which is actually dilute hydrochloric acid, according to the following equation:

$$CaCO_3(s) + 2HCl(aq) \rightarrow CaCl_2(aq) + CO_2(g) + H_2O(l)$$

We'll use exactly this reaction to observe and quantify the effect of concentration of reaction rates.

PROCEDURE

1. If you have not already done so, put on your splash goggles, gloves, and protective clothing.
2. Label three foam cups 4 M, 2 M, and 1 M.
3. In the first foam cup, make up 100 mL of 4 M hydrochloric acid by adding 33.3 mL of 12 M HCl to 66.7 mL of water. Again, make sure that the combined mass of the foam cup, antacid tablet, and 100 mL of hydrochloric acid is less than the maximum capacity of your balance. If not, reduce the quantity of hydrochloric acid accordingly.)
4. In the second foam cup, make up 100 mL of 2 M hydrochloric acid by adding 16.7 mL of 12 M HCl to 83.3 mL of water.

REQUIRED EQUIPMENT AND SUPPLIES

- ☐ goggles, gloves, and protective clothing
- ☐ balance and weighing papers
- ☐ graduated cylinder, 100 mL
- ☐ thermometer
- ☐ timer
- ☐ foam cup (3)
- ☐ marking pen
- ☐ calcium carbonate antacid tablets (3)
- ☐ hydrochloric acid, 4 M (100 mL)
- ☐ hydrochloric acid, 2 M (100 mL)
- ☐ hydrochloric acid, 1 M (100 mL)
- ☐ water

SUBSTITUTIONS AND MODIFICATIONS

- You may substitute small (~1 cm) pieces of chalk (calcium carbonate) for the calcium carbonate antacid tablets.
- In making up the dilute hydrochloric acid solutions, you may substitute muriatic acid from the hardware store, which is typically 31.45% (10.3 M), for reagent grade hydrochloric acid, which is typically 37% (12 M). Adjust quantities accordingly.

CAUTIONS

Hydrochloric acid is corrosive and emits strong fumes. Wear splash goggles, gloves, and protective clothing at all times.

5. In the third foam cup, make up 100 mL of 1 M hydrochloric acid by adding 8.3 mL of 12 M HCl to 91.7 mL of water.

CAUTIONS
ADD ACID TO WATER AND WAIT

Always add acid to water. Adding water to a concentrated acid can cause splattering. Diluting acid produces heat. Because we want concentration to be the only variable, it's important to ensure that all three acid solutions are at the same initial temperature. Make your dilute acid solutions ahead of time and allow all three cups to cool to room temperature before proceeding.

OPTIONAL ACTIVITIES

If you have time and the required materials, consider performing these optional activities:

• Graph your results for the trials using 4 M, 2 M, and 1 M HCl. Determine whether the reaction rate is linear over time and whether the concentration of the HCl affects linearity.

• In the first laboratory, we mentioned that a rule of thumb says that an increase of 10°C in temperature should approximately double reaction rate. (Did your experimental data verify or disprove this speculation?) Repeat this experiment, using 100 mL of 1 M HCl at a temperature 10°C higher than the temperature of the acid in your original run. Compare the results to your data for the 1 M and 2 M HCl at the lower temperature.

6. Weigh one of the antacid tablets to 0.01 g and record its mass on line A of Table 12-3.
7. Place the first foam cup with 4 M hydrochloric acid on the balance and record the mass to 0.01 g on line B of Table 12-3.
8. Measure and record the temperature of the HCl solution on line C of Table 12-3.
9. With the cup and its contents still on the balance, drop the antacid tablet into the cup.
10. Note the combined mass of the cup, acid, and tablet every five seconds and record each mass in Table 12-3. (It may be helpful to have one person watching the clock while another calls out the mass reading at each 5-second milestone.)
11. Continue recording the changing mass until you reach one minute or until the reaction completes, as evidenced by the cessation of bubbling.
12. When the reaction completes, record the final mass of the cup, water, and tablet on line Q of Table 12-3.
13. Dispose of the spent solution and rinse out the cup.
14. Repeat steps 6 through 9, using 2 M hydrochloric acid.
15. Repeat steps 6 through 9, using 1 M hydrochloric acid.

DISPOSAL:
Neutralize the spent acid solutions with sodium bicarbonate or another base and flush the neutralized solutions down the drain with plenty of water.

TABLE 12-3: *Effect of concentration on reaction rate—observed and calculated data*

Item	Trial A (4 M HCl)	Trial B (2 M HCl)	Trial C (1 M HCl)
A. Mass of tablet	_____.__ g	_____.__ g	_____.__ g
B. Mass of cup + hydrochloric acid	_____.__ g	_____.__ g	_____.__ g
C. Temperature of hydrochloric acid	_____.__ °C	_____._ °C	_____._ °C
D. Mass at 0:00 (A + B)	_____.__ g	_____.__ g	_____.__ g
E. Mass at 0:05	_____.__ g	_____.__ g	_____.__ g
F. Mass at 0:10	_____.__ g	_____.__ g	_____.__ g
G. Mass at 0:15	_____.__ g	_____.__ g	_____.__ g
H. Mass at 0:20	_____.__ g	_____.__ g	_____.__ g
I. Mass at 0:25	_____.__ g	_____.__ g	_____.__ g
J. Mass at 0:30	_____.__ g	_____.__ g	_____.__ g
K. Mass at 0:35	_____.__ g	_____.__ g	_____.__ g
L. Mass at 0:40	_____.__ g	_____.__ g	_____.__ g
M. Mass at 0:45	_____.__ g	_____.__ g	_____.__ g
N. Mass at 0:50	_____.__ g	_____.__ g	_____.__ g
O. Mass at 0:55	_____.__ g	_____.__ g	_____.__ g
P. Mass at 1:00	_____.__ g	_____.__ g	_____.__ g
Q. Mass at completion of reaction	_____.__ g	_____.__ g	_____.__ g
R. Mass loss (D − Q)	_____.__ g	_____.__ g	_____.__ g
S. Mass loss percentage [(R/A)·100]	_____._ %	_____._ %	_____._ %

REVIEW QUESTIONS

Q1: What effect did you observe concentration to have on reaction rate?

Q2: Based on the data you recorded in Table 12-3, is the effect of concentration on reaction rate linear?

Q3: Based on the data you recorded in Table 12-3, at any particular concentration does reaction rate appear to be approximately linear over time? If you noticed an increase or decrease in reaction rate over time, propose at least one possible explanation.

LABORATORY 12.4:

DETERMINE THE EFFECT OF A CATALYST ON REACTION RATE

A *catalyst* is a substance that increases the rate of a chemical reaction, but is not consumed or changed by the reaction. A catalyst works by reducing the activation energy needed to initiate and sustain the reaction. For example, two molecules of hydrogen peroxide can react to form two molecules of water and one molecule of molecular oxygen gas by the following reaction:

$$2\ H_2O_2(aq) \rightarrow 2\ H_2O(l) + O_2(g)$$

At room temperature, this reaction occurs very slowly because few of the collisions between hydrogen peroxide molecules have sufficient energy to activate the reaction. Furthermore, commercial hydrogen peroxide solutions, such as the 3% hydrogen peroxide solution sold in drugstores and the 6% solution sold by beautician supply stores, are treated with stabilizers (sometimes called *negative catalysts*) that *increase* the activation energy for the reaction, further inhibiting it from occurring.

If you add a catalyst to a solution of hydrogen peroxide, the effect is immediately evident. The solution begins bubbling, as oxygen gas is evolved. Numerous substances can catalyze the reaction of hydrogen peroxide to water and oxygen gas, including many metal oxides such as manganese dioxide, but the efficiency of catalysts varies. One of the most efficient catalysts for hydrogen peroxide is the enzyme *catalase* that is contained in blood. (Catalase functions in the body as a peroxide scavenger, destroying peroxide molecules that would otherwise damage cells.)

One catalase molecule can catalyze the reaction of millions of hydrogen peroxide molecules per second. Immediately after each pair of hydrogen peroxide molecules reacts, catalyzed by the catalase molecule, that catalase molecule is released unchanged and becomes available to catalyze the reaction of another pair of hydrogen peroxide molecules. When all of the hydrogen peroxide has reacted to form water and oxygen gas, you end up with as many catalase molecules remaining as you started with.

In this lab session, we'll measure the reaction rate of the catalyzed reaction of hydrogen peroxide by adding a fixed amount of catalase enzyme to measured samples of hydrogen peroxide. After allowing the reaction to continue for measured periods of time, we'll stop the reaction by adding sulfuric acid to denature (deactivate) the catalase and then titrate the resulting solutions

REQUIRED EQUIPMENT AND SUPPLIES

☐ goggles, gloves, and protective clothing

☐ beaker, 150 mL (6)

☐ graduated cylinder, 10 mL

☐ graduated cylinder, 100 mL

☐ pipette, Mohr or serological, 10.0 mL

☐ pipette, Beral (or other disposable)

☐ burette, 50 mL

☐ funnel (for filling burette)

☐ ring stand

☐ burette clamp

☐ stopwatch

☐ catalase enzyme solution (see Substitutions and Modifications)

☐ hydrogen peroxide, 3% (50 mL)

☐ potassium permanganate solution, 0.1 M (~150 mL)

☐ sulfuric acid, 0.1 M (~250 mL)

☐ distilled or deionized water

with a dilute solution of potassium permanganate to determine how much unreacted hydrogen peroxide remains in each sample. In acidic solution, the intensely purple permanganate (MnO_4^-) ion reacts with hydrogen peroxide to form the light-brown Mn^{2+} ion according to the following equation:

$$5\ H_2O_2(aq) + 2\ MnO_4^-(aq) + 6\ H^+(aq)$$
$$\rightarrow 2\ Mn^{2+}(aq) + 8\ H_2O(l) + 5\ O_2(g)$$

Because the purple color of the permanganate ion is so intense, the titrant can serve as its own indicator for this titration. As long as hydrogen peroxide remains in excess, MnO_4^- ions are quickly reduced to Mn^{2+} ions, and the solution remains a light brown color. As soon as permanganate ions are slightly in excess, the solution assumes a purple color. By determining the amount of permanganate titrant required, we can calculate the amount of hydrogen peroxide that remained in the original samples.

CAUTIONS

Sulfuric acid is corrosive. Hydrogen peroxide is a strong oxidizer and bleach. Potassium permanganate is a strong oxidizer and stains skin and clothing. Wear splash goggles, gloves, and protective clothing at all times.

PROCEDURE

1. If you have not already done so, put on your splash goggles, gloves, and protective clothing.
2. Transfer about 40 mL of the sulfuric acid to the 100 mL graduated cylinder.
3. Fill the 10 mL graduated cylinder with the catalase solution.
4. Label six beakers or other containers from A through F.
5. Use the pipette to transfer as closely as possible 10.00 mL of the hydrogen peroxide solution to each of the beakers. Record the amount of hydrogen peroxide in each beaker to 0.01 mL on the corresponding line in Table 12-4.
6. Beaker A is the control, to which we will not add any catalase enzyme. Set it aside for now.
7. Use the Beral pipette to withdraw 1.0 mL of catalase solution from the 10 mL graduated cylinder.
8. As close to simultaneously as possible, start the stopwatch or timer and squirt the catalase solution into beaker B. Swirl the beaker to mix the solutions.

TOO FAST OR TOO SLOW

If the reaction is so vigorous that the solution foams over the top of the container, repeat step 8 using a more dilute catalase solution, and use that more dilute solution for the following steps as well. If the reaction rate is too slow—not enough bubbles forming quickly enough—use a more concentrated catalase solution or more of it.

9. With the graduated cylinder of sulfuric acid held ready, when the timer reaches the 15.0 second mark, dump the sulfuric acid quickly into beaker B and swirl to stop the reaction. Record the elapsed reaction time as closely as possible on the corresponding line in Table 12-4.
10. Refill the 100 mL graduated cylinder with 40 mL of sulfuric acid solution.
11. Repeat steps 7 through 10 for beakers C, D, E, and F using reaction times of 30 seconds, 60 seconds, 120 seconds, and 240 seconds.

SUBSTITUTIONS AND MODIFICATIONS

- If you don't have six 150 mL beakers, you may substitute Erlenmeyer flasks, foam cups, or other containers of similar capacity. Alternatively, you can complete this lab session using only one beaker or other container by doing the timed runs sequentially instead of simultaneously.

- You may substitute a 10 mL graduated cylinder for the 10 mL pipette, at the expense of accuracy.

- If you do not have a burette, you may use the alternative titration procedure described in Chapter 5.

- If you do not have a stopwatch, you may substitute a timer or watch with a second hand.

- Rather than purchase catalase enzyme, you may substitute a dilute solution of animal blood, which contains catalase. The liquid that leaks from a package of raw ground beef suffices, or you can obtain a small amount of animal blood from a butcher. The exact concentration of catalase in unimportant, as long as that concentration is the same for all of your test runs. One or two mL of animal blood diluted with distilled water to 10 mL works well.

12. Set up your burette, rinse it with the 0.1 M potassium permanganate titrant, and refill it to near the 0.00 mL line.
13. Titrate the solution in beaker A by adding titrant to the beaker until a slight purple coloration just persists.
14. Record the volume of titrant required to 0.01 mL on the corresponding line in Table 12-4. (Assuming nominal concentrations, about 35 mL of titrant should be required for beaker A, and correspondingly less for beakers B, C, D, E, and F.)
15. Repeat step 13 for beakers B, C, D, and E. For each titration, calculate the number of millimoles (mM) of potassium permanganate required to neutralize the remaining hydrogen peroxide. (One millimole is 0.001 mole. Using mM locates the decimal point more conveniently for calculations.) Enter the number of millimoles of titrant required for each titration in Table 12-4.
16. Using the balanced equation provided in the introduction, calculate the number of millimoles of unreacted hydrogen peroxide in each beaker, and enter that value in Table 12-4.
17. For each beaker, calculate the reaction rate in millimoles/second (mM/s) and enter that value in Table 12-4.

TABLE 12-4: *Effect of a catalyst on reaction rate—observed and calculated data*

	Hydrogen peroxide	Reaction time	Titrant volume	Titrant millimoles	Peroxide remaining	Reaction rate
A.	_____.___ mL	0.0 s	_____.___ mL	_____.___ mM	_____.___ mM	n/a
B.	_____.___ mL	_____.__ s	_____.___ mL	_____.___ mM	_____.___ mM	_____.___ mM/s
C.	_____.___ mL	_____.__ s	_____.___ mL	_____.___ mM	_____.___ mM	_____.___ mM/s
D.	_____.___ mL	_____.__ s	_____.___ mL	_____.___ mM	_____.___ mM	_____.___ mM/s
E.	_____.___ mL	_____.__ s	_____.___ mL	_____.___ mM	_____.___ mM	_____.___ mM/s
F.	_____.___ mL	_____.__ s	_____.___ mL	_____.___ mM	_____.___ mM	_____.___ mM/s

(Beaker — brackets grouping rows A–F)

OPTIONAL ACTIVITIES

If you have time and the required materials, consider performing these optional activities:

- Graph your results for the trials to determine whether the reaction rate is linear.

- Repeat the trials, using more concentrated catalase solution (or more of it) and determine the effect of additional catalyst on the reaction rate.

- Boil a solution of catalase for 30–60 seconds and then use the boiled catalase solution to determine whether the catalase enzyme is denatured by heat.

- Test other materials such as manganese dioxide, copper(II) oxide, and zinc oxide to determine whether they function as catalysts for the decomposition of hydrogen peroxide and, if so, how efficient each is compared with each other and with catalase.

DISPOSAL:
All waste solutions from this laboratory session can be disposed of by flushing them down the drain with plenty of water.

REVIEW QUESTIONS

Q1: What effect did you observe the catalase catalyst to have on reaction rate?

_____

_____

_____

_____

Q2: Based on the data you recorded in Table 12-4, is the effect of the catalyst on reaction rate linear? If not, propose an explanation.

_____

_____

_____

_____

Q3: Would you expect the reaction rate to increase, decrease, or remain the same if you increased the amount of catalyst? Why?

Q4: When you begin a titration of a reacted hydrogen peroxide solution, you find that the first drop of potassium permanganate titrant causes the solution to assume a purple color. What has happened?

Laboratory:
Chemical Equilibrium and Le Chatelier's Principle

13

In a chemical reaction, *chemical equilibrium* is the state in which the concentrations of the reactants and products exhibit no net change over time. This state occurs when the *forward reaction rate* (conversion of reactants to products) is equal to the *reverse reaction rate* (conversion of products to reactants). These reaction rates are generally greater than zero, but because the rates are equal, reactants are converted to products at the same rate that products are converted to reactants, yielding zero net change in the concentrations of reactants and products. A system in this state is said to have achieved *dynamic chemical equilibrium*.

The equilibrium is referred to as "dynamic" because any change to the reaction environment in concentrations, temperature, volume, or pressure forces a corresponding change in the equilibrium state. For example, if a chemical reaction has achieved dynamic equilibrium and you add reactants, the state of the system changes to achieve a new dynamic equilibrium. Some (but not all) of the additional reactants react to form additional products, changing the concentrations of both the reactants and the products. Similarly, if you remove products from the reaction vessel, the dynamic equilibrium changes because additional reactants are converted to products.

The nineteenth-century French chemist Henry Louis Le Chatelier explained the effect of environmental changes on dynamic chemical equilibrium in a short statement that is known to every chemist as *Le Chatelier's Principle*:

"If a chemical system that has achieved dynamic chemical equilibrium experiences an imposed change in concentration, temperature, volume, or pressure, the equilibrium state shifts to counteract, insofar as is possible, that imposed change."

It's important to understand two things about Le Chatelier's Principle. First, the equilibrium change can (almost) never return the system to its original equilibrium state. For example, if you add reactants to a system that has achieved dynamic chemical equilibrium, the concentration of reactants at the new equilibrium state will be higher than it was at the original equilibrium state (although lower than it would have been without the change in equilibrium state). Second, Le Chatelier's Principle is qualitative, rather than quantitative. That is, applying Le Chatelier's Principle tells you whether the reaction will be forced to the left (more reactants) or to the right (more products), but it cannot tell you how far the reaction will be forced or what the eventual concentrations of reactants and products will be.

In this chapter, we'll examine chemical reactions and Le Chatelier's Principle.

EVERYDAY CHEMICAL EQUILIBRIUM

The importance of chemical equilibrium and Le Chatelier's Principle isn't limited to chemistry labs. These concepts are important in everyday life, including at least one application that we literally couldn't live without. Here are just a few of them.

- Human blood is a complex chemical system that is constantly in a finely balanced equilibrium. As you burn calories, your cells produce carbon dioxide, which enters your bloodstream, forming carbonic acid and lowering the pH. Biological pH meters located in your carotid artery, called *carotid bodies*, detect the lowered pH and cause you to breathe faster. As the blood circulates through your lungs, it is exposed through the permeable membranes that make up your lungs' alveoli to atmospheric air, which contains less carbon dioxide. Carbon dioxide migrates from the bloodstream to the outside air. By Le Chatelier's Principle, the equilibrium shifts to the left. The concentration of carbonic acid in your blood is reduced, and its pH returns to its normal level of about 7.4.

- Hundreds of industrial chemicals used in thousands of everyday products from plastics to drugs, gasoline, and coffee are produced in continuous-mode processes that depend upon chemical equilibrium. In these processes, products are continuously removed from the reaction vessel to force the reaction to the right, giving higher yields of the desired products.

- Carbonated beverages are a perfect illustration of Le Chatelier's Principle. These beverages are bottled under carbon dioxide gas at high pressure to ensure that they'll still be fizzy when you open them. Under high pressure, the carbon dioxide gas reacts with water to form carbonic acid, as illustrated by the equilibrium equation: $CO_2 + H_2O \Leftrightarrow H_2CO_3$. When you remove that pressure by opening the bottle or can, the equilibrium shifts to the left, converting the carbonic acid back into water and carbon dioxide gas.

LABORATORY 13.1:
OBSERVE LE CHATELIER'S PRINCIPLE IN ACTION

Le Chatelier's Principle states that a forced change in concentration, temperature, volume, or pressure results in change to the dynamic equilibrium state. In this lab, we'll examine various reactions that are in a state of dynamic chemical equilibrium, force changes to each of these environmental characteristics, and observe the results.

PROCEDURE
This lab has four parts, each of which examines one of the effects of concentration, temperature, volume, and pressure on dynamic equilibrium states. As you observe each experiment, record your observations in the Observations section that follows the procedure sections.

CAUTIONS

With the exception of concentrated hydrochloric acid, the chemicals used in this laboratory are reasonably safe. Hydrochloric acid is toxic, corrosive, and produces irritating fumes. This experiment uses an open flame, so use caution and have a fire extinguisher handy. Wear splash goggles, gloves, and protective clothing.

REQUIRED EQUIPMENT AND SUPPLIES

☐ goggles, gloves, and protective clothing

☐ dropper or Beral pipette (3)

☐ test tubes (6) and rack

☐ beaker, 150 mL (2)

☐ alcohol lamp or other heat source

☐ ring stand

☐ support ring

☐ wire gauze

☐ stirring rod

☐ tap water

☐ sodium chloride (~40 g)

☐ sodium carbonate (~1.6 g)

☐ magnesium sulfate (~1.3 g)

☐ hydrochloric acid, concentrated (~5 mL)

☐ small bottle of carbonated soft drink (2, one chilled and one at room temperature)

SUBSTITUTIONS AND MODIFICATIONS

- You may substitute ordinary table salt for the sodium chloride.

- You may substitute washing soda or soda ash for the sodium carbonate.

- You may substitute Epsom salts for the magnesium sulfate.

- You may substitute muriatic acid (from the hardware store) of 30% or higher concentration for the hydrochloric acid.

DISPOSAL: All of the solutions used in this laboratory can safely be flushed down the drain with copious water.

PART I: EXAMINE THE EFFECT OF CONCENTRATION ON EQUILIBRIUM
In this part of the lab, we examine the effect of changing the concentration of the reactant ions on a saturated solution of sodium chloride (NaCl). That solution contains sodium ions (Na^+) and chloride ions (Cl^-). To saturated sodium chloride solution samples, we'll add concentrated hydrochloric acid (which contains Cl^- ions but not Na^+ ions), saturated sodium carbonate solution (which contains Na^+ ions but not Cl^- ions), and saturated magnesium sulfate solution (which contains neither), and observe the effects on the saturated sodium chloride solution.

1. If you have not already done so, put on your splash goggles, gloves, and protective clothing.
2. Prepare a saturated solution of sodium carbonate by adding about 1.6 g of sodium carbonate to 5 mL of water in a test tube. Stir or shake the solution to ensure that the sodium carbonate dissolves completely, and continue

adding sodium carbonate until some of it remains undissolved at the bottom of the test tube.

3. Prepare a saturated solution of magnesium sulfate by adding about 1.3 g of magnesium sulfate to 5 mL of water in a test tube. Stir or shake the solution to ensure that the magnesium sulfate dissolves completely, and continue adding magnesium sulfate until some of it remains undissolved at the bottom of the test tube.

4. Prepare a saturated solution of sodium chloride by adding about 40 g of sodium chloride to 100 mL of water in a small beaker or flask. Stir or shake the solution to ensure that the sodium chloride dissolves completely, and continue adding sodium chloride until some of it remains undissolved at the bottom of the beaker or flask.

5. Transfer about 5.0 mL of the saturated sodium chloride solution to each of four test tubes, labeled A through D, one of which will serve as the control. The exact amount of solution is not critical, but make sure that each test tube contains the same amount.

6. Place the test tubes adjacent to each other in a test tube rack under a strong light. (It may be easier to observe the reactions if you place a sheet of black construction paper or a similar material behind the test tube rack.)

7. Add concentrated hydrochloric acid dropwise to test tube A, observing any change that occurs as each drop is added. Continue adding hydrochloric acid until you have added about 5 mL.

8. Add saturated sodium carbonate solution dropwise to test tube B, observing any change that occurs as each drop is added. Continue adding the saturated sodium carbonate solution until you have added about 5 mL.

9. Add saturated magnesium sulfate solution dropwise to test tube C, observing any change that occurs as each drop is added. Continue adding the saturated magnesium sulfate solution until you have added about 5 mL.

10. Add water dropwise to test tube D, the control test tube, observing any change that occurs as each drop is added. Continue adding water until you have added about 5 mL.

FIGURE 13-1: *Saturated sodium chloride solutions with (from left to right) hydrochloric acid, sodium carbonate, magnesium sulfate, and water added*

PART II:
EXAMINE THE EFFECT OF TEMPERATURE ON EQUILIBRIUM
In our saturated solution of sodium chloride, solid sodium chloride exists in equilibrium with aqueous sodium ions and chloride ions. Le Chatelier's Principle tells us that changing the temperature of the reaction environment will cause a change in that dynamic chemical equilibrium, but we have no way of knowing the direction of that change. Intuitively, we expect that more sodium chloride will dissolve in a given volume of water at a higher temperature, so let's test that hypothesis:

1. If you have not already done so, put on your splash goggles, gloves, and protective clothing.

2. Transfer about half of the remaining saturated sodium chloride solution (~40 mL) to a small beaker or flask.

3. Add a small amount of sodium chloride to the beaker or flask; just enough so that undissolved sodium chloride crystals are visible on the bottom of the beaker or flask.

4. Set up your tripod stand, wire gauze, and alcohol burner.

5. With stirring, gently heat the beaker or flask that contains the saturated sodium chloride solution, keeping an eye on the undissolved sodium chloride.

PART III:
EXAMINE THE EFFECT OF VOLUME ON EQUILIBRIUM
In our saturated solution of sodium chloride, solid sodium chloride exists in equilibrium with aqueous sodium ions and chloride ions. Le Chatelier's Principle tells us that changing the volume of the solvent will cause a change in that dynamic chemical equilibrium. Intuitively, we expect that if X g of sodium chloride dissolves in Y mL of water, increasing the amount of water will increase the amount of sodium chloride that dissolves. Let's test that hypothesis:

1. If you have not already done so, put on your splash goggles, gloves, and protective clothing.

2. Make sure that the remaining saturated sodium chloride solution (~40 mL) has some undissolved sodium chloride visible in the bottom of the beaker or flask. Note the approximate amount of undissolved sodium chloride.

3. Add 10 drops of water to the beaker or flask, and stir or swirl the solution to determine whether any additional sodium chloride dissolves.

4. Continue adding water 10 drops at a time until all of the visible undissolved sodium chloride has dissolved.

PART IV: EXAMINE THE EFFECT OF PRESSURE (AND TEMPERATURE) ON EQUILIBRIUM
Le Chatelier's Principle says that changing the pressure of the reaction environment will change the dynamic chemical equilibrium. Carbonated soft drinks contain dissolved carbon dioxide gas, which is more soluble at higher pressures. If we reduce the pressure, we expect the soft drink solution

to reduce the concentration of dissolved carbon dioxide by liberating carbon dioxide gas. (You may want to do this experiment outdoors to avoid making a mess indoors.) Let's test that hypothesis:

1. If you have not already done so, put on your splash goggles, gloves, and protective clothing.
2. Leaving it capped, shake the chilled sealed bottle of carbonated soft drink. (Agitating the sealed bottle mixes small bubbles of the carbon dioxide gas in the "empty" part of the bottle into the liquid, providing loci for the dissolved carbon dioxide to come out of solution.)
3. Immediately after you give the carbonated soft drink bottle a thorough shaking, point the mouth in a safe direction and remove the cap.
4. Repeat steps 2 and 3 with the bottle of carbonated soft drink that's at room temperature.
5. Clean up.

OBSERVATIONS

If you are not maintaining a lab notebook, record your observations narratively here.

Part I:

Part II:

Part III:

Part IV:

REVIEW QUESTIONS

Q1: In Part I, what substance was formed as you added concentrated hydrochloric acid to the saturated solution of sodium chloride? Why?

_____

_____

_____

_____

Q2: In Part I, what substance was formed as you added saturated sodium carbonate solution to the saturated solution of sodium chloride? Why?

_____

_____

_____

_____

Q3: In Part I, explain what occurred when you added the saturated magnesium sulfate solution.

_____

_____

_____

_____

Q4: If you repeated Part I in a pressure chamber at 10 atmospheres, would you expect different results? Why or why not?

_____

_____

_____

_____

_____

Q5: In Part II, we observed the qualitative effect of temperature on the dynamic equilibrium of a saturated sodium chloride solution. If you were charged with designing an experiment to obtain quantitative data, how would you proceed?

Q6: In Part III, we observed the effect of increasing the volume of solvent on the dynamic equilibrium of a saturated sodium chloride solution. Is it possible that, for some compounds, increasing the volume of solvent would reduce the amount of solute that dissolved? Why or why not?

Q7: In Part IV, we observed the effect of pressure on the dynamic equilibrium of a carbonated soft drink at two different temperatures. The differing results at the two temperatures illustrate the concept of *retrograde solubility*, which is the chemist's way of referring to a compound that is less soluble at higher temperature. What mechanism explains the lower solubility of carbon dioxide gas in water at higher temperatures?

LABORATORY 13.2:
QUANTIFY THE COMMON ION EFFECT

In Part I of the preceding laboratory, we examined the effect on equilibrium of changing the concentration of one of the reactant ions. We found that increasing the concentration of either reactant ion (Na^+ or Cl^-) by adding sodium carbonate or hydrochloric acid, respectively, forced solid sodium chloride to precipitate from the saturated solution. This phenomenon is called the *common ion effect*.

In addition to affecting solubility, the common ion effect reduces the ionization of weak acids and weak bases in solution. Acetic acid is a weak organic acid that dissociates only partially in aqueous solutions, yielding hydronium (H_3O^+) ions and acetate (CH_3COO^-) ions:

$$CH_3COOH(l) + H_2O \Leftrightarrow CH_3COO^-(aq) + H_3O^+(aq)$$

Sodium acetate is an ionic compound that dissociates essentially completely in aqueous solution, yielding sodium (Na^+) ions and acetate (CH_3COO^-) ions.

$$CH_3COONa(s) \rightarrow CH_3COO^-(aq) + Na^+(aq)$$

The common ion effect tells us that acetic acid should dissociate less completely in sodium acetate solution than in pure water, because the sodium acetate increases the concentration of acetate ions. Less dissociation means fewer protons, and an accordingly higher pH. In this lab, we'll test that hypothesis by measuring the pH of a fixed concentration of acetic acid in solutions that contain various concentrations of sodium acetate.

REQUIRED EQUIPMENT AND SUPPLIES

- ☐ goggles, gloves, and protective clothing
- ☐ balance and weighing paper
- ☐ dropper or beral pipette (1)
- ☐ graduated cylinder, 10 mL
- ☐ test tubes (6) and rack
- ☐ pH meter
- ☐ tap water
- ☐ acetic acid, glacial (6 mL)
- ☐ sodium acetate, anhydrous (4.0 g)

SUBSTITUTIONS AND MODIFICATIONS

- You may substitute narrow-range pH test paper for the pH meter, but the accuracy and precision of your data will be lower.
- You may substitute white vinegar for glacial acetic acid, but the lower concentration of acetic acid reduces the pH variation between samples.
- You may substitute 6.7 g of sodium acetate trihydrate for the 4.0 g of anhydrous sodium acetate.
- You may make your own sodium acetate by neutralizing baking soda with white vinegar and evaporating the solution to dryness.

CAUTIONS

Glacial acetic acid is corrosive, toxic, and has a very strong, biting odor. Wear splash goggles, gloves, and protective clothing.

PROCEDURE

1. If you have not already done so, put on your splash goggles, gloves, and protective clothing.
2. Label six test tubes A through F.
3. Prepare a concentrated (nearly saturated) solution of sodium acetate by adding 4.0 g of anhydrous sodium acetate to 7 mL of water in the graduated cylinder and then adding water to bring the total solution volume to 10 mL. Stir or shake the solution to ensure that the sodium acetate dissolves completely. We'll define this concentration of sodium acetate as 100%.
4. Transfer 5.0 mL of the 100% sodium acetate solution to test tube A and 5.0 mL to test tube B.
5. Add 5.0 mL of water to test tube B and mix thoroughly.
6. Transfer 5.0 mL of the solution in test tube B to test tube C, add 5.0 mL of water to test tube C, and mix thoroughly.
7. Repeat these steps until the five test tubes labeled A through E contain, respectively, 5.0 mL each of sodium acetate solutions of 100%, 50%, 25%, 12.5%, and 6.25%.
8. Transfer 5.0 mL of water to test tube F.
9. Add 1.0 mL of glacial acetic acid to test tube A and mix thoroughly.
10. Following the directions supplied with your pH meter, test the pH of the solution and record the data in Table 13-1.

> **OPTIONAL ACTIVITIES**
>
> If you have time and the required materials, consider performing these optional activities:
>
> - Graph pH versus concentration of sodium acetate to determine whether the change in pH is linear with respect to concentration of sodium acetate.
>
> - Repeat the experiment using 2 mL of glacial acetic acid per test tube to determine how (or whether) the change in pH varies with the concentration of acetic acid.
>
> - Calculate the initial concentrations of acetate ion in the samples. The acid dissociation constant (pK_a) of acetic acid is 4.76 at 25°C. Calculate the expected pH value for each of the samples. How closely do the theoretical values correspond to the observed values?

11. Repeat steps 9 and 10 for test tubes B through F.
12. After 15 minutes have passed, retest the pH of each of the six test tubes. Repeat the test again after 60 minutes. Record the data in Table 13-1.

TABLE 13-1: *pH values of acetic acid in sodium acetate solutions*

Test tube	Concentration	pH (immediate)	pH (15 minutes)	pH (60 minutes)
A	100%			
B	50%			
C	25%			
D	12.5%			
E	6.25%			
F	0.00%			

> DISPOSAL: All of the solutions used in this laboratory can safely be flushed down the drain with copious water.

REVIEW QUESTIONS

Q1: What effect does a higher concentration of sodium acetate have on the pH of solution after the acetic acid has been added? Propose an explanation based on your understanding of Le Chatelier's Principle.

Q2: Do the pH values of the various samples change after 15 minutes and/or 60 minutes? Why or why not?

Q3: If you repeated this experiment, substituting sodium chloride for sodium acetate and concentrated hydrochloric acid for acetic acid, would you expect the pH changes in different concentrations of sodium chloride solution to be greater than or less than the corresponding changes in the sodium acetate solution. Why?

LABORATORY 13.3:
DETERMINE A SOLUBILITY PRODUCT CONSTANT

The coexistence of an ionic solid and its component ions in solution is one example of a chemical equilibrium. This equilibrium can be quantified using the *solubility product principle*, which states that in a saturated solution of an ionic compound, the product of the molar activities, called the *solubility product constant* or K_{sp}, has a constant value at any particular temperature and pressure.

Because ionic activities are very difficult to determine accurately and molarities are easy to determine, solubility product calculations are normally performed using molarity values. For dilute solutions, using molarities rather than activities introduces only tiny errors, because molarity and activity are nearly identical in dilute solutions. For concentrated solutions, molarity and activity may differ significantly, so solubility product calculations, although useful for sparingly soluble ionic compounds, are of limited use for very soluble ionic compounds.

Consider the sparingly soluble salt silver chloride. In a saturated aqueous solution of silver chloride, the silver and chlorine are present primarily as solvated silver ions and chlorine ions. (*Solvated ions* are ions that have bound to solvent molecules.) Undissociated molecular silver chloride exists in the solution, but it is present in such a tiny amount that it can be ignored. The following equilibrium equation describes the saturated silver chloride solution:

$AgCl(s) \Leftrightarrow Ag^+(aq) + Cl^-(aq)$

The solubility product constant for this equilibrium is:

$K_{sp} = [Ag^+] \cdot [Cl^-]$

Square brackets in equilibrium expressions indicate molar concentration, so this expression is a short way to say that the solubility product constant is equal to the molar concentration of the silver ions multiplied by the molar concentration of the chloride ions. Because one molecule of silver chloride dissociates into one silver ion and one chloride ion, the concentrations of the silver ions and chloride ions are identical. Assigning that concentration the value **x** simplifies the expression to:

$K_{sp} = x \cdot x = x^2$

REQUIRED EQUIPMENT AND SUPPLIES

☐ goggles, gloves, and protective clothing

☐ balance and weighing paper

☐ beaker, 250 mL (2)

☐ volumetric flask, 100 mL

☐ eye dropper or disposable pipette

☐ thermometer

☐ stirring rod

☐ funnel

☐ filter paper

☐ ring stand

☐ support ring

☐ wire gauze

☐ burette clamp

☐ gas burner

☐ sodium chloride (~50 g)

☐ potassium hydrogen tartrate (~2 g)

☐ sodium hydroxide, 0.1000 M, standardized (~50 mL)

☐ phenolphthalein indicator solution (a few drops)

which means that the concentration of either ion is the square root of the K_{sp}. That means that if we know the K_{sp}, we can determine the molar concentration, and vice versa. For example, if we know that the K_{sp} of silver chloride at a particular temperature and pressure is $1.8 \cdot 10^{-10}$, we can rewrite the equilibrium expression as:

$1.8 \cdot 10^{-10} = x^2$

Solving for x (taking the square root of the K_{sp}) tells us that the molar concentrations of the silver ions and chloride ions are both $1.3 \cdot 10^{-5}$, so a saturated solution of silver chloride at this temperature and pressure is 0.000013 M. The gram molecular mass of silver chloride is 143.32 g/mol, so a saturated silver chloride solution contains about 0.0019 g/L.

Conversely, if we know the molar solubility, we can calculate the K_{sp}. For example, if we know that a saturated solution of silver bromide at a particular temperature and pressure is $7.2 \cdot 10^{-7}$ M, we can calculate the K_{sp} of silver bromide as follows:

$$K_{sp} = (7.2 \cdot 10^{-7})^2 = 5.2 \cdot 10^{-13}$$

Until now, we've considered only the simplest case, compounds in which a molecule dissociates into two ions. For ionic compounds that dissociate into three or more ions, the calculations are a bit more involved. Consider silver chromate, which dissociates as follows:

$$Ag_2CrO_4(s) \Leftrightarrow 2\ Ag^+(aq) + CrO_4^{2-}(aq)$$

We can no longer use simple squares and square roots, because each molecule of silver chromate dissociates into three ions—two silver ions and one chromate ion. We can generalize the solubility product constants for a salt A_xB_y that dissociates into the ions A^{y+} and B^{x-} in the form:

$$K_{sp} = [A^{y+}]^x \cdot [B^{x-}]^y$$

so, for silver chromate, $Ag_2(CrO_4)_1$, the solubility product constant can be expressed as:

$$K_{sp} = [Ag^+]^2 \cdot [Cr_2O_4^{2-}]^1$$

If we assign the molar concentration of the chromate ion as **x**, and knowing that the concentration of the silver ion is twice that of the chromate ion, the expression becomes:

$$K_{sp} = [2x]^2 \cdot [x]^1 = 4x^3$$

If we make up a saturated solution of silver chromate, we might determine by titration that the concentration of silver ion is $1.3 \cdot 10^{-4}$ M, from which it follows that the concentration of the chromate ion must be $0.65 \cdot 10^{-4}$ M, which can also be expressed as $6.5 \cdot 10^{-5}$ M. Plugging these values into the equation yields:

$$K_{sp} = (1.3 \cdot 10^{-4})^2 \cdot (6.5 \cdot 10^{-5})^1 = 1.1 \cdot 10^{-12}$$

If the K_{sp} is known, we can determine the molar solubility of silver chromate as follows:

$$K_{sp} = 1.1 \cdot 10^{-12} = 4x^3$$

$$x^3 = (1.1 \cdot 10^{-12}/4) = 2.8 \cdot 10^{-13}$$

$$x = 6.5 \cdot 10^{-5}$$

In principle, you can use solubility product constants to calculate molar solubilities of ionic salts. In practice, observed solubilities often differ significantly from such calculated values. Electrostatic interaction between ions can affect the results, particularly in any but very dilute solutions, and other equilibria occurring simultaneously in the solution can have a major impact. For example, the solubility of magnesium hydroxide varies greatly with the pH of the solution because of the common ion effect. Magnesium hydroxide is much more soluble in acid solutions, which have a very low $[OH^-]$ and correspondingly less soluble in basic solutions, which have a high $[OH^-]$. Finally, the assumption that ionic substances fully dissociate in solution into simple solvated ions is not always true. For example, magnesium fluoride (MgF_2) does not dissociate completely into Mg^{2+} and F^- ions. Instead, some partially dissociated magnesium fluoride is present as MgF^+ ions.

In this lab, we'll determine solubility product constants for two compounds, using two different methods.

CAUTIONS

Glacial acetic acid is corrosive, toxic, and has a very strong, biting odor. Wear splash goggles, gloves, and protective clothing.

SUBSTITUTIONS AND MODIFICATIONS

- You may substitute uniodized table salt (popcorn or kosher salt) for the sodium chloride.

- You may substitute cream of tartar from the grocery store for the potassium hydrogen tartrate.

PROCEDURE

This lab session is in two parts. In Part I, we'll determine the solubility product constant for sodium chloride gravimetrically, by evaporating a known volume of a saturated sodium chloride solution and determining the mass of the dry solid remaining. In Part II, we'll determine the solubility product constant for potassium hydrogen tartrate titrimetrically (volumetrically) by titrating a known volume of saturated potassium hydrogen tartrate solution with sodium hydroxide titrant of known molarity.

PART I: DETERMINE THE K_{SP} OF SODIUM CHLORIDE GRAVIMETRICALLY

1. If you have not already done so, put on your splash goggles, gloves, and protective clothing.
2. Weigh a clean, dry, empty 250 mL beaker and record its mass to 0.01 g on line A of Table 13-2.
3. Add about 125 mL of room temperature distilled or deionized water and about 50 g of sodium chloride to the 250 mL beaker. Allow the salt to dissolve for at least 15 minutes, stirring periodically, to ensure that the solution is saturated. Record the temperature of the solution on line B of Table 13-2.
4. Carefully decant 100.00 mL of the saturated sodium chloride solution into the 100 mL volumetric flask, making sure that none of the undissolved sodium chloride is transferred to the volumetric flask. Fill the flask to the index line, using a dropper or disposable pipette to add the last mL or so, and record the volume on line C of Table 13-2. (If you've calibrated the volumetric flask, as described in Chapter 5, record the actual volume that the flask contains rather than the nominal 100.00 mL volume.)
5. Discard the remaining sodium chloride solution and undissolved sodium chloride and rinse the beaker thoroughly, first with tap water and then with distilled water.
6. Pour the contents of the 100 mL volumetric flask into the 250 mL beaker. Do a quantitative transfer, rinsing the volumetric flask several times with a few mL of distilled water and adding the rinse water to the beaker to make sure all of the sodium chloride is transferred to the beaker.
7. Place a stirring rod in the beaker to prevent boil-over, and heat the beaker until the solution comes to a gentle boil. Continue heating the beaker gently until nearly all of the water has evaporated. When most of the water has vaporized, remove the stirring rod from the beaker, and use a few mL of distilled water to rinse any sodium chloride into the beaker.

8. Continue heating the beaker to drive off all remaining water. When you have evaporated all of the remaining water, remove the heat and allow the beaker to cool to room temperature.
9. Weigh the beaker and record the mass of the beaker plus sodium chloride to 0.01 g on line D of Table 13-2.
10. Subtract the empty mass of the beaker from the mass of the beaker plus sodium chloride, and record the mass of sodium chloride on line E of Table 13-2.
11. Calculate the solubility of sodium chloride in g/L and enter that value on line F of Table 13-2.
12. The gram molecular mass of sodium chloride is 58.442 g/mol. Use that value to calculate the molar solubility of sodium chloride and enter that value on line G of Table 13-2.
13. The K_{sp} of sodium chloride is the product of the activities (concentrations) of the sodium ion and the chloride ion: $K_{sp} = [Na^+] \cdot [Cl^-]$. Those concentrations are the same, and are equal to the concentration of sodium chloride, so we can simplify the equation to: $K_{sp} = [NaCl]^2$. Therefore, the solubility product constant of sodium chloride equals the square root of the molar solubility of sodium chloride. Calculate that value and enter it on line H of Table 13-2.

TABLE 13-2: *Determine a solubility product constant, Part I—observed and calculated data*

Item	Data
A. Mass of empty beaker	_____.____ g
B. Temperature of solution	_____.____ °C
C. Volume of sodium saturated sodium chloride solution	_____.____ mL
D. Mass of beaker + sodium chloride	_____.____ g
E. Mass of sodium chloride (D − A)	_____.____ g
F. Mass solubility of sodium chloride ([1000/C] · E)	_____.____ g/L
G. Molar solubility of sodium chloride (F/58.442 g/mol)	_____.____ mol/L
H. K_{sp} of sodium chloride (square root of G)	_____.____

PART II: DETERMINE THE K_{SP} OF POTASSIUM HYDROGEN TARTRATE TITRIMETRICALLY (VOLUMETRICALLY)

1. If you have not already done so, put on your splash goggles, gloves, and protective clothing.
2. To about 150 mL of distilled or deionized water in the 250 mL beaker, add about 2 g of potassium hydrogen tartrate. Allow the salt to dissolve for at least 15 minutes, stirring periodically, to ensure that the solution is saturated. Record the temperature of the solution on line A of Table 13-3.
3. Allow the solution to sit undisturbed for several minutes until most of the undissolved material has settled to the bottom of the beaker. Carefully filter the supernatant liquid through a dry filter paper into a second 250 mL beaker, trying to keep as much as possible of the undissolved solid in the first beaker.
4. Transfer 100.00 mL of the saturated potassium hydrogen tartrate solution into the 100 mL volumetric flask. Fill the flask to the index line, using a dropper or disposable pipette to add the last mL or so, and record the volume on line B of Table 13-3. (If you've calibrated the volumetric flask, as described in Chapter 5, record the actual volume that the flask contains rather than the nominal 100.00 mL volume.)
5. Discard the remaining potassium hydrogen tartrate solution and undissolved potassium hydrogen tartrate. Rinse the beaker thoroughly, first with tap water and then with distilled water.
6. Pour the contents of the 100 mL volumetric flask into the 250 mL beaker. Do a quantitative transfer, rinsing the volumetric flask several times with a few mL of distilled water and adding the rinse water to the beaker to make sure all of the potassium hydrogen tartrate is transferred to the beaker.
7. Rinse your 50 mL burette twice with 0.1000 M sodium hydroxide, running the rinse solution through the tip and into a waste container.
8. Clamp the 50 mL burette in a burette clamp and fill it to above the 0.00 mL index line with 0.1000 M sodium hydroxide solution.
9. Run solution through the burette until the level drops to or slightly below the 0.00 mL index line. Make sure that there are no bubbles in the body or tip of the burette. Record the initial burette reading as accurately as possible on line C of Table 13-3. Interpolate the reading to 0.05 mL or better.
10. Add a few drops of phenolphthalein indicator solution to the beaker of potassium hydrogen tartrate solution, and swirl slightly to mix.
11. Titrate the potassium hydrogen tartrate solution until you reach the endpoint, when the phenolphthalein gives the solution a distinct pink coloration that persists for at least 30 seconds. Record the final burette reading as accurately as possible on line D of Table 13-3. Interpolate the reading to 0.05 mL or better.
12. Subtract the initial burette reading from the final burette reading to determine the volume of 0.1000 M sodium hydroxide titrant that was required to neutralize the potassium hydrogen tartrate aliquot. Record that volume on line E of Table 13-3.
13. The gram molecular mass of sodium hydroxide is 39.9971 g/mol. Use that value and the actual molarity of the titrant solution to calculate the number of moles of sodium hydroxide required to neutralize the potassium hydrogen tartrate aliquot, and enter that value on line F of Table 13-3.
14. Using the number of moles of sodium hydroxide needed to neutralize the aliquot of potassium hydrogen tartrate and the volume of that aliquot, calculate the molar solubility of potassium hydrogen tartrate and enter that value on line G of Table 13-3.
15. Calculate the K_{sp} of potassium hydrogen tartrate, and enter that value on line H of Table 13-3.

TABLE 13-3: *Determine a solubility product constant, Part II— observed and calculated data*

Item	Data
A. Temperature of solution	_____._____ °C
B. Volume of sodium saturated potassium hydrogen tartrate solution	_____._____ mL
C. Initial burette reading	_____._____ mL
D. Final burette reading	_____._____ mL
E. Volume of 0.1000 M sodium hydroxide titrant used (D – C)	_____._____ mL
F. Moles of sodium hydroxide required	_____._____ mol
G. Molar solubility of potassium hydrogen tartrate	_____._____ mol/L
H. K_{sp} of potassium hydrogen tartrate	_____

REVIEW QUESTIONS

Q1: Why did we record the temperatures of the saturated solutions?

Q2: The solubility of silver chromate, Ag_2CrO_4, is $1.3 \cdot 10^{-4}$ mol/L. What is the K_{sp} of silver chromate?

Q3: The K_{sp} of lead chloride ($PbCl_2$) is $1.6 \cdot 10^{-5}$. What is solubility of lead chloride in g/L?

Q4: What is solubility of lead chloride in mol/L in 0.1 M hydrochloric acid? (Assume that the HCL is fully dissociated.)

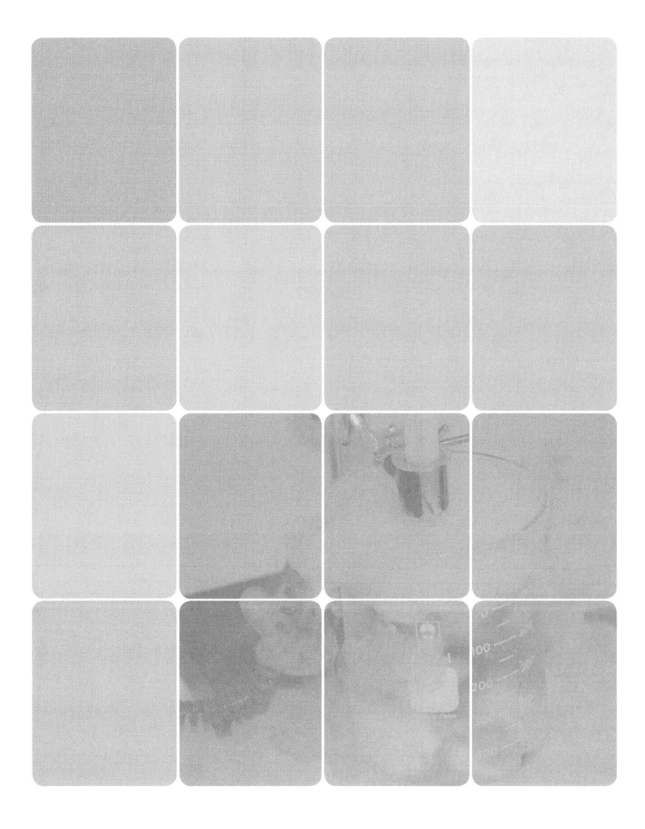

Laboratory: Gas Chemistry

14

The study of gases and their chemistry, composition, properties, and interactions is an important part of general chemistry. By their nature, gases are more difficult to study than solids and liquids. Unlike solids and liquids, which tend to stay put in their containers, gases dissipate rapidly into the surrounding atmosphere. (Imagine studying liquids if you had to work under water.) Their low density relative to solids and liquids makes it difficult to weigh small volumes of gases accurately. Many gases are colorless and impossible to discriminate visually or by odor.

Despite these difficulties, gases have been studied intensively since chemistry evolved as a separate science in the late eighteenth century. Pioneers like Antoine-Laurent de Lavoisier, Carl Wilhelm Scheele, Joseph Priestley, and Joseph Louis Gay-Lussac built on the earlier work of the seventeenth-century natural philosopher Robert Boyle to discover and explore the basic principles that govern the behavior of gases.

Naturally enough, early chemists thought of all gases as forms of air. As they isolated various gases experimentally—such as hydrogen, oxygen, carbon dioxide, chlorine, and others—in relatively pure forms, it soon became obvious that these new forms of "air" had properties much different from plain old air-air. Immersing a candle in oxygen, for example, caused it to burn fiercely, and carbon dioxide snuffed it out instantly. Accordingly, the early chemists named gases as airs with particular properties, not recognizing that each of these special "airs" was in fact a discrete chemical element or compound. It was not until the seminal works of Scheele, Lavoisier, and Priestley were published that scientists recognized that these newfangled gases were in fact discrete chemical elements and compounds.

In addition to studying their chemical properties, these early gas chemists also studied the physical properties of gases; in particular, the relationships between volume, pressure, and temperature. Anyone who has used a manual pump to inflate a tire or a basketball intuitively understands the relationship of these three characteristics of gases. As the plunger is depressed, the pressure increases, the volume decreases, and the pump gets noticeably warmer. When you used canned air, the opposite occurs. When you press the trigger, gas is released, increasing its volume and reducing its pressure, and the can gets noticeably colder, a phenomenon known as the *Joule-Thompson effect*.

In 1662, Boyle quantified the relationship between the volume of a gas and its pressure, a discovery that became known as **Boyle's Law**. In 1802, Gay-Lussac quantified the relationship between the volume of a gas and its temperature. That relationship became known as **Charles' Law**, chiefly because Gay-Lussac was kind enough to credit some preliminary unpublished work done about 15 years earlier by the French chemist Jacques Charles. In 1809, Gay-Lussac set in place the third of the fundamental gas laws by quantifying the relationship between the pressure of a gas and its temperature, a relationship known as **Gay-Lussac's Law**. These three gas laws are inextricably related. All of them specify the relationship between volume, pressure, and temperature, with one of those held constant.

Boyle's Law states that, at constant temperature, the volume of a gas is inversely proportional to its pressure. For example, if you double the pressure of the gas you halve its volume and vice versa. Boyle's Law can be expressed as the equation:

$$V_1 \cdot P_1 = V_2 \cdot P_2$$

where the variables on the left side refer to the values before a change in volume or pressure and those on the right refer to the new values after the volume or pressure is changed. If any three of those values are known, the fourth can be determined algebraically.

Charles' Law states that, at constant pressure, the volume of a gas is proportional to its absolute temperature, specified in kelvins (see note in Lab 14.2). For example, if you double the temperature of a gas, you double its volume, and vice versa. Charles' Law can be expressed as the equation:

$$V_1 \cdot T_2 = V_2 \cdot T_1$$

where the variables subscripted with 1 refer to the values before a change in volume or temperature and those subscripted with 2 refer to the new values after the volume or temperature is changed. Again, if any three of those values are known, the fourth can be determined algebraically.

Knowing Boyle's Law and Charles' Law, you can derive Gay-Lussac's Law mathematically by substitution (as indeed you can derive any of these three laws if the other two are known). Gay-Lussac's Law states that, at constant volume, the pressure of a gas is proportional to its absolute temperature, specified in kelvins. For example, if you double the temperature of a gas, you double its pressure, and vice versa. Gay-Lussac's Law can be expressed as the equation:

$$P_1 \cdot T_2 = P_2 \cdot T_1$$

where again, the variables subscripted with 1 refer to the values before a change in volume or temperature and those subscripted with 2 refer to the new values after the pressure or temperature is changed. As before, if any three of those values are known, the fourth can be determined algebraically.

These three related gas laws are sometimes referred to as the **Combined Gas Law**, but although true individually and in combination, they are insufficient to define a generalized law of gases. In 1811, the Italian scientist Amedeo Avogadro postulated the fourth and final fundamental law of gases, referred to as **Avogadro's Principle**, which states that equal volumes of gases at the same temperature and pressure contain the same number of particles.

Avogadro's Principle was the final piece in the puzzle. With the Combined Gas Law and Avogadro's Principle, it's possible to define a generalized law of gases, called the **Ideal Gas Law**. First stated in 1834 by the French scientist Benoît Paul Émile Clapeyron, the Ideal Gas Law can be expressed as the equation:

$$PV = nRT$$

where

- P = the absolute pressure of the gas
- V = the volume of the gas
- n = the number of moles of the gas
- R = the ideal gas constant
- T = the temperature of the gas in kelvins (K)

The value for R, the ideal gas constant, depends on the units used for volume and pressure. In SI units, the value of R is 8.314472 joules per mole per kelvin ($J \cdot mol^{-1} \cdot K^{-1}$)

The name of this law reflects the fact that it is completely accurate only for an "ideal" gas, which is to say one that comprises monoatomic particles of infinitely small volume, at very high temperature and very low pressure, and in which no attractions or repulsions exist between particles. Real gases depart from values calculated using the Ideal Gas Law, because their atoms or molecules have finite volumes and do interact.

In this chapter, we'll examine the characteristics of gases and experimentally verify these fundamental laws of gases.

EVERYDAY GAS CHEMISTRY

The importance of gas laws isn't limited to chemistry labs. Applied gas laws are important in everyday life, including at least one application that we literally couldn't live without. Here are just a few of them:

- Every breath you take illustrates Boyle's Law. When you inhale, your muscles expand the volume of your chest and lungs, reducing the gas pressure inside. Outside air, at constant atmospheric pressure, enters your lungs until the pressure equalizes. When you exhale, the volume of your chest and lungs decreases, increasing the gas pressure above ambient atmospheric pressure. Air exits your lungs until the pressure equalizes.

- Diesel engines use the physical manifestations of gas laws to produce power. During the intake stroke, the engine cylinder fills with a mixture of air and vaporized diesel fuel at relatively low pressure. During the compression stroke, the piston compresses the fuel-air mixture to a volume that's only a few percent of its original volume, which increases the temperature (Charles' Law) sufficiently to ignite the fuel-air mixture without a spark. Because one molecule of hydrocarbon fuel reacts with oxygen to form many molecules of carbon dioxide and water vapor, the total number of moles of gas increases (Avogadro's Principle). The combustion of the fuel-air mixture produces a great deal of heat, which increases pressure in the cylinder (Gay-Lussac's Law). Because the piston is free to move within the cylinder, the increased pressure forces it outward, producing power. That power is transferred via the drive train to the wheels, moving the vehicle.

- Refrigerators and air conditioners are closed systems that use the physical manifestations of gas law to transfer heat. The compressor compresses a gas (sometimes so far that the gas becomes a liquid), during which the temperature of the gas increases according to Charles' Law. The compressed gas is passed through a heat exchanger, where it gives up most of its heat to the outside environment. The cool compressed gas is then passed through cooling coils, where it is permitted to expand, reducing its pressure and (again according to Charles' Law) its temperature by the Joule-Thompson effect. The cold gas absorbs heat from the outside environment, cooling it. The cycle is then repeated, resulting in a net transfer of heat from one place (such as the inside of the refrigerator or the inside of your home) to another place (such as the outside of the refrigerator or your home). The same process is used in reverse by heat pumps, which transfer heat from the cold outside air to the warmer air inside your home.

- Automobile air bags illustrate Avogadro's Principle dramatically (and loudly). An air bag is simply a large balloon constructed of very strong material and with all gas exhausted from it. The air bag mechanism includes a small canister that contains 50 g to 200 g of solid sodium azide (NaN_3) or a similar propellant. When the air bag deploys, the sodium azide reacts almost instantly to form metallic sodium and nitrogen gas. In about 50 milliseconds (0.05 second), the airbag that originally contained almost no gas is filled with sufficient nitrogen gas to inflate it completely. (Sodium azide is extremely toxic, more so than potassium cyanide, so don't mess with it at home.)

SI SYSTEM

SI = International System of Units from the French, Le Système International d'Unités. SI is the modern form of the now-obsolete metric system. Some metric system units (e.g. centimetre, litre) are deprecated in SI and replaced with the SI standard units (metre and cubic decimetre, respectively).

LABORATORY 14.1:

OBSERVE THE VOLUME-PRESSURE RELATIONSHIP OF GASES
(Boyle's Law)

Boyle's Law states that, at constant temperature, the volume of a gas is inversely proportional to its pressure. In this laboratory session, we'll verify Boyle's Law experimentally, using the setup shown in Figure 14-1. (You can buy a ready-made Boyle's Law apparatus from Home Science Tools or other vendors, but $10 or $12 is a pretty high price to pay for a disposable syringe and two blocks of wood.) The apparatus shown in Figure 14-1 uses only standard lab equipment, and is at least as accurate as a ready-made apparatus.

FIGURE 14-1:

Measuring gas volume as compressed by a measured mass

REQUIRED EQUIPMENT AND SUPPLIES

- ☐ goggles, gloves, and protective clothing
- ☐ balance and weighing cup
- ☐ caliper
- ☐ barometer (optional)
- ☐ ring stand
- ☐ burette or utility clamp (to fit syringe)
- ☐ 4" (100 mm) support ring
- ☐ plastic syringe, 10 mL to 50 mL, graduated, with cap
- ☐ mineral oil or petroleum jelly (1 drop)
- ☐ plastic cup (to fit support ring)
- ☐ lead shot (10 pounds or 5 kilograms)

SUBSTITUTIONS AND MODIFICATIONS

- If you do not have a caliper, you may substitute a metric ruler with millimeter markings, although you will sacrifice significant accuracy.

- If you do not have a barometer, you may use the barometric pressure broadcast by a local TV or radio station or the Weather Channel web site for your zip code, but see the note on the next page.

- You may substitute any sturdy lightweight container of similar size for the plastic cup, including an aluminum beverage can with the top removed. The container should fit loosely within the support ring; not closely enough to bind, but closely enough to keep the container centered over the syringe plunger when mass is added to the container.

- You may substitute any dense material for the lead shot, such as old wheel weights, fishing sinkers, spools of solder, and so on. Ideally, the material should be dense enough to allow the container to hold at least 3 kilograms of mass. Syringes with large bores require more mass for equivalent compression.

Using this apparatus, we'll begin with the syringe containing a known volume of gas under normal atmospheric pressure. We'll then add mass incrementally to the container above the syringe to increase the pressure on the gas contained in the syringe and record the gas volume under differing amounts of pressure.

Gas pressure is specified in units of mass, weight, or force per unit area. For example, in traditional units, standard atmospheric pressure is about 14.7 pounds per square inch. Chemists use the SI unit of pressure, the *pascal* (**Pa**), which equals one newton per square metre ($N \cdot m^{-2}$ or $kg \cdot m^{-1} \cdot s^{-2}$). In SI units, standard atmospheric pressure is about 101,325 Pa, which may also be stated as 101.325 kilopascal.

The total pressure exerted on the gas in the syringe is the sum of the atmospheric pressure, the pressure exerted by the mass of the syringe plunger and the container above it, and the pressure exerted by the mass added to the container. But there's one more piece of the puzzle. Because pressure is specified per unit area, we need to know the area of the syringe bore. With all of those data known, we can calculate the pressure of the gas contained within the syringe and correlate that pressure with the observed volume of the gas.

ATMOSPHERIC PRESSURE VERSUS BAROMETRIC PRESSURE

Although most people assume that *atmospheric pressure* and *barometric pressure* mean the same thing, they don't. Atmospheric pressure is the actual pressure of the atmosphere. Barometric pressure, the value given by TV and radio stations (and indicated by your barometer if you adjusted it to the pressure given in a local weather report), is an adjusted value that reflects the elevation of the reporting station. Barometric pressure is what the atmospheric pressure would be *if the reporting station were located at sea level*. For a reporting station located at sea level, barometric pressure is equal to atmospheric pressure. But for a reporting station located above sea level, the actual atmospheric pressure is lower—sometimes significantly lower—than the barometric pressure reported by the station.

To obtain an accurate atmospheric pressure reading, you can adjust the barometric pressure reported by a station for that station's altitude. Alternatively, you can check with your local airport, which reports both barometric pressure and atmospheric pressure.

CAUTIONS

The only real hazard in this lab is that the apparatus may topple or collapse if you add too much mass to the container. Wear splash goggles, gloves, and protective clothing.

PROCEDURE

1. If you have not already done so, put on your splash goggles, gloves, and protective clothing.
2. Use the caliper to measure the inside diameter of the syringe bore in millimeters. Calculate the surface area of the syringe bore by dividing the diameter by 2 to determine the radius, squaring the radius, and multiplying by pi (π, 3.14159). Divide by 1,000,000 to convert the result from square millimeters to square meters and enter that value on line A of Table 14-1. (The number you obtain will be very small. For example, my syringe has an internal bore of 15.0 mm, which corresponds to an area of $0.000177\ m^2$.)
3. Weigh the container and syringe plunger (just the plunger—not the body of the syringe) together, and record their combined mass on line B of Table 14-1.
4. Spread a tiny amount (~1 drop) of mineral oil or petroleum jelly on the seal at the bottom of the plunger. Use just enough to lubricate the seal so that it slides freely within the body of the syringe.
5. With the cap removed, slide the plunger in and out of the syringe several times to distribute the lubricant. There should be little discernible resistance when moving the plunger.
6. Adjust the plunger to the top gradation line on the syringe. If the plunger seal has multiple wipes (as are visible in Figure 14-2), the bottom wipe is the one that matters. Once the plunger is in the correct position, reinstall the cap on the syringe.

FIGURE 14-2: *Measure the gas volume at the lowest bearing surface of the plunger*

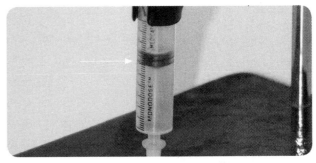

7. Clamp the syringe as shown in Figure 14-2. Make sure that the cap of the syringe is pressing directly against the base of the ring stand. Otherwise, when you add mass to the container, the cap might pop off the syringe.

8. Obtain the atmospheric pressure using a barometer or from a local radio or TV broadcast. Convert that value to pascals and enter it on line D of Table 14-1.

9. Verify the initial volume reading on the syringe, and record that value as accurately as possible on line D of Table 14-1. (Your syringe may use cc graduations, which are the same as mL graduations for our purposes.)

10. Weigh 500 g of lead shot and add it to the container above the syringe. If your balance does not have a 500 g capacity, do repeated weighings and accumulate the shot until you reach 500 g. Record the volume reading of the syringe on line E of Table 14-1. The first 500 g mass addition should cause a noticeable but not excessive reduction in volume. For example, with my setup, the first 500 g mass addition reduced the volume from 10.0 mL to 8.2 mL. Depending on the bore size and capacity of your syringe, you may need to modify the mass increments from 500 g to some other value. If 500 g causes too much volume reduction, as it may with a narrow-bore syringe, reduce

the incremental mass additions to some smaller value, such as 250 g. If the first 500 g addition causes too little volume reduction, as it may with a large-bore syringe, increase the incremental mass additions to some larger value, such as 1,000 g.

11. Repeat step 9, adding 500 g of lead shot each time and recording each volume reading on the syringe in Table 14-1. Continue adding mass until the container is nearly full or the syringe fails to compress any more, whichever comes first. As you add mass, keep an eye on the stability of the apparatus, or you may find yourself picking up 5 kilograms of lead shot from the floor, piece by piece.

12. The atmospheric pressure, the mass of the empty container and plunger, the bore area of the syringe, and the masses added to the container are all known. Force equals mass times acceleration (F = ma). Gravitational acceleration (a) is 9.81 m · s^{-2}. Pressure equals force divided by area (P = F/A), where the area is the cross-sectional area of the syringe. Using these data, calculate the pressures and expected volumes for each of the mass/volume readings, and record those values in the right columns of Table 14-1.

TABLE 14-1: *Observe the volume-pressure relationship of gases—observed and calculated data.*

Item	Observed value	Calculated pressure	Calculated volume
A. Area of syringe bore	_____ m^2	n/a	n/a
B. Mass of empty container and plunger	_____.___ g	n/a	n/a
C. Atmospheric pressure	_____ Pa	n/a	n/a
D. Initial volume reading	_____.___ mL	n/a	n/a
E. Volume reading (+500 g)	_____.___ mL	_____ Pa	_____.___ mL
F. Volume reading (+1,000 g)	_____.___ mL	_____ Pa	_____.___ mL
G. Volume reading (+1,500 g)	_____.___ mL	_____ Pa	_____.___ mL
H. Volume reading (+2,000 g)	_____.___ mL	_____ Pa	_____.___ mL
I. Volume reading (+2,500 g)	_____.___ mL	_____ Pa	_____.___ mL
J. Volume reading (+3,000 g)	_____.___ mL	_____ Pa	_____.___ mL
K. Volume reading (+3,500 g)	_____.___ mL	_____ Pa	_____.___ mL
L. Volume reading (+4,000 g)	_____.___ mL	_____ Pa	_____.___ mL
M. Volume reading (+4,500 g)	_____.___ mL	_____ Pa	_____.___ mL
N. Volume reading (+5,000 g)	_____.___ mL	_____ Pa	_____.___ mL

DISPOSAL: No waste products are produced in this
laboratory session. Save the lead shot for future use.

REVIEW QUESTIONS

Q1: Graph pressure (x-axis) versus volume (y-axis) using your observed and calculated (predicted) volumes. What general form do the graphs take (linear, exponential, etc.)? How closely do your observed values correspond to your calculated values?

Q2: Calculate percent error for your observed data. What experimental errors might explain the differences between the observed and calculated values?

Q3: Could you use this apparatus to determine the mass of an unknown sample? If so, how would you go about determining the mass?

Q4: The initial pressure in a Boyle's Law apparatus is known to be 29,084 Pa and the initial volume 10.0 mL. Assuming a perfect apparatus (no friction, etc.), how much total pressure must be applied to reduce the volume to 5.0 mL?

LABORATORY 14.2:

OBSERVE THE VOLUME-TEMPERATURE RELATIONSHIP OF GASES
(Charles' Law)

Charles' Law states that, at constant pressure, the volume of a gas is proportional to its absolute temperature, specified in kelvins. According to Charles' Law, then, doubling the temperature of a gas—say, from 200 K to 400 K—doubles its volume, and halving the temperature halves the volume. This proportionality holds true regardless of the percentage change. For example, if we increase the temperature of a 7.5 mL gas sample from 293.15 K (20.00°C) to 373.15 K (100.00°C), we can calculate the volume at the higher temperature by substituting the known values in the equation for Charles' Law:

$$V_1 \cdot T_2 = V_2 \cdot T_1$$

or

$$(7.5 \text{ mL}) \cdot (373.15 \text{ K}) = (x \text{ mL}) \cdot (293.15 \text{ K})$$

Solving for x, we find that the gas volume at the higher temperature is about 9.5 mL.

Similarly, if we decrease the temperature of a 7.5 mL gas sample from 293.15 K (20.00°C) to 194.65 K (−78.50°C), we can calculate the volume at the lower temperature, again by substituting the known values in the Charles' Law equation:

$$(7.5 \text{ mL}) \cdot (194.65 \text{ K}) = (x \text{ mL}) \cdot (293.15 \text{ K})$$

Again solving for x, we find that the gas volume at the lower temperature is about 5.0 mL.

I didn't choose that volume and those temperatures arbitrarily. The syringe I used in my apparatus has a full-scale reading of 10.0 mL, so an initial volume of 7.5 mL at 20°C (about room temperature) allowed me to get near (but not exceed) the full-scale 10.0 mL graduation when heating the syringe to 100°C. That's the approximate temperature of boiling water, which was the highest temperature I was comfortable using for this experiment. I could have substituted vegetable oil or another higher-boiling liquid and gotten up to 150°C—the upper limit of my thermometer's scale—but doing that would add little to the

REQUIRED EQUIPMENT AND SUPPLIES

☐ goggles, gloves, and protective clothing

☐ thermometer

☐ beaker, 150 mL or larger (2)

☐ ring stand

☐ burette or utility clamp (to fit syringe)

☐ 4" (100 mm) support ring

☐ wire gauze

☐ alcohol lamp, gas burner, or other heat source

☐ plastic syringe, 10 mL to 50 mL, graduated, with cap

☐ mineral oil or petroleum jelly (1 drop)

☐ ethanol, isopropanol, or acetone (sufficient to fill beaker; needed only if you use dry ice)

☐ ice

☐ dry ice (optional, see Substitutions and Modifications)

WHAT'S A KELVIN?

Note that the SI unit of temperature is not "degrees kelvin" or "kelvin degrees" but just "kelvins" all by itself. The Kelvin temperature scale uses the same incremental units of temperature as the familiar Celsius scale. In other words, increasing or decreasing the temperature by one degree Celsius (1°C) also increases or reduces the temperature by one kelvin (1 K).

The difference between the Kelvin scale and the Celsius scale is the baselines. The Kelvin scale assigns a temperature of 0 K to absolute zero, which is the coldest possible temperature, where even atomic vibrations cease. The Celsius scale assigns a temperature of 0°C to the melting point of pure water, which can also be specified as 273.15 K. It follows, therefore, that on the Celsius scale, absolute zero is −273.15°C.

SUBSTITUTIONS AND MODIFICATIONS

- If you use dry ice, the thermometer needs to be accurate down to at least −80°C. If you use a standard freezer as your cold bath, the thermometer needs to be accurate down to −20°C or so. Typical inexpensive digital thermometers are accurate down to −40 or −50°C, and typical inexpensive glass thermometers are accurate down to −20°C.

- If you do not have a ring stand, clamp, and so on, you may substitute beaker tongs and a large steel bolt or other weight. Tie the bolt securely to the body of the syringe (not the plunger) and use the weight to keep the syringe fully immersed in the hot and cold baths. Use the beaker tongs to add and remove the syringe from the baths.

value of the experiment, increase the danger level, and use up a lot of vegetable oil for no good reason.

I wanted the temperature of my cold bath as low as possible. In my first pass, I put the apparatus in a full-size freezer, which got the temperature down to about −20°C. I was able to do a bit better than that, though. As I was working on this chapter, I happened to receive a FedEx shipment of food packed in dry ice. Putting chunks of dry ice in a beaker full of alcohol allowed us to reach a temperature considerably colder than our freezer. You can purchase small amounts of dry ice locally. Check the Yellow Pages for "dry ice."

In this laboratory session, we'll use an apparatus similar to the one we used in the preceding lab. The difference is that in the previous lab we held temperature constant and changed the pressure, and in this lab we hold pressure constant and change the temperature. Thus, we won't need the container and the masses we used in the preceding lab. Instead, we'll record the volume contained by the syringe at room temperature, and then immerse the syringe in liquids at various temperatures and record the changes in volume.

CAUTIONS

Use extreme care with dry ice. At atmospheric pressure, dry ice sublimates (changes directly from a solid to a gas) at −78.5°C, cold enough to cause severe frostbite almost instantly if it contacts your skin. Handle dry ice only with tongs. Make sure the alcohol lamp or gas burner is widely separated from the beaker of alcohol—across the room is best. Wear splash goggles, heavy-duty gloves, and protective clothing.

PROCEDURE

1. If you have not already done so, put on your splash goggles, gloves, and protective clothing.
2. Spread a tiny amount (~1 drop) of mineral oil or petroleum jelly on the seal at the bottom of the plunger. Use just enough to lubricate the seal so that it slides freely within the body of the syringe.
3. With the cap removed, slide the plunger in and out of the syringe several times to distribute the lubricant. There should be little discernible resistance when moving the plunger.
4. Adjust the plunger to about 75% of full scale. For example, for my 10 mL syringe, I adjusted the plunger to 7.5 mL. Once the plunger is in the correct position, reinstall the cap on the syringe, and record the initial volume on line A of Table 14-2.
5. Fill the beaker about 90% full of ethanol, isopropanol, or acetone and add several chunks of dry ice. Insert the thermometer and observe the temperature until it stabilizes at its lowest point. (Continue adding chunks of dry ice as necessary, as the dry ice is consumed.) Once the temperature has stabilized, immerse the syringe, making sure that the portion filled with gas is below the surface of the alcohol.
6. Allow the syringe to remain in the dry ice bath for two to three minutes to ensure that the gas has equilibrated at the temperature of the bath, and then withdraw the syringe and note the volume indicated. (Bump the plunger once or twice to ensure that it is actually at equilibrium position.) Record the temperature of the bath and the volume of the syringe on line B of Table 14-2.

FIGURE 14-3: *Measuring gas volume in a bath of alcohol and dry ice*

7. Remove the syringe from the dry ice bath and set it aside to return to room temperature. When it has done so, verify that the volume has returned to the original reading. (Press the plunger gently once or twice to ensure that it can move freely; the readings should be the same each time.)

8. Set the beaker aside with a watch glass or similar item loosely covering it. (Make sure that it is vented to allow gas to escape.) Allow the dry ice to be consumed and the alcohol to return to room temperature, which may require an hour or more. Once the alcohol is again at room temperature, you may return it to its container. The dry ice does not contaminate the alcohol any more than storing it in the freezer would. Allowing it to return to room temperature also allows dissolved carbon dioxide gas to dissipate.

9. Fill the second beaker with a mixture of tap water and ordinary ice, and allow the temperature to stabilize at 0°C. Make sure that some ice remains unmelted in the beaker.

10. Verify that the syringe is at the original volume—75% of full scale, or whatever you used initially—and then immerse the syringe in the ice-water bath, making sure that the portion filled with gas is below the surface of the ice water.

11. Allow the syringe to remain in the ice-water bath for two to three minutes to ensure that the gas has equilibrated at the temperature of the bath, and then withdraw the syringe and note the volume indicated. (Bump the plunger.) Record the temperature of the bath and the volume of the syringe on line C of Table 14-2.

12. Remove the syringe from the ice-water bath and set it aside to return to room temperature. When it has done so, verify that the volume has returned to the original reading. (Bump the plunger.)

13. Repeat steps 9 through 12 using baths of hot tap water and boiling water. Record the temperatures and volumes on lines D and E of Table 14-2, respectively.

14. Calculate the expected volume at each temperature, and enter the results on lines B through E in the right column of Table 14-2.

A DRY ICE EXPLOSION

Dr. Paul Jones, one of our technical advisors, suggests an amusing use for any dry ice you have left over when you finish this lab session. Place a chunk of dry ice in a disposable latex glove and knot the top of the glove. As the dry ice warms, it sublimates from solid to a gas and inflates the rubber glove. Eventually, the glove can stretch no farther, and it explodes with a loud bang. Our dogs may never forgive me.

Just don't use anything more robust than a thin latex glove. Confining dry ice in a glass jar, plastic soda bottle, or a similar container can have tragic results. When the container inevitably shatters, shards of glass or plastic moving at high velocity will puncture anything or anyone in the area. People have been blinded and even killed by doing stupid things with dry ice.

DISPOSAL:
No waste products are produced in this laboratory session. Save the alcohol for use in your alcohol lamp or similar noncritical purposes.

TABLE 14-2: *Observe the volume-temperature relationship of gases—observed and calculated data*

Trial	Temperature	Actual volume	Calculated volume
A. Room temperature	_____.__ K	____.____ mL	n/a
B. Dry ice in alcohol	_____.__ K	____.____ mL	____.____ mL
C. Ice water	_____.__ K	____.____ mL	____.____ mL
D. Hot tap water	_____.__ K	____.____ mL	____.____ mL
E. Boiling water	_____.__ K	____.____ mL	____.____ mL

REVIEW QUESTIONS

Q1: Graph kelvin temperature (x-axis) versus volume (y-axis) using your observed and calculated (predicted) volumes? What general form do the graphs take (linear, exponential, etc.)? How closely do your observed values correspond to your calculated values?

_____

_____

_____

_____

_____

_____

_____

Q2: What volume would your gas sample have if you were able to cool it to absolute zero? Extrapolate your experimental data to estimate the kelvin temperature of absolute zero. How closely does your extrapolation correspond to the actual temperature of absolute zero (0 K)?

_____

_____

_____

_____

_____

_____

Q3: A sample of gas has a volume of 10.00 mL at 26.85°C. With pressure held constant, what volume would that sample occupy at −173.15°C? At 126.85°C?

_____

_____

_____

Q4: A sample of gas has a volume of 10.00 mL at 26.85°C. With pressure held constant, at what temperature would that sample occupy 15.00 mL?

_____

_____

_____

_____

LABORATORY 14.3:

OBSERVE THE PRESSURE-TEMPERATURE RELATIONSHIP OF GASES (Gay-Lussac's Law)

Gay-Lussac's Law states that, at constant volume, the pressure of a gas is proportional to its absolute temperature, specified in kelvins. For example, if you double the temperature of a gas, you double its pressure, and vice versa. Gay-Lussac's Law can be expressed as the equation:

$$P_1 \cdot T_2 = P_2 \cdot T_1$$

For example, if we increase the temperature of a 7.5 mL gas sample at atmospheric pressure from 293.15 K (20.00°C) to 373.15 K (100.00°C), we can calculate the pressure at the higher temperature by substituting the known values in the Gay-Lussac's Law equation:

$$(101{,}325 \text{ Pa}) \cdot (373.15 \text{ K}) = (x \text{ Pa}) \cdot (293.15 \text{ K})$$

Solving for x, we find that the gas pressure at the higher temperature is about 128,976 Pa.

In this lab, we'll verify Gay-Lussac's Law experimentally by using the apparatus shown in Figure 14-4, along with some of the measurements and data we recorded in the first lab session in this chapter. We'll start with a known volume of air in the syringe, with the syringe in a water bath at room temperature. We'll then heat the water bath to boiling, which will cause the volume of air in the syringe to increase. With the gas sample held at constant temperature in the boiling water bath, we'll add mass to the container until the syringe is depressed to its original 7.5 mL volume reading. Knowing the mass required to compress the hotter gas sample to its original volume, we can calculate the pressure of the 7.5 mL gas sample at that higher temperature.

SUBSTITUTIONS AND MODIFICATIONS

- If you are using the same syringe, container, and other components you used in Laboratory 14.1, you may use the measurements and calculations you did for that laboratory session rather than repeating them here.
- Other substitutions and modifications are as listed in Laboratory 14.1.

REQUIRED EQUIPMENT AND SUPPLIES

- ☐ goggles, gloves, and protective clothing
- ☐ balance and weighing cup
- ☐ caliper
- ☐ barometer (optional)
- ☐ ring stand
- ☐ burette or utility clamp (to fit syringe)
- ☐ 4" (100 mm) support ring (2)
- ☐ wire gauze
- ☐ alcohol lamp, gas burner, or other heat source
- ☐ beaker, 150 mL
- ☐ plastic syringe, 10 mL to 50 mL, graduated, with cap
- ☐ mineral oil or petroleum jelly (1 drop)
- ☐ plastic cup (to fit support ring)
- ☐ lead shot (10 pounds or 5 kilograms)

CAUTIONS

The real hazard in this lab is that the apparatus may topple or collapse as you add too much mass to the container, splashing boiling water everywhere. Take extreme care with the hot water, make sure that the container remains centered over the syringe plunger as you add mass, and use the smallest beaker that allows the gas-filled portion of the syringe to be fully immersed. If you have a third support ring, use two rings to surround the mass container to prevent it from tipping. Wear splash goggles, gloves, and protective clothing.

FIGURE 14-4: *Our apparatus for verifying Gay-Lussac's Law*

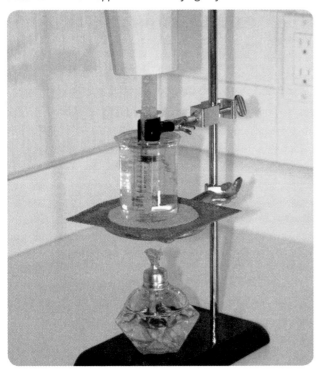

PROCEDURE

1. If you have not already done so, put on your splash goggles, gloves, and protective clothing.
2. Use the caliper to measure the inside diameter of the syringe bore in millimeters. Calculate the surface area of the syringe bore by dividing the diameter by 2 to determine the radius, squaring the radius, and multiplying by pi (π, 3.14159). Divide by 1,000,000 to convert the result from square millimeters to square meters and enter that value on line A of Table 14-3. (The number you obtain will be very small. For example, my syringe has an internal bore of 15.0 mm, which corresponds to an area of 0.000177 m^2.)
3. Weigh the container and syringe plunger (just the plunger—not the body of the syringe) together, and record their combined mass on line B of Table 14-3.
4. Spread a tiny amount (~1 drop) of mineral oil or petroleum jelly on the seal at the bottom of the plunger. Use just enough to lubricate the seal so that it slides freely within the body of the syringe.

5. With the cap removed, slide the plunger in and out of the syringe several times to distribute the lubricant. There should be little discernible resistance when moving the plunger.
6. Adjust the plunger to about 75% of the maximum capacity of the syringe. For example, with my 10 mL syringe, I set the initial volume to 7.5 mL. If the plunger seal has multiple wipes (as are visible in Figure 14-2), the bottom wipe is the one that matters. Once the plunger is in the correct position, reinstall the cap on the syringe.
7. Clamp the syringe as shown in Figure 14-4. Make sure that the cap of the syringe is pressing directly against the bottom of the beaker. Otherwise, when you add mass to the container, the cap may pop off the syringe.
8. Obtain the atmospheric pressure using a barometer or from a local radio or TV broadcast. Convert that value to Pa and enter it on line C of Table 14-3.
9. Verify the initial volume reading on the syringe, and record that value as accurately as possible on line D of Table 14-3. (Your syringe may use cc graduations, which are the same as mL graduations for our purposes.)
10. Fill the beaker with tap water at room temperature, making sure that the gas-filled portion of the syringe body is as far below the water line as possible. Record the temperature on line E of Table 14-3.
11. Bring the water to a boil, and allow it to continue boiling for at least two or three minutes to give the gas in the syringe a chance to equilibrate. Record the temperature on line F of Table 14-3. (Do not assume 100°C; the actual boiling point of water varies with atmospheric pressure.)
12. Carefully add lead shot to the container while watching the volume reading on the syringe. When the volume approach the initial volume, "bump" the plunger carefully by raising the container once or twice and then allowing it to come to rest on the plunger again. Continue adding lead shot until the volume is as close as possible to the original volume.
13. Remove heat from the beaker and allow it to cool. While it cools, weigh the lead shot in the container and record its mass on line G of Table 14-3.
14. Using the combined masses of the empty container, plunger, and lead shot, calculate the pressure as described in Laboratory 14.1. Enter that value on line H of Table 14-3.

DISPOSAL: No waste products are produced in this laboratory session. Save the lead shot for future use.

TABLE 14-3:

Observe the volume-pressure relationship of gases—observed and calculated data.

Item	Value
A. Area of syringe bore	_____ m^2
B. Mass of empty container and plunger	_____.__ g
C. Initial (atmospheric) pressure	_____ Pa
D. Initial volume reading	____.____ mL
E. Initial temperature reading	_____.__ °C
F. Final temperature reading	_____.__ °C
G. Mass of lead shot	_____.__ g
H. Final pressure	._____ Pa

REVIEW QUESTIONS

Q1: A gas sample at a pressure of 101,325 Pa and 26.85°C has a volume of 10.00 mL. At constant volume, at what pressure would that gas sample have a temperature of 126.85°C?

Q2: A sample of gas at a pressure of 101,325 Pa and 26.85°C has a volume of 10.00 mL. If the temperature is increased to 126.85°C and the volume decreased to 5.00 mL, what is the pressure of the gas sample?

LABORATORY 14.4:
USE THE IDEAL GAS LAW TO DETERMINE THE PERCENTAGE OF ACETIC ACID IN VINEGAR

The Ideal Gas Law states that $PV = nRT$, where P is absolute pressure, V is volume, n is the number of moles, R is the ideal gas constant, and T is the temperature in kelvins. If the values for any four of these are known, the fifth can be determined by solving algebraically. The value of R, the ideal gas constant, depends on the units used for volume and pressure. In SI units, the value of R is 8.314472 joules per mole per kelvin ($J \cdot mol^{-1} \cdot K^{-1}$), which can also be stated as $8.314472 \ L \cdot kPa \cdot mol^{-1} \cdot K^{-1}$.

In this lab, we'll use the ideal gas constant to determine the actual percentage of acetic acid in distilled white vinegar. Most distilled white vinegars are nominally 5% acetic acid by weight, but industry standards permit that percentage to vary from 4.5% to 5.5%. To determine the actual mass percentage of acetic acid in our sample, we'll react a known mass of vinegar with an excess of sodium hydrogen carbonate (sodium bicarbonate or baking soda) to produce water, sodium acetate, and carbon dioxide gas by the following balanced equation:

$$CH_3COOH(aq) + NaHCO3(aq)$$
$$\rightarrow H_2O(l) + CH_3COONa(aq) + CO_2(g)$$

Stoichiometrically, one mole of acetic acid reacts with one mole of sodium hydrogen carbonate to produce one mole each of water, sodium acetate, and carbon dioxide. Because the value for the ideal gas constant is known and the pressure and temperature can be determined experimentally, if we capture the carbon dioxide gas evolved by this reaction and measure its volume, we can apply the Ideal Gas Law to determine the number of moles of carbon dioxide produced and therefore the number of moles of acetic acid that were present in the vinegar sample. With that value, the mass percentage of acetic acid in the original sample can be derived by a simple calculation.

Although you can purchase a gas-generating bottle, gas-collection trough, and gas-collection bottle from a laboratory supplies vendor, there's no need to do so. I built my own gas-generating bottle with a 250 mL Erlenmeyer flask, a one-hole rubber stopper, and a short length of glass tubing. I used one of our large plastic storage tubs that is ordinarily used for chemical

storage for the gas-collection trough and an empty 2-liter soft drink bottle for the gas-collection bottle. Figure 14-5 shows my setup.

REQUIRED EQUIPMENT AND SUPPLIES

☐ goggles, gloves, and protective clothing

☐ balance

☐ paper towel (quarter sheet)

☐ barometer (optional)

☐ thermometer

☐ graduated cylinder, 100 mL

☐ funnel

☐ Erlenmeyer flask, 250 mL

☐ stopper, 1-hole (to fit flask)

☐ glass tube, 75 mm (to fit stopper)

☐ plastic/rubber tubing (about 500 mm; to fit glass tube)

☐ soft drink bottle, 2-liter, with cap

☐ large plastic storage tub (Rubbermaid "S" 17-liter or similar)

☐ vinegar (75.0 g)

☐ sodium hydrogen carbonate (7.5 g)

SUBSTITUTIONS AND MODIFICATIONS

• If you do not have a barometer, you may use the current barometric pressure broadcast by a local TV or radio station, but see the note on atmospheric pressure versus barometric pressure in Laboratory 14.1.

• You may use a commercial gas-generation bottle, gas-collection trough, and gas-collection bottle instead of the apparatus shown in Figure 14-5.

FIGURE 14-5: *My apparatus for gas generation and collection*

My first thought was to use an inverted 100 mL graduated cylinder as the gas collection vessel, which would have made it easy to measure directly the volume of gas produced. A few quick calculations convinced me to use a larger gas collection vessel. I reasoned as follows:

1. At standard temperature and pressure (STP), one mole of gas occupies about 22.4 liters or 22,400 mL.
2. At full capacity, our 100 mL graduated cylinder could therefore contain about [100 mL]/[22,400 mL/mol] moles of gas, or about 0.0045 moles.
3. The gram-molecular mass of acetic acid is 60.05 g/mol, so 0.0045 mol would be about 0.27 g of acetic acid.
4. If the vinegar has a mass percentage of 5%, 0.27 g of acetic acid would be contained in about 5.36 g of vinegar.
5. In order to make sure I didn't overshoot the capacity of my graduated cylinder, I should reduce that amount by 25% or so, say to 4.00 g.
6. Using such small masses of reactants might well introduce significant experimental error in weighing and determining volume.

So I decided to use a larger gas-collection vessel. A 2-liter soft drink bottle seemed perfect for the job. It has more than 20 times the capacity of our 100 mL graduated cylinder, is easy to manipulate, and comes with a screw cap that makes it easy to seal. (The reason for that will become obvious shortly.) The larger size means we can use much larger amounts of the reactants, which minimizes measuring errors.

PROCEDURE

1. If you have not already done so, put on your splash goggles, gloves, and protective clothing.
2. Assemble the gas-generating bottle, as shown in Figure 14-5. (Be careful when you insert the glass tubing into the stopper; lubricate it with glycerol or another lubricant,

CAUTIONS

The only real hazard in this lab is making a mess. (Dr. Mary Chervenak says, "Yay!") Wear splash goggles, gloves, and protective clothing.

and turn the tubing constantly as you slide it into the hole. Use a heavy towel or similar item to protect your fingers if the glass breaks.)

3. Determine the volume of the soft drink bottle by filling the 100 mL graduated cylinder repeatedly with water and emptying it into the bottle until the bottle is full to the brim. (I cheated by using a 500 mL volumetric flask to put 2,000 mL of water into the bottle and then topping off the bottle with the 100 mL graduated cylinder. My soft drink bottle held 2,128.4 mL when filled to the brim.) Record the volume in liters as accurately as possible on line A of Table 14-4.
4. Without spilling any of the water from the bottle, replace the cap and set the bottle aside.
5. Place the 100 mL graduated cylinder on your balance, tare the balance, and weigh 75.00 g of vinegar. Record the mass of the vinegar to 0.01 g on line B of Table 14-4.
6. Transfer the vinegar to the gas-generating bottle. To make sure you've done a quantitative transfer, rinse the graduated cylinder two or three times with several mL of water and transfer the rinse water to the gas-generating bottle.
7. Place a quarter-sheet of paper towel on the balance, tare the balance, and weigh out about 7.5 g of sodium hydrogen carbonate. The exact amount is not critical, as long as the sodium hydrogen carbonate is stoichiometrically in excess relative to the amount of acetic acid.
8. Center the sodium hydrogen carbonate on the paper towel, and roll up the paper towel such that several layers of paper towel cover the chemical. Secure the package with a rubber band or small piece of adhesive tape.
9. Place the soft drink bottle inverted in the plastic tub, which should be about 80% full of water at room temperature. With the mouth of the bottle submerged, remove the cap from the bottle and lean the bottle against the corner of the tub. (If the bottle won't stay put, you can either hold it in place or use a mouse pad or similar nonslip item to prevent the mouth of the bottle from sliding on the bottom of the tub. Alternatively, put a brick in the tub and use it to keep the bottle upright.)
10. Insert the free end of the plastic tubing well up into the bottle.

11. With the stopper in one hand, drop the package of sodium hydrogen carbonate into the gas-generating bottle and immediately insert the stopper securely.

12. It takes a moment for the vinegar to penetrate the paper towel package. As soon as the vinegar penetrates the package, the reaction commences and vigorous fizzing begins as carbon dioxide is generated. The increased pressure in the flask forces carbon dioxide through the plastic tubing and into the gas-collection bottle, where it begins to displace the water in the bottle.

13. Hold the gas-collection bottle to make sure that it doesn't tip over as it fills with gas. When the reaction ceases and bubbles are no longer being produced, remove the free end of the plastic tubing from the gas-collection bottle and recap the bottle while keeping its mouth under water.

14. Remove the gas-collection bottle from the tub and dry its exterior.

15. Determine the volume of water remaining in the bottle by repeatedly filling the 100 mL graduated cylinder. Record the total volume of the water remaining in the gas-collection bottle in liters on line C of Table 14-4.

16. Determine the volume of gas produced by subtracting the volume of water remaining (line C) from the total volume of the gas-collection bottle (line A). Record the volume of gas produced on line D of Table 14-4.

17. Determine the local atmospheric pressure (not barometric pressure) in kilopascals (kPa) and enter that value on line E of Table 14-4.

18. Use the thermometer to determine room temperature in kelvins, and enter that value on line F of Table 14-4.

19. Use the ideal gas law, $P \cdot V = n \cdot R \cdot T$ to calculate n, the number of moles of gas produced by the reaction (which equals the number of moles of acetic acid consumed). We have already determined pressure (in kPa), volume (in L), and temperature (in K). The ideal gas constant, R, for these units is $8.314472 \ L \cdot kPa \cdot mol^{-1} \cdot K^{-1}$. Solve for n, the number of moles, and enter that value on line G of Table 14-4.

20. The gram-molecular mass of acetic acid is 60.05 g/mol. Determine the reactant mass of acetic acid by multiplying the number of moles (line G) by 60.05 g/mol, and enter that value on line H of Table 14-4.

21. Calculate the mass-percentage of acetic acid in the vinegar by dividing the mass of acetic acid (line H) by the mass of the vinegar sample (line B) and multiplying that result by 100 to convert it to a percentage. Enter that value on line I of Table 14-4.

TABLE 14-4: *Use the Ideal Gas Law to determine the percentage of acetic acid in vinegar—observed and calculated data*

Item	Value
A. Volume of gas-collection bottle	___._____ L
B. Mass of vinegar	_____.____ g
C. Volume of water remaining	___._____ L
D. Volume of gas produced (A − C)	___._____ L
E. Atmospheric pressure	_____.____ kPa
F. Room temperature	_____.___ K
G. Moles of gas (and acetic acid)	___._____ mol
H. Mass of acetic acid (G · 60.05 g/mol)	_____.____ g
I. Mass percentage of acetic acid (H · 100/B)	_____.____ %

DISPOSAL:
The gas-generating bottle contains only a dilute solution of sodium acetate and sodium hydrogen carbonate, and may be flushed down the drain with plenty of water.

OPTIONAL ACTIVITIES

If you have time and the required materials, consider performing these optional activities:

- Verify your experimental results for this lab by standardizing the acetic acid concentration in an aliquot of vinegar via titration, as described in Laboratory 11.4.

- There are two possibly significant sources of experimental error that we ignored for this lab. Carbon dioxide is soluble in water, which reduces the amount of gas collected. Offsetting this, as the carbon dioxide bubbles through the water in the gas-collection trough, it absorbs water in the form of vapor, which increases the amount of gas collected. Use a standard reference source to look up the solubility of carbon dioxide in water and the vapor pressure of water (both are temperature-sensitive) and decide how you might take these variables into account.

REVIEW QUESTIONS

Q1: If a reaction between acetic acid and sodium hydrogen carbonate produces 1.585 liters of carbon dioxide gas at a temperature of 297.0 K and a pressure of 100.000 kPa, how many moles of carbon dioxide were produced?

Q2: Use the Ideal Gas Law to determine how many liters one mole of gas occupies at a temperature of 273.15 K and a pressure of 101.325 kPa.

LABORATORY 14.5:
DETERMINE MOLAR MASS FROM VAPOR DENSITY

According to the Ideal Gas Law, PV = nRT. Rearranging this equation to put n, the number of moles, on one side gives us:

$$n = (PV)/(RT)$$

For a sample of a gas in a container, R (the ideal gas constant) is known, and the pressure (P), volume (V), and temperature (T) are easy to determine experimentally. With these four values known, determining n, the number of moles of gas in the container, is a simple calculation. Because the number of moles in the sample (n) equals the mass of the sample divided by the molar mass of the substance, the only additional datum we need to calculate the molar mass of the substance is the mass of the sample.

In 1826, the French chemist Jean Baptiste André Dumas developed a method and an apparatus for determining molar mass from vapor density. The original apparatus from his illustration in his 1826 report is shown in Figure 14-6.

FIGURE 14-6:
Apparatus used by Dumas in 1826 for determining molar mass from vapor density

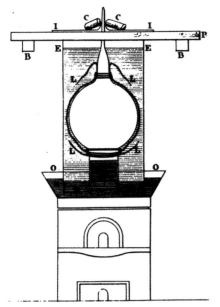

REQUIRED EQUIPMENT AND SUPPLIES

☐ goggles, gloves, and protective clothing

☐ balance

☐ barometer (optional)

☐ thermometer

☐ graduated cylinder, 100 mL

☐ Erlenmeyer flask, 250 mL or larger (see Substitutions and Modifications)

☐ beaker, 250 mL

☐ ring stand

☐ support clamp (for flask)

☐ warm water bath (see Substitutions and Modifications)

☐ pin or needle

☐ aluminum foil

☐ acetone (~ 5 mL)

SUBSTITUTIONS AND MODIFICATIONS

- If you do not have a barometer, you may use the current barometric pressure broadcast by a local TV or radio station, but see the note on atmospheric pressure versus barometric pressure in Laboratory 14.1.

- Use the largest Erlenmeyer flask you have that weighs less than the maximum capacity of your balance. If you use a larger flask, increase the amount of acetone proportionally. For example, with a 500 mL flask, use 10 mL of acetone.

- This experiment requires slowly increasing the temperature of a small amount of acetone contained in a large Erlenmeyer flask until the acetone boils. The easiest and most consistent way to do this is to use an old kitchen pot or similar container as a warm water bath. At standard pressure, acetone boils at about 56.5°C. Suspend the flask inside the water bath container, and add sufficient water at about 50°C to immerse as much as possible of the flask under the surface of the water. Allow a few minutes for the flask and its contents to stabilize at the temperature of the water bath, and then begin adding small amounts of boiling water with stirring until the acetone in the flask begins to boil.

FIGURE 14-7:

My apparatus for determining molar mass from vapor density

We use a simpler apparatus, shown in Figure 14-7, to determine the molar mass of acetone by the Dumas method. First, we accurately determine the mass of an empty Erlenmeyer flask and a piece of aluminum foil used to stopper it. We then introduce a small amount of liquid acetone and immerse the flask in a warm water bath to boil the acetone, converting it to vapor. As the acetone vaporizes, it displaces the air in the flask through a small pinhole in the aluminum foil. When all of the acetone has vaporized, we record the temperature (T), and then cool the flask, causing the vapor to condense to liquid acetone. We reweigh the flask to determine by difference the mass of acetone vapor it contained. We then determine the volume (V) of the flask, and the atmospheric pressure (P). Using those data, we calculate the molar mass of acetone.

PROCEDURE

1. If you have not already done so, put on your splash goggles, gloves, and protective clothing.
2. Loosely crimp a piece of aluminum foil around the mouth of the flask.
3. Weigh the flask and aluminum foil, and record the mass to 0.01 g on line A of Table 14-5.
4. Transfer about 5 mL of acetone to the flask. The exact amount is not critical.
5. Crimp the aluminum foil tightly around the neck of the flask, covering the mouth, and use the pin or needle to poke one tiny hole in the center of the foil.
6. Assemble the vapor density apparatus, as shown in Figure 14-6.
7. Immerse the flask in the water bath, with the flask tilted slightly to make the liquid more visible, and *gradually*

CAUTIONS

Although only a small amount is used in this lab, acetone is extremely flammable. Be careful not to expose acetone liquid or vapor to an open flame. Use an exhaust hood, or work outdoors or in a well-ventilated area. Water at 50°C is hot enough to burn you badly. Wear splash goggles, gloves, and protective clothing.

heat the flask until the liquid acetone boils. As the acetone boils, acetone vapor displaces the air in the flask.

8. Just as the final drop of acetone boils away, note the temperature of the water bath (which, if you've been heating it very gradually, is the same as the temperature inside the flask). Record the temperature in kelvins on line B of Table 14-5.
9. Immediately remove the flask from the water bath, and run cold water over it. As the flask cools, the acetone vapor condenses inside the flask.
10. Carefully dry the outside of the flask with a paper towel.
11. Reweigh the flask, aluminum foil, and condensed acetone, and record that mass to 0.01 g on line C of Table 14-5.
12. Remove the aluminum foil from the flask, pour out as much as possible of the condensed acetone, and then replace the flask in the warm water bath. Allow the flask to warm for a minute or two while you complete the following calculations.
13. Calculate the mass of condensed acetone, and enter that value on line D of Table 14-5.
14. Calculate the number of moles of condensed acetone, and enter that value on line E of Table 14-5.
15. Remove the flask from the warm water bath and make sure that no liquid remains in the flask. Use the graduated cylinder to fill the flask with water, noting the total value of water held by the flask when it is full to the brim. Record that volume as accurately as possible on line F of Table 14-5.
16. Use the barometer (or a local weather report) to determine the atmospheric pressure. (See the note in Lab 14.1 about atmospheric pressure versus barometric pressure.) Record the atmospheric pressure on line G of Table 14-5.
17. With the temperature, mass of acetone, volume, and pressure known, you have all the information you need to calculate the molar mass of acetone. Make that calculation, and enter the value on line H of Table 14-5.

TABLE 14-5: *Determine molar mass from vapor density—observed and calculated data*

Item	Value
A. Mass of flask + aluminum foil	_____.____ g
B. Temperature of water bath	_____.____ K
C. Mass of flask + aluminum foil + condensed acetone	_____.____ g
D. Mass of condensed acetone (C – A)	_____.____ g
E. Moles of condensed acetone (D/58.09 g/mol)	___._____ mol
F. Volume of flask	_____.____ mL
G. Atmospheric pressure	_____.____ kPa
H. Molar mass of acetone	_____.____ g/mol

> DISPOSAL:
> No waste is produced in
> this lab session.

REVIEW QUESTIONS

Q1: The actual molar mass of acetone is 58.09 g/mol. How closely did your experimental result correspond to the actual value? Propose at least two possible explanations for any variation in your result.

_____

_____

_____

_____

_____

Q2: Could this method be used unmodified to determine the molar mass of trichloroacetic acid? If not, why not, and what modification to the apparatus would be required to determine the molar mass of trichloroacetic acid?

_____

_____

_____

_____

_____

_____

_____

Q3: At 25°C, the vapor pressure of acetone is about 26.67 kilopascals. Can this information be used to increase the accuracy of your calculated molar mass of acetone? If so, how?

Q4: What modifications could you make to this procedure to reduce experimental error?

Q5: Propose a method using only the standard laboratory equipment listed in Chapter 3 for determining the molar mass of carbon dioxide gas. What change to your proposed method would you make if you had a vacuum pump available?

Q6: If your sample had been unknown, would you have been able to identify the substance from the molar mass you determined experimentally? What other datum you obtained could be used to confirm the identity of the sample?

Laboratory:
Thermochemistry and Calorimetry

15

Thermochemistry is the study of the heat evolved or absorbed during chemical reactions, physical transformations (such as mixing), and phase changes (such as melting). *Calorimetry*, from the Latin words *calor* for heat and *metrum* for measure, is a technique used to measure the heat changes that occur in such processes.

All of the laboratories in this chapter require a *calorimeter*, a device that thermally isolates the process being investigated, preventing gain or loss of heat from the external environment. We used an inexpensive purpose-built calorimeter, which is available from Home Science Tools (*http://www.homesciencetools.com*) and similar vendors for $15 or so. If you don't want to buy a formal calorimeter, you can achieve results nearly as accurate with a home-built calorimeter. Both types of calorimeter are shown in Figure 15-1.

FIGURE 15-1: *Store-bought (left) and homemade calorimeters*

HEAT VERSUS TEMPERATURE

Heat and temperature are related but not identical concepts. A sample may contain a small amount of heat, but the temperature may be relatively high, or vice versa. For example, the sparks coming off a grinding wheel are tiny particles of burning metal. Their temperature is very high—perhaps 1,000°C or higher—but the particles contain little heat, because the mass of each is a fraction of a milligram. If one of those particles were to contact your skin, you might not even notice it. Conversely, a liter of boiling water is at a relatively low temperature, only 100°C, but the kilogram mass of water stores a considerable amount of heat. If that water comes into contact with your skin, you will be burned badly.

All you need for a homemade calorimeter are a couple of 8- or 16-ounce (250 or 500 mL) foam cups, a lid, and a beaker or similar heavy container to provide stability. Place one cup inside the other to provide better insulation. Insert your thermometer or temperature probe through the center hole in the lid, and use it both for measuring temperatures and as a stirring rod. You can significantly improve the accuracy of your homemade calorimeter by replacing the snap-on plastic lid with a lid constructed of foam. Using the top (smallest diameter part) of the plastic lid as a template, mark and cut a circular piece of foam just large enough to fit inside the mouth of the cup. Using the top of the cup as a template, mark and cut a second circle of foam just a bit too large to fit inside the mouth of the cup. Center the smaller circle of foam on the larger one, and glue the two pieces together. Poke a small hole through the center of the new lid assembly to allow the insertion of a stirring rod or thermometer.

Measuring temperature changes accurately and precisely is the foundation of thermochemistry. We used an inexpensive digital temperature probe (thermometer) to measure temperatures to 0.1°C. You can substitute a standard glass-tube thermometer with some loss of accuracy and precision. If you do use a glass-tube thermometer, try to interpolate your temperature measurements by estimating values between the markings. With a standard 300 mm glass-tube lab thermometer, it's usually possible to interpolate values accurate to 0.2°C. With longer thermometers, or those with narrower ranges, it may be possible to interpolate to 0.1°C.

In this chapter, we'll use calorimetry to examine several aspects of thermochemistry.

EVERYDAY THERMOCHEMISTRY AND CALORIMETRY

- The chemical cold packs sold in drugstores illustrate heat of solution (as defined on the next page). These packs have two compartments. One contains water and the second a solid chemical (usually ammonium nitrate or sodium thiosulfate) with a positive (endothermic) heat of solution. When you break the divider between the compartments, the solid chemical dissolves, absorbing heat from the environment.

- The reusable chemical heat packs sold in drugstores illustrate a different aspect of heat of solution. These packs contain a very concentrated solution of sodium acetate. Placing the pack in boiling water or heating it in a microwave oven for a few minutes dissolves the solid sodium acetate. When the pack has cooled to room temperature, it contains a supersaturated, supercooled solution of sodium acetate. When the pack is initiated, normally by bending a the small embedded metal disc that contains tiny slots with microscopic crystals of sodium acetic, the sodium acetate immediately begins recrystallizing, producing heat as it does so.

- If you've ever made ice cream in an old-fashioned ice cream maker, you've used a practical example of the heat of fusion of ice (as defined in Lab 15.2). Adding salt to water ice reduces the freezing point, converting some of the ice to liquid water. The heat of fusion required to melt the ice must come from somewhere, and the only available source of heat is the ice itself. As the salt dissolves, the temperature of the ice/salt bath decreases enough to freeze the milk and cream, producing ice cream.

LABORATORY 15.1:
DETERMINE HEAT OF SOLUTION

Heat of solution, also called *enthalpy change of solution*, is heat that is absorbed or released when a solute is dissolved in a solvent. Dissolution is a complex process that involves both absorption and release of energy. Energy is absorbed (*endothermic* energy) to break the attractions between the solute molecules and the attractions between the solvent molecules. Conversely, energy is released (*exothermic* energy) as new attractions form between solute molecules and solvent molecules. The net difference between the energy absorbed and the energy released per mole of solute is defined as the heat of solution for that solute.

For a particular solvent, some solutes have a positive heat of solution, because the energy that must be absorbed to break the solute-solute bonds and the solvent-solvent bonds is greater than the energy that is released by the formation of solute-solvent bonds. Dissolving such a solute reduces the temperature of the solution relative to the original temperature of the solvent. Other solutes have a negative heat of solution, because less energy is needed to break the solute-solute and solvent-solvent attractions than is released by the forming of solute-solvent bonds. (Remember, heat of solution is defined as endothermic energy minus exothermic energy, so dissolving a compound with negative heat of solution causes the temperature of the solution to increase.)

Heat of solution (represented as $\Delta H_{solution}$) is properly quantified using SI units of kJ/mol (kiloJoules/mole), but many older sources (and many older chemists . . .) continue to use the old-style C/mol (kilocalories/mole), where one kJoule equals 2.390 x 10^{-1} kilocalories (0.2390 C) and one kilocalorie (*kcal* or *C*) equals 4.1841 kJ. For any particular solute/solvent combination, a value for heat of solution can be determined experimentally from data derived by dissolving a known mass of the solute in a known mass of the solvent and measuring the temperature change. That's what we do in this lab.

REQUIRED EQUIPMENT AND SUPPLIES

- ☐ goggles, gloves, and protective clothing
- ☐ balance and weighing papers
- ☐ calorimeter
- ☐ thermometer
- ☐ graduated cylinder, 100 mL
- ☐ ammonium nitrate (40.0 g = 0.5 mol)
- ☐ sodium chloride (29.2 g = 0.5 mol)
- ☐ sodium hydroxide (20.0 g = 0.5 mol)
- ☐ water (at room temperature)

SUBSTITUTIONS AND MODIFICATIONS

- At the risk of some loss of accuracy, you can run this lab on a semi-micro scale by reducing the quantities of solutes and water by a factor of 10 and substituting a test tube for the calorimeter. (This is not ideal, because the glass of the test tube has much higher thermal mass than a foam cup and is a poor insulator.)

- You may substitute pure ammonium nitrate fertilizer for laboratory- or reagent-grade ammonium nitrate.

- You may substitute table salt for sodium chloride.

- You may substitute crystal drain opener that lists its contents as 100% sodium hydroxide for laboratory- or reagent-grade sodium hydroxide.

- If you are using a calorimeter in which the solutions will come into contact with aluminum or glass, substitute calcium chloride for the sodium hydroxide. In areas that have snowy winters, relatively pure calcium chloride is sometimes available as ice-melt crystals, although many different formulations are used, so it's necessary to check the label.

PROCEDURE

To achieve the highest accuracy, it's best to have the solutes, solvent (water), and calorimeter at room temperature. The temperature of cold tap water is usually lower than room temperature. If you have time to prepare beforehand, fill a two-liter soft drink bottle or similar container with tap water and allow it to sit for several hours to equilibrate to room temperature.

1. If you have not already done so, put on your splash goggles, gloves, and protective clothing.
2. Weigh 40.0 g of ammonium nitrate and record the mass in Table 15-1.
3. Use the graduated cylinder to measure 100.0 mL (100.0 g) of water.
4. Add the water to the calorimeter and replace the cover.
5. Measure the temperature of the water in the calorimeter as accurately as possible and record the temperature in Table 15-1. See Figure 15-2.
6. Add the ammonium nitrate to the calorimeter and replace the cover.
7. Stir the solution (or swirl the calorimeter) to dissolve the ammonium nitrate.
8. Watch the thermometer. When the temperature change reaches its maximum, record that temperature in Table 15-1. Calculate the temperature difference and record it in Table 15-1.
9. Dispose of the spent solution and rinse out the calorimeter.
10. Repeat steps 1 through 9, substituting 29.2 g of sodium chloride for the 40.0 g of ammonium nitrate.
11. Repeat steps 1 through 9, substituting 20.0 g of sodium hydroxide for the 40.0 g of ammonium nitrate.

OPTIONAL ACTIVITIES

If you have time and the required materials, consider performing these optional activities:

- Repeat the experiment using 10.0 mL of concentrated (98%) sulfuric acid as the solute. (Caution: concentrated sulfuric acid is extremely corrosive.) Calculate the heat of solution for concentrated sulfuric acid.

- Repeat the experiment using 10.0 mL of 35% sulfuric acid (battery acid) as the solute. (Caution: 35% sulfuric acid is extremely corrosive.) Calculate the heat of solution for 35% sulfuric acid. Does this value differ from the value for concentrated sulfuric acid on a per mole basis? If so, propose an explanation.

CAUTIONS

Ammonium nitrate is a strong oxidizer, and may detonate if heated strongly. Sodium hydroxide is corrosive, reacts strongly with aluminum (some commercial calorimeters have aluminum bodies), and in high concentrations etches or dissolves glass. Wear splash goggles, gloves, and protective clothing at all times.

FIGURE 15-2:
Measuring the temperature of the contents of the calorimeter

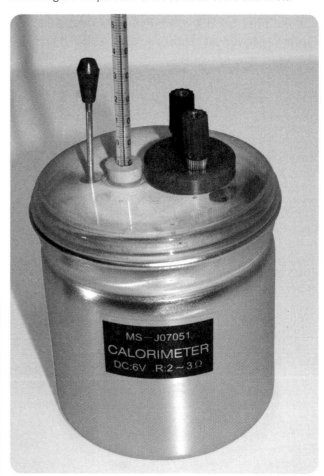

DISPOSAL: All of the solutions from this laboratory can be flushed down the drain with plenty of water.

TABLE 15-1: *Heat of solution—observed and calculated data*

Solute	Mass	Water temperature	Solution temperature	Temperature difference	Calculated heat of solution
A. Ammonium nitrate	_____.___ g	_____.__ °C	_____.__ °C	(+/-) _____.__ °C	(+/−) _____.___ kJ/mol
B. Sodium chloride	_____.___ g	_____.__ °C	_____.__ °C	(+/-) _____.__ °C	(+/−) _____.___ kJ/mol
C. Sodium hydroxide	_____.___ g	_____.__ °C	_____.__ °C	(+/-) _____.__ °C	(+/−) _____.___ kJ/mol

REVIEW QUESTIONS

Q1: Using your experimental data, calculate the heats of solution for ammonium nitrate, sodium chloride, and sodium hydroxide. Enter your calculated values in Table 15-1.

Q2: Look up published values for the heats of solution of those three compounds on the Internet or in a printed reference. How closely do the values that you obtained experimentally correspond to the published values? If your values are significantly different, propose possible explanations.

LABORATORY 15.2:
DETERMINE THE HEAT OF FUSION OF ICE

If you add heat to a mixed-phase system that comprises a substance in its solid and liquid phases, both at the melting point of the solid, the temperature of the system does not increase until all of the solid is converted to liquid. The additional heat energy that is absorbed without causing a temperature increase is required to give the atoms or molecules that make up the solid sufficient energy to break the attraction that holds them in solid form. The additional heat energy that is required to change a solid to a liquid at its melting point is called *heat of fusion*, *enthalpy of fusion*, or *specific melting heat*.

In this laboratory, we'll measure the heat of fusion of ice, which has a high heat of fusion because of the strong intermolecular attraction of hydrogen atoms. Heat of fusion is (officially) expressed as Joules per mole (J/mol) in the SI system, but many older and alternate forms are still used. Among the most common of those are calories per gram (cal/g), kiloJoules per kilogram (kJ/kg), and even British Thermal Units per pound (BTU/lb). In particular, cal/g is still very widely used, and is the format we use in this laboratory.

CAUTIONS

Take care with the hot water and hotplate to avoid burns. Wear splash goggles, gloves, and protective clothing.

SUBSTITUTIONS AND MODIFICATIONS

- You may substitute a sauce pan or similar container for the 600 mL beaker.
- You may substitute any heat source for the hotplate.

REQUIRED EQUIPMENT AND SUPPLIES

- ☐ goggles, gloves, and protective clothing
- ☐ calorimeter
- ☐ thermometer
- ☐ graduated cylinder, 100 mL
- ☐ ice bucket or similar wide-mouth container
- ☐ beaker, 600 mL
- ☐ beaker tongs
- ☐ hotplate
- ☐ strainer
- ☐ ice, crushed or chipped
- ☐ hot tap water

PROCEDURE

1. If you have not already done so, put on your splash goggles, gloves, and protective clothing.
2. Fill one of the 600 mL beakers about half full of crushed or chipped ice and allow the ice to begin melting while you perform the following steps.

COLD AS ICE

Ice in a typical home freezer is typically at a temperature of −15°C to −20°C, well below the melting point of ice. To determine the heat of fusion of ice accurately, we need to start with ice at 0°C. Otherwise, some of the heat absorbed will be needed to warm the ice to 0°C, giving a false high value for the heat of fusion of ice. You can avoid this by crushing or chipping the ice into small pieces, allowing it to melt partially into slush, and using the strainer to separate solid ice from the water.

3. Add about 400 mL of hot tap water to the second 600 mL beaker and heat the water to 65°C ± 5°C.
4. Using the beaker tongs, pour about 100 mL of the heated water into the 100 mL graduated cylinder to preheat it. Wait 30 seconds and then dump the water from the graduated cylinder into the sink.

5. Repeat the preceding step with another 100 mL of heated water from the large beaker. After 30 seconds, dump the water into the sink.
6. While you are preheating the graduated cylinder, use the strainer to separate solid ice from the ice-water slush mixture and fill the calorimeter about half full with ice. Replace the cover on the calorimeter to prevent the ice from absorbing heat from the room air.
7. Using the beaker tongs, add about 25 mL of the heated water to the graduated cylinder. Measure the volume of the water to 0.1 mL and record the volume on line A of Table 15-2.
8. Measure the temperature of the water in the graduated cylinder to 0.1 °C and record that temperature on line B of Table 15-2.
9. Carefully pour off any water that has accumulated in the calorimeter, and add the 25 mL of warm water from the graduated cylinder to the calorimeter. Replace the calorimeter cover and stir the ice-water mixture. (Some ice *must* remain in the calorimeter at the conclusion of

this step. If all of the ice melts, remove the calorimeter cover and quickly add more ice.)
10. Measure the temperature of the ice-water mixture, and record the value to within 0.1°C on line C of Table 15-2. (Ideally, that temperature should closely approach 0°C, but under experimental conditions, a temperature of 2°C to 3°C is acceptable. Simply watch the temperature until it stabilizes at its lowest value and record it.)
11. Pour the contents of the calorimeter through the strainer and into the graduated cylinder. Try to make sure that all of the water in the calorimeter is transferred to the graduated cylinder without spillage and that none of the remaining ice is transferred to the graduated cylinder. Measure the volume of the water in the graduated cylinder and record that value to 0.1 mL on line D of Table 15-2.
12. Repeat steps 1 through 10 at least once, and record the observed values in Table 15-2. If your results are poor (that is, they vary widely), run additional trials with the same steps and record the observed values in Table 15-2.

TABLE 15-2: *Heat of fusion of ice—observed and calculated data*

Item	Trial #1	Trial #2	Trial #3	Trial #4	Trial #5
A. Volume, initial	___.__ mL	___.__ mL	___.__ mL	___.__ mL	___.__ mL
B. Temperature, initial	___.__ °C	___.__ °C	___.__ °C	___.__ °C	___.__ °C
C. Temperature, final	___.__ °C	___.__ °C	___.__ °C	___.__ °C	___.__ °C
D. Volume, final	___.__ mL	___.__ mL	___.__ mL	___.__ mL	___.__ mL
E. Temperature change (B – C)	___.__ °C	___.__ °C	___.__ °C	___.__ °C	___.__ °C
F. Volume change (D – A)	___.__ mL	___.__ mL	___.__ mL	___.__ mL	___.__ mL
G. Heat of fusion of ice	___.__ cal/g	___.__ cal/g	___.__ cal/g	___.__ cal/g	___.__ cal/g

REVIEW QUESTIONS

Q1: Using your experimental data for each of the first two trials, calculate the heat of fusion of ice in cal/g and record your calculated values on line G of Table 15-2. The actual value for the heat of fusion of ice is 79.72 cal/g. If the value you obtained experimentally differs significantly, propose possible reasons for this variation.

LABORATORY 15.3:

DETERMINE THE SPECIFIC HEAT OF A METAL

When material at one temperature comes into contact with material at another temperature, heat flows from the warmer material to the cooler material until thermal equilibrium is attained when both materials are at the same temperature. In a controlled environment, such as our calorimeter, no heat is gained from or lost to the environment, so the heat gained by the (originally) cooler material equals the heat lost by the hotter material. (In practice, of course, there are minor heat gains or losses, even with the calorimeter, but these gains or losses are quite minor relative to the large changes that we'll measure in this laboratory.)

In this laboratory, we'll determine the specific heat of two metals, lead and iron, by heating a known mass of each metal to a known temperature, and then adding the hot metal to a known mass of cooler water at a known temperature and allowing the system to come to thermal equilibrium. Heat flows from the hot metal, warming the cooler water. Because the amount of heat lost by the metal is equal to the amount of heat gained by the water, we can calculate the specific heat of the metal by measuring the temperature increase of the water.

When the unit quantity is specified in mass, the heat transfer equation is:

$$Q = mc\Delta T$$

where Q is the amount of heat transferred, m is the mass of the sample, c is the specific heat of the substance, and ΔT is the change in temperature. Because the heat transfer for the metal and the water is equal—although with different signs, because heat is lost by the metal and gained by the water—we can express this equivalence as:

$$Q_{water} = -Q_{metal}$$
or
$$(mc\Delta T)_{water} = -(mc\Delta T)_{metal}$$

The specific heat of water is known to be 4.181 Joules per gram per Kelvin ($J/g \cdot K$ or $J \cdot g^{-1} \cdot K^{-1}$). Because the masses and temperatures of the metals and water will be measured and are therefore known, the only unknown is the value for the specific

REQUIRED EQUIPMENT AND SUPPLIES

☐ goggles, gloves, and protective clothing

☐ balance and weighing boat

☐ calorimeter

☐ thermometer

☐ graduated cylinder, 100 mL

☐ test tube (2)

☐ test tube holder

☐ beaker, 600 mL

☐ hotplate

☐ lead and iron shot (see Substitutions and Modifications)

☐ water

SUBSTITUTIONS AND MODIFICATIONS

• You may substitute a small Petri dish or similar container for the weighing boat.

• You may substitute a sauce pan or similar container for the 600 mL beaker.

• You may substitute any appropriate heat source for the hotplate.

• You may substitute other forms of lead and iron, as long as the samples fits inside the test tube. Lead shot may be obtained from laboratory supply houses or from sporting goods stores that sell shotgun reloading supplies. (Lead shot used in shotgun shells is an alloy, but contains a high percentage of lead; some shotgun shells use copper-plated steel shot instead of lead shot.) At the expense of some accuracy, you may substitute wheel weights or fishing sinkers for the lead shot. Because these items are often made from a lead alloy rather than pure lead, your experimental results may differ significantly from the values obtained with a pure lead sample. You may substitute small steel nuts or bolts for the iron shot. Sporting goods stores sell lead airgun pellets, which are almost pure lead, and BBs, which are copper-plated steel shot.

CAUTIONS

Take care with the hot water and hotplate to avoid burns. Wear splash goggles, gloves, and protective clothing.

DEGREES CELSIUS VERSUS KELVINS

The SI unit of temperature is the kelvin—not the degree Kelvin, just the kelvin. For all practical purposes one kelvin, abbreviated K, equals one degree Celsius, or 1°C. (I'm so old that I still think of the "C" as meaning centigrade rather than Celsius, so please excuse me if "centigrade" slips in occasionally.) For our purposes, J/g · K and J/g · °C can be considered equivalent.

heat of the metal samples. We can calculate that value by rearranging the preceding formula to:

$$c_{metal} = [(mc\Delta T)_{water} \div -(m\Delta T)_{metal}]$$

Plugging in the known and experimentally determined values on the right side of the equation gives us the unknown value for the specific heat of the metal on the left side. This works both ways. If we happened to have a known value for the metal but not a known value for the specific heat of water, we could rearrange the equation slightly and solve for the specific heat of water.

If you look up published values for specific heat, keep in mind that specific heat can be specified on a mass basis, as we are doing in this laboratory, or on a mole basis. Mass-based specific heat values are represented by a lowercase c; mole-based specific heat values are represented by an uppercase C. These values can differ greatly. For example, the C (mole-based) specific heats of lead and iron are quite similar, at 26.4 and 25.1 J/mol · K, respectively, while the mass-based specific heat of lead, at 0.127 J/g · K, is less than one third the mass-based specific heat of iron, at 0.450 J/g · K.

You may also see specific heat values represented as c_p, c_v, C_p, or C_v. The subscript indicates whether the specific heat value is specified at constant pressure or constant volume, which is particularly important if you are measuring the specific heat of a gas. The difference between specific heat values at constant pressure versus constant volume is small for solids and liquids, and can be ignored for our purposes.

PROCEDURE

1. If you have not already done so, put on your splash goggles, gloves, and protective clothing.
2. Fill the 600 mL beaker nearly full of hot tap water, add a boiling chip, place the beaker on the hotplate, and bring the water to a boil. Once the water boils, turn down the temperature of the hotplate until the water just maintains a gentle boil.
3. Pour lead shot into one test tube until it is about half full. We'll be immersing the test tube in the boiling water, and it's important that the level of the lead shot in the test tube be well below the surface of the boiling water bath.
4. Place the weighing boat on the balance and tare the balance to read 0.00 g.
5. Transfer the lead shot from the test tube into the weighing boat and determine its mass. Record the mass of the lead shot on line A of Table 15-3.
6. Transfer the lead shot back into the test tube and immerse the test tube in the boiling water bath, making sure that none of the water is transferred into the test tube. If necessary, clamp the test tube in place to secure it.
7. Repeat steps 3 through 6 with the iron shot.
8. Measure the temperature of the boiling water bath to 0.1°C and record that value on line B of Table 15-3.
9. Keep the test tubes in the boiling water bath for at least 10 to 15 minutes, by which time the lead and iron shot will have reached the temperature of the boiling water bath. While you wait for the shot in the two test tubes to equilibrate to the temperature of the boiling water bath, set up your calorimeter.
10. Use the graduated cylinder to measure about 50.0 mL of cold tap water to 0.1 mL, and add the water to the calorimeter. Record the volume on line C of Table 15-3.
11. Measure the temperature of the water in the calorimeter to 0.1°C and record that value on line D of Table 15-3.
12. Using the test tube holder, remove the test tube of lead shot from the boiling water bath, transfer the lead shot into the calorimeter, and replace the cover of the calorimeter.
13. Swirl or stir the contents of the calorimeter, and measure the temperature of the water inside it. When the temperature peaks, record that value to 0.1°C on line E of Table 15-3.
14. Empty the calorimeter, and repeat steps 10 through 13 with the steel shot.

DISPOSAL: Dry the lead and iron shot and save them for future use.

OPTIONAL ACTIVITIES

If you have time and the required materials, consider performing these optional activities:

- Repeat the experiment using different types of metal shot, such as aluminum, copper, magnesium, tin, and zinc.

- Repeat the experiment using alloyed metals rather than pure metals. For example, you might try solder (lead/tin), brass (copper/zinc), nickel silver (copper/nickel), bronze (copper/tin), and so on. Solder is a good choice, because the label usually gives the percentages of lead and tin in the solder. Measure the specific heat for the alloy and compare it to the specific heats for the component metals to determine whether the specific heat of the alloy is directly proportional to the specific heats of the component metals.

- Repeat the experiment using mineral oil, vegetable oil, motor oil, or a similar liquid in place of water. Using either published or your experimental values for the specific heat of lead, calculate the specific heat of the liquid.

REVIEW QUESTIONS

Q1: Using your experimental data for the lead shot and iron shot, calculate the specific heat of lead and iron in J/g · °C and record your calculated values on line G of Table 15-3.

TABLE 15-3: *Determine the specific heat of a metal—observed and calculated data*

Item	Lead	Iron
A. Mass of shot	_____.__ g	_____.__ g
B. Temperature of water bath	_____.__ °C	_____.__ °C
C. Volume of water	_____.__ mL	_____.__ mL
D. Temperature of water (initial)	_____.__ °C	_____.__ °C
E. Temperature of water (final)	_____.__ °C	_____.__ °C
F. Temperature change (E − D)	_____.__ °C	_____.__ °C
G. Specific heat	___._____ J/(g · °C)	___._____ J/(g · °C)

Q2: The actual values for the specific heat of lead and iron are 0.127 J/g · °C and 0.450 J/g · °C, respectively. Calculate the percent error of the value you obtained experimentally.

Q3: Propose several possible explanations for experimental error.

Q4: After completing this laboratory, a student learned that her digital thermometer consistently reads 0.8°C high. What effect did this error have on the specific heat values that she determined experimentally? Why?

Q5: After completing this laboratory, a student learned that his graduated cylinder consistently delivered 5% less liquid than indicated. What effect did this error have on the specific heat values that he determined experimentally? Why?

LABORATORY 15.4:
DETERMINE THE ENTHALPY CHANGE OF A REACTION

Chemical reactions absorb or release energy, usually in the form of heat (also called **thermal energy**). Endothermic reactions absorb heat; exothermic reactions release heat. If the reaction occurs in a solution in a calorimeter, the heat absorbed (or released) by the reaction reduces (or increases) the temperature of the solvent. This heat transfer to or from the solvent can be quantified with the familiar formula:

$$Q = mc_\Delta T$$

or, if water is the solvent:

$$Q = (mc_\Delta T)_{water}$$

where Q is the amount of heat transferred, m is the mass of the water, c is the specific heat of water, and ΔT is the change in temperature of the water. With known values for the mass and specific heat of water and ΔT determined experimentally, the value of Q can be calculated. (Remember that Q for an exothermic reaction is a negative value.) Q may be expressed in the traditional units of calories (cal) or in SI units of joules (J). The calculation is the same in either case. Only the units for the specific heat of water (c) are different:

$$c_{water} = 1.00 \ cal/(g \cdot °C)$$

$$c_{water} = 4.18 \ J/(g \cdot °C)$$

In this laboratory, we neutralize 50.0 mL of 1.0 M sodium hydroxide solution with an equal volume of 1.2 M hydrochloric acid solution. (We use a slight excess of HCl to ensure that all of the sodium hydroxide is consumed by the reaction.) The balanced equation for this reaction is:

$$HCl_{(aq)} + NaOH_{(aq)} \rightarrow H_2O + NaCl_{(aq)}$$

This reaction is exothermic, so the final temperature of the solution is higher than the starting temperature, and the value of Q is negative. Because the starting and final solutions are dilute, we can as a working approximation assume that their density is the same as that of water, 1.00 g/mL, which simplifies calculations. Filling in the known values gives us:

$$Q = [100 \ g] \cdot [1.00 \ cal/(g \cdot °C)] \cdot [\Delta T]$$

REQUIRED EQUIPMENT AND SUPPLIES

☐ goggles, gloves, and protective clothing

☐ calorimeter

☐ thermometer

☐ graduated cylinder, 100 mL

☐ sodium hydroxide (NaOH) solution, 1.0 M (50 mL)

☐ hydrochloric acid (HCl) solution, 1.2 M (50 mL)

SUBSTITUTIONS AND MODIFICATIONS

• You can prepare 50 mL of 1.0 M sodium hydroxide solution by dissolving 2.00 g of sodium hydroxide in about 40 mL of water in a small beaker and then making it up to 50.0 mL in your graduated cylinder. Allow the sodium hydroxide solution to cool to room temperature before using it.

• You can prepare 50 mL of 1.2 M hydrochloric acid solution by adding 5.0 mL of concentrated (37%, 12 M) hydrochloric acid to about 35 mL of water in a small beaker and then making it up to 50.0 mL in your graduated cylinder. If you use hardware-store muriatic acid (typically 31.45% or 10.3 M HCl), add 5.8 mL of the acid to about 35 mL of water in a small beaker and then make it up to 50.0 mL in your graduated cylinder. Sodium hydroxide is the limiting reagent in this experiment, so the exact molarity of the hydrochloric acid is not critical, as long as HCl is in excess. Allow the hydrochloric acid solution to cool to room temperature before using it.

• If you already have 1.0 M stock solutions of sodium hydroxide and hydrochloric acid made up, you can use them. Make sure the HCl is slightly in excess. For example, use 45.0 mL of 1.0 M sodium hydroxide and 55.0 mL of 1.0 M hydrochloric acid.

Once we determine ΔT experimentally, plugging that value into the equation gives us the value of Q—the amount of thermal energy transferred. Determining the enthalpy change requires one more step. Q is denominated in calories or Joules, and the enthalpy change of reaction, $\Delta H°_{reaction}$, is denominated in

CAUTIONS

Sodium hydroxide and hydrochloric acid solutions are corrosive and toxic, including the dilute solutions used in this laboratory. Wear splash goggles, gloves, and protective clothing.

OPTIONAL ACTIVITIES

If you have time and the required materials, consider performing these optional activities:

- Repeat the experiment using 1.2 M sulfuric acid in place of the 1.2 M hydrochloric acid. Look up the published value for the enthalpy change of reaction in the CRC handbook or an online reference and compare the known value to your experimentally determined value.

- Repeat the experiment using 1.2 M acetic acid in place of the 1.2 M hydrochloric acid. Look up the published value for the enthalpy change of reaction in the CRC handbook or an online reference and compare the known value to your experimentally determined value.

calories per mole or Joules per mole. To determine the enthalpy change of reaction, $\Delta H°_{reaction}$, we need to divide Q, the heat of reaction we observe experimentally, by the number of moles that yielded that value of Q. (If we had a large calorimeter and reacted 1.0 mole of sodium hydroxide, the values of Q and $\Delta H°_{reaction}$ would be equal. Because we're reacting only a small fraction of a mole of sodium hydroxide, the absolute value of $\Delta H°_{reaction}$ will be much larger than Q.)

PROCEDURE

1. If you have not already done so, put on your splash goggles, gloves, and protective clothing.
2. If you have just made up the sodium hydroxide solution and/or the hydrochloric acid solution, make sure that both are at room temperature before proceeding.
3. Use a clean, dry graduated cylinder to accurately measure 50.0 mL of 1.2 M hydrochloric acid, transfer it to the calorimeter, and replace the lid of the calorimeter. Record the volume of HCl solution on line A of Table 15-4. (Although the exact molarity of the hydrochloric acid is unimportant, it is important to record the volume of the HCl solution that you use as accurately as possible.)
4. Measure the temperature of the HCl solution to 0.1°C and record the temperature on line B of Table 15-4.
5. Rinse and dry the graduated cylinder, and then use it to accurately measure 50.0 mL of 1.0 M sodium hydroxide solution. Record the volume on line C of Table 15-4.
6. Rinse and dry the thermometer and then use it to measure the temperature of the sodium hydroxide solution to 0.1°C. Record the temperature on line D of Table 15-4. (The temperature of both solutions should be the same, but if they are not, you can average the two temperatures to arrive at a valid initial temperature. For example, if you start with 50.0 mL of hydrochloric acid at 22.3°C and 50.0 mL of sodium hydroxide solution at 21.1°C, your actual starting point is 100.0 mL of solution at 21.7°C.)
7. Remove the lid of the calorimeter, quickly add the sodium hydroxide solution, and replace the lid.
8. Stir the contents (or swirl the calorimeter gently) to mix the solutions thoroughly. Observe the temperature of the mixed solutions for several minutes until the temperature reaches its highest point. Record that temperature to 0.1°C on line E of Table 15-4.

TABLE 15-4: *Determine the enthalpy change of a reaction—observed and calculated data*

Item	Data
A. Volume of HCl solution	_____.___ mL
B. Temperature of HCl solution	_____.___ °C
C. Volume of NaOH solution	_____.___ mL
D. Temperature of NaOH solution	_____.___ °C
E. Temperature of solution (final)	_____.___ °C
F. Temperature change [E − (B+D)/2]	_____.___ °C
G. Heat of reaction	_____.___ cal
H. Moles of sodium hydroxide	___._____ mol
I. Enthalpy change of reaction	_____.___ cal/mol

REVIEW QUESTIONS

Q1: Calculate the heat of reaction and enter the value on line G of Table 15-4. (Hint: remember to use the proper sign.)

Q2: Calculate the number of moles of sodium hydroxide that reacted and enter that value on line H of Table 15-4.

Q3: Calculate the enthalpy change of reaction and enter that value on line I of table 15-4.

Q4: Using the known value for enthalpy of reaction, calculate the percent error in your experimental results. (Use the proper sign.)

Q5: Propose several possible reasons for the experimental error that you calculated in the preceding question.

DISPOSAL: The waste solution in the calorimeter contains only sodium chloride (table salt) and a small amount of hydrochloric acid. Dump it down the sink and flush with plenty of water.

Laboratory: Electrochemistry

16

Electrochemistry is the study of the relationship between chemical energy and electrical energy. Electrochemical processes involve two types of redox reactions. In the first type of electrochemical process, applying an electric current causes a chemical reaction to occur. For example, applying an electric current to water causes a chemical reaction that splits the water into hydrogen gas and oxygen gas, an example of a process called *electrolysis*. In the second type of electrochemical process, a chemical reaction produces an electric current. For example, the chemical reaction that occurs in a standard alkaline AA cell produces electric current that can be converted to light, heat, or mechanical energy.

Normally, a redox reaction occurs when an oxidizing agent and a reducing agent are brought into contact. For example, placing an iron nail in a solution of copper sulfate causes a spontaneous redox reaction. Iron atoms lose electrons (are oxidized) and enter solution as iron ions, while copper ions gain electrons (are reduced) and are deposited as copper metal on the nail.

In such redox reactions, electrons are transferred directly from one species to another. However, if you physically separate the oxidizing agent (copper, in this example) and the reducing agent (iron), the electron transfer that occurs during this redox reaction can be redirected to flow through an external conductor, such as a wire. This is the basis of a *voltaic cell*. Because the electric current is flowing through an external conductor rather than contained within the cell, it is accessible to do useful work, such as lighting a light bulb or powering an MP3 player.

Physical work (W) can be quantified as the product of the force (F) required to move an object and the distance (d) the object is to be moved, or, mathematically, $F = W \cdot d$. For example, if you lift a box from the floor and place it on a shelf, the amount of work required is the product of the force needed to lift the box times the distance the box is lifted. In its new position on the shelf, the box has higher potential energy than it did when sitting on the floor. (The work you did in lifting the box is converted to potential energy.) If the box falls from the shelf to the floor, that higher potential energy is converted to kinetic energy, and the potential energy of the box returns to its old, lower value.

Electrical work is similar to physical work, but instead of the physical work of moving a box from the floor to a shelf, electrical work is required to move electrons through a conductor or move ions through a solution to an electrode. And, just as the position of the box determines its physical potential energy, the type and state of charge of ions determine their electric potential energy (or *electric potential* for short).

Just as air moves (as wind) from a point of higher air pressure to a point of lower air pressure, an electric charge moves (as an electric current) from a point of higher electric pressure to a point of lower electric pressure. Electric pressure is denominated in volts, so an electric charge moves from an area at higher voltage to an area at lower voltage. To determine the electric work needed to move a charge through a conductor, multiply the charge by the potential difference:

Charge (C) · Potential Difference (V) = Electric work (J)

Electric charge is specified in Coulombs (C), the potential difference in volts (V), and electric work in Joules (J).

In this chapter, we'll examine both types of electrochemical processes by using an electric current to initiate and sustain a chemical reaction, and by using chemical reactions to produce electric current.

EVERYDAY ELECTROCHEMISTRY

- All electric batteries and cells are electrochemical in nature, from the AA alkaline cells in your flashlight to the nickel metal hydride (NiMH) battery in your digital camera to the lead-acid battery in your car.

- Many pure gases, such as hydrogen, the oxygen used in hospitals, and the chlorine used for water purification and many other purposes, are produced electrochemically, by passing electric current through water or an aqueous solution of a salt that contains the gaseous element in ionic form.

- Many commercially important elements, including aluminum, bromine, chlorine, and sodium, are produced exclusively or primarily electrochemically.

- Commercially important pure forms of many metals, including copper, silver, gold, and platinum, are produced electrochemically.

LABORATORY 16.1:
PRODUCE HYDROGEN AND OXYGEN BY ELECTROLYSIS OF WATER

Electrolysis is the process of forcing a redox reaction to occur by supplying an external electric current. For example, electrolysis can be used to plate an iron object with chromium by reducing chromium ions from a solution, replacing them with iron ions. Ordinarily, this reaction could not occur, because chromium is higher in the activity series of metals than iron. The application of an external electric current overcomes the difference in activity (potentials) between these two metals, allowing the chromium ions to be reduced to chromium metal.

Electrolysis is commonly used industrially and in laboratories to produce pure hydrogen and oxygen by splitting otherwise stable water molecules into their component gases, a process that requires providing an external electrical current at 1.23V or higher. This voltage is applied to two inert electrodes immersed in a dilute aqueous solution of an ionic compound. (Pure water cannot be used, because it is a very poor conductor of electricity.) In an *electrolytic cell* (as opposed to a *galvanic cell*, where the direction is reversed), the negative electrode, or *cathode*, provides the electrons needed to reduce H^+ ions to hydrogen gas. Conversely, the positive electrode, or *anode*, accepts the electrons needed to oxidize O^- ions to oxygen gas.

We can balance this redox reaction using half-reactions. To do so, we must identify which species is being oxidized and which reduced by looking at the change in oxidation states:

$$2 \, H^{+1}_2 O^{-2}(l) \rightarrow 2 \, H^0_2(g) + O^0_2(g)$$

From this balanced equation, we see that hydrogen is being reduced and oxygen is being oxidized, which allows us to set up the following two half-reactions:

(oxidation) $2 \, O^{-2} \rightarrow O_2 + 4 \, e^-$

(reduction) $2 \, H^+ + 2 \, e^- \rightarrow H_2$

Doubling the reduction half-reaction to put the same number of electrons on each side—everything, including electrons, must balance—and then adding the two half-reactions gives us the final balanced reaction:

REQUIRED EQUIPMENT AND SUPPLIES

- ☐ goggles, gloves, and protective clothing
- ☐ balance and weighing papers
- ☐ ring stand
- ☐ burette clamp
- ☐ test tube, with stopper (4)
- ☐ test tube rack
- ☐ graduated cylinder, 10 mL
- ☐ beaker, 600 mL
- ☐ 9V transistor battery
- ☐ marking pen
- ☐ matches, strike–anywhere
- ☐ rubber band (2)
- ☐ magnesium sulfate heptahydrate (~15 g)
- ☐ distilled or deionized water

SUBSTITUTIONS AND MODIFICATIONS

- You may substitute any container of similar size for the beaker.
- You may substitute wood splints or toothpicks for the strike–anywhere matches.
- Magnesium sulfate heptahydrate is sold in drugstores as Epsom salts. You may also substitute sodium bicarbonate (baking soda) for the magnesium sulfate.

$$[2 \, O^{-2} \rightarrow O_2 + 4 \, e^-] + [4 \, H^+ + 4 \, e^- \rightarrow 2 \, H_2]$$
$$= 2 \, H_2O \rightarrow 2 \, H_2 + O_2$$

In this laboratory session, we'll use electrolysis to produce hydrogen and oxygen from water.

PROCEDURE

1. If you have not already done so, put on your splash goggles, gloves, and protective clothing.
2. Weigh about 15 g of magnesium sulfate heptahydrate and transfer it to the beaker.
3. Fill the beaker nearly full with distilled water and stir until the magnesium sulfate dissolves completely.

WHY MAGNESIUM SULFATE?

Pure water does not conduct electricity. Dissolving magnesium sulfate—which is cheap, readily available, and electrochemically inert—produces a solution of charged magnesium ions and sulfate ions that allows the solution to conduct electricity. Some other ionic salts, such as sodium bicarbonate, can be substituted for the magnesium sulfate, but not all ionic salts can be used. For example, if you use sodium chloride (common table salt), chlorine gas is produced at one electrode and sodium hydroxide (lye) at the other electrode. The electrochemical potentials of the magnesium and sulfate ions are such that they act only to increase the conductivity of the solution, and otherwise remain unaffected by the electrolysis.

4. Fill one of the test tubes completely with the magnesium sulfate solution, place your finger or thumb over the tube, invert it, and immerse it in the beaker. Make sure the tube is completely filled with the solution and contains no air bubble. Repeat this procedure for the second test tube, and then use the rubber band to secure the two tubes together. Make sure the mouths of the test tubes are at least several centimeters below the surface of the solution.
5. Repeat step 4 with the second pair of test tubes.
6. Place the beaker on the base of the ring stand, with its center under the clamp.
7. Gently lower the 9V transistor battery into the beaker, with its electrodes up. You'll see a trail of bubbles begin to rise immediately from each electrode.
8. Clamp the first pair of test tubes and position them over the electrodes, placed so that each tube catches the trail of bubbles from one of the electrodes.
9. Note which tube is over the cathode of the battery and which is over the anode. Use the marking pen to make a minus sign on the tube over the cathode and a plus sign on the tube over the anode.
10. As the reaction proceeds, you'll notice that the gas in one tube accumulates about twice as fast as the gas in the other tube. When most of the water has been displaced from one of the tubes, which may require 10 to 30 minutes, depending on the size of the test tubes,

CAUTIONS

Although the quantities produced are very small, this experiment produces hydrogen gas, which is very flammable, and oxygen gas, which causes vigorous combustion of any burning substance. Wear splash goggles, heavy-duty gloves, and protective clothing.

insert a stopper into each of the two tubes while keeping their mouths below the surface of the solution.
11. Unclamp the first set of tubes, remove them from the beaker, rinse them with tap water, and place them in the test tube rack.
12. Clamp the second set of tubes into position, and allow gas to begin accumulating in them.
13. Remove the tube that was over the anode from the rack. Strike a match, allow it to catch fire completely, and then blow it out. While the burned end of the match is still glowing, remove the stopper from the test tube and thrust the glowing end of the match into the tube. Record your observations on line A of Table 16-1.
14. Remove the tube that was over the cathode from the rack. Strike a match, hold the burning match over the mouth of the tube, and remove the stopper. Record your observations on line B of Table 16-1.
15. Pour the solution that remains in the anode test tube into the graduated cylinder, and record the volume of the solution to 0.1 mL on line C of Table 16-1.
16. Pour the solution that remains in the cathode test tube into the graduated cylinder, and record the volume of the solution to 0.1 mL on line D of Table 16-1.
17. Fill one of the test tubes to the brim with water, and use the graduated cylinder to measure the volume of water the test tube contains when full. (You will probably have to fill and empty the graduated cylinder two or three times.) Record the volume of the test tube to 0.1 mL on line E of Table 18–1.
18. Note the level of the gases in the second set of test tubes. When the tube with the larger volume of gas has had about two-thirds of the solution displaced by the gas, reposition the test tubes so that the tube that was initially over the anode is now over the cathode and vice versa.
19. Allow the reaction to proceed until both tubes are nearly full of gas and the volume of gas in each tube is about equal.
20. Stopper the tubes while keeping their mouths under the surface of the solution.
21. Unclamp the second set of tubes, remove them from the beaker, rinse them with tap water, and place them in the test tube rack.

22. Light a match, remove the stopper from one of the tubes, and hold the lit match just over the mouth of the tube. Record your observations on line H of Table 16-1.

TABLE 16-1: *Produce hydrogen and oxygen by electrolysis of water—observed and calculated data*

Item	Result
A. Glowing match + anode gas	
B. Burning match + cathode gas	
C. Volume of solution remaining in anode tube	_____._____ mL
D. Volume of solution remaining in cathode tube	_____._____ mL
E. Volume of test tube	_____._____ mL
F. Volume of anode gas (E − C)	_____._____ mL
G. Volume of cathode gas (E − D)	_____._____ mL
H. Burning match + mixed anode and cathode gases	

FIGURE 16-1: *A homemade version of a commercial Brownlee electrolysis setup*

DISPOSAL: The waste solutions from this lab can be flushed down the drain with plenty of water.

REVIEW QUESTIONS

Q1: From looking at the balanced equation and your own observations during this lab, which gas was produced at the anode and which at the cathode? What evidence supports your conclusions?

Q2: Using the values you recorded on lines F and G of Table 16-1, calculate the volume ratio of the two gases produced and the percentage error from the theoretical yield.

Q3: Comparing your observations for igniting the cathode gas versus igniting the mixed cathode and anode gases, why was the combustion of the mixed gases so much more vigorous?

Q4: At standard temperature and pressure, one mole of a gas occupies about 22.4 L. The density of water is 1.00 g/mL and its formula weight is 18.02 g/mol. If your test tube of mixed anode and cathode gases contains a total of 25.0 mL of gases in a 2:1 hydrogen:oxygen ratio, how much liquid water is produced when they combust?

LABORATORY 16.2:
OBSERVE THE ELECTROCHEMICAL OXIDATION OF IRON

In the preceding lab session, we used an external electrical current to force a non-spontaneous redox reaction to occur. But many redox reactions occur spontaneously, without requiring an external energy source to initiate or sustain them. In this lab, we'll examine a very familiar spontaneous redox reaction, the oxidation of iron-to-iron oxide, otherwise known as rusting.

Although rusting occurs spontaneously, it happens only when specific conditions exist; namely, the presence of oxygen and water. The balanced equation for this electrochemical reaction shows why:

$$4\ Fe(s) + 6\ H_2O(l) + O_2(g) \rightarrow 2\ Fe_2O_3 \cdot 3H_2O(s)$$

In the absence of oxygen or water, the reaction cannot proceed. And, although these conditions are both necessary and sufficient for the reaction to occur, the reaction rate varies with the presence or absence of electrolytes. Because the current flow inherent to an electrochemical reaction involves the migration of ions, the presence of an electrolyte, such as sodium chloride (common salt) or another ionic salt, increases the reaction rate. That's why, for example, iron and steel rust much faster near the ocean than they do in drier areas far from salt air, and why automotive rust is a major problem in areas where roads are salted during winter weather.

In this lab session, we'll examine the rusting of iron. We'll expose clean iron surfaces to oxygen, water, and salt, alone or in any combination, and observe the effects of each of these environments on the reaction rate.

CAUTIONS

This experiment uses flame. Be careful with the flame, have a fire extinguisher readily available, and use care in handling hot objects (a hot beaker looks exactly like a cold beaker). Wear splash goggles, gloves, and protective clothing.

REQUIRED EQUIPMENT AND SUPPLIES

☐ goggles, gloves, and protective clothing

☐ test tubes (6)

☐ beaker, 150 mL

☐ stoppers or corks for test tubes (2)

☐ test tube rack

☐ ring stand

☐ support ring

☐ wire gauze

☐ alcohol lamp or burner

☐ beaker tongs

☐ spatula

☐ steel wool or sandpaper

☐ marking pen

☐ common iron or steel nails (ungalvanized), 6d to 12d (6)

☐ sodium chloride, solid (~5 to 10 g)

☐ sodium chloride, 1 M (~100 mL)

☐ distilled or deionized water (~100 mL)

PROCEDURE

1. If you have not already done so, put on your splash goggles, gloves, and protective clothing.
2. Label six clean, dry test tubes A through F and place them in the test tube rack.
3. Use the steel wool or sandpaper to polish six iron or steel nails until they are bright and shiny. Make sure to clean the entire nail, including the head, shank, and tip.
4. Place one of the nails in test tube A and return the test tube to the rack.
5. Fill test tube B with tap water to just below the rim and insert the stopper. The goal is to determine the water level needed to allow the stopper to seat, leaving only water in the test tube, with no air bubble. When you have determined how much water is needed, mark the level on the outside of the tube and empty the tube.

SUBSTITUTIONS AND MODIFICATIONS

- You may substitute a hotplate, kitchen stove burner, or similar heat source for the alcohol lamp or burner, which also eliminates the need for the tripod stand or ring stand and the wire gauze. If you use a hotplate or kitchen stove burner, place a large tin can lid or similar item between the burner and the beaker.

- You may substitute bolts, nuts, or other hardware items for the nails. Make sure that whatever items you use are iron or steel and not galvanized or otherwise plated with a substance other than iron or steel.

- You may substitute ordinary table salt for the sodium chloride.

- If you do not have a 1 M solution of sodium chloride on hand, weigh 5.85 g of sodium chloride, dissolve it in distilled or deionized water, and make up the solution to 100 mL.

6. Repeat step 5 for test tube E.
7. Set up the beaker on the tripod stand or ring stand, using the wire gauze to protect the beaker from direct flame.
8. Use test tube B to transfer about two test tubes worth of distilled or deionized water to the beaker.
9. Light the alcohol lamp or burner, and bring the water to a gentle boil. Continue boiling it gently for several minutes. The goal is to drive off all of the oxygen dissolved in the water.
10. Place the second clean nail in test tube B.

11. Use the beaker tongs to fill test tube B with the boiled water to the line you previously made on the test tube and insert the stopper gently. You needn't force the stopper into place. The idea is simply to prevent the contents of the test tube from being exposed to atmospheric oxygen.
12. Place the third clean nail in test tube C, and fill the tube to near the rim with distilled or deionized water that has not been boiled. Leave this test tube unstoppered.
13 Place the fourth clean nail in test tube D, and fill the tube with solid sodium chloride until about half the length of the nail is covered by the sodium chloride.
14. Use test tube E to transfer about two test tubes worth of 1 M sodium chloride solution to the beaker. Bring the sodium chloride solution to a gentle boil and continue boiling it gently for several minutes to drive out dissolved oxygen.
15. Place the fifth clean nail in test tube E, use the beaker tongs to fill test tube E with the boiled sodium chloride solution to the line you previously made on the test tube, and insert the stopper gently. Again, seat the stopper gently. All we want to do is prevent the contents of the test tube from being exposed to atmospheric oxygen.
16. Place the sixth clean nail in test tube F, fill that test tube with unboiled 1M sodium chloride solution, and return the test tube to the rack. Leave it unstoppered.
17. Set the rack of six test tubes aside; check it under a strong light after several hours or overnight to see if the nail in one or more of the test tubes has begun rusting. Continue checking the tubes periodically until it is evident that rusting is or is not occurring in each of the tubes. Record your observations for each of the tubes in Table 16-2.

TABLE 16-2: *Observe the electrochemical oxidation of iron—observed data*

Test tube	O_2	H_2O	NaCl	Amount of oxidation
A	•	○	○	none···light···moderate···heavy···extreme
B	○	•	○	none···light···moderate···heavy···extreme
C	•	•	○	none···light···moderate···heavy···extreme
D	•	○	•	none···light···moderate···heavy···extreme
E	○	•	•	none···light···moderate···heavy···extreme
F	•	•	•	none···light···moderate···heavy···extreme

DISPOSAL: The solid waste material from this lab can be disposed of with household waste and the liquid waste material may be flushed down the drain with plenty of water.

REVIEW QUESTIONS

Q1: Which of the nails showed the greatest amount of rust and which the least? Why?

_____

_____

_____

_____

Q2: What might explain some rust appearing on the nails in test tubes A or D?

_____

_____

_____

Q3: What might explain some rust appearing on the nails in test tubes B or E?

_____

_____

_____

Q4: Did a chemical reaction occur between the iron and the sodium chloride in test tube F? Explain your reasoning.

_____

_____

_____

_____

Q5: Is the presence of sodium chloride alone sufficient to initiate or accelerate rusting? Explain your reasoning.

_____

_____

_____

_____

LABORATORY 16.3:
MEASURE ELECTRODE POTENTIALS

In the first lab of this chapter, you may have wondered how we knew that the electrolysis of water would require at least 1.23V. A glance at the standard reduction potentials listed in Table 16-3 provides the answer. (Table 16-3 lists standard reduction potentials for only a handful of half-reactions. Comprehensive tables that list reduction potentials for thousands of half-reactions are available online and in printed reference works.) Of the half-reactions involved in the electrolysis of water, the reduction of oxygen gas and hydrogen ions (protons) to water:

$$O_2(g) + 4 H^+(aq) + 4 e^- \rightarrow 2 H_2O(l)$$

has the greatest reduction potential, +1.23V. Of course, oxidation and reduction are two sides of the same coin, so the reverse reaction is equally valid:

$$2 H_2O(l) \rightarrow O_2(g) + 4 H^+(aq) + 4 e^-$$

So, looking at the oxidation reaction rather than the reduction reaction, the oxidation of water to oxygen gas and hydrogen ions has the greatest oxidation potential, which means that external current must be supplied at a minimum voltage of −1.23V for the half-reaction to occur.

Half-reactions that appear near the top of the table—those with high negative values—represent strong reducers. Those listed near the bottom of the table—those with high positive values—represent strong oxidizers. It's important to understand that a species in isolation cannot be said to be a reducer or an oxidizer, because the species with which it reacts determines whether the first species functions as a reducer or an oxidizer.

For example, consider the pairs of ions and elements, Zn^{2+}/Zn, Fe^{2+}/Fe, and Cu^{2+}/Cu, which have reduction potentials of −0.76V, −0.44V and +0.34V, respectively. If you place solid iron in a solution of Cu^{2+} ions, those ions are reduced to solid copper while the solid iron is oxidized to Fe^{2+} ions. With this combination of species, iron serves as the reducer and copper as the oxidizer. Conversely, if you place solid zinc in a solution of Fe^{2+} ions, those ions are reduced to solid iron while the solid

TABLE 16-3: *Standard reduction potentials at 25°C.*

Half-Reaction	E^0 (V)
$Li^+(aq) + e^- \rightarrow Li(s)$	−3.05
$K^+(aq) + e^- \rightarrow K(s)$	−2.93
$Ba^{2+}(aq) + 2 e^- \rightarrow Ba(s)$	−2.90
$Ca^{2+}(aq) + 2 e^- \rightarrow Ca(s)$	−2.87
$Na^+(aq) + e^- \rightarrow Na(s)$	−2.71
$Mg^{2+}(aq) + 2 e^- \rightarrow Mg(s)$	−2.37
$Al^{3+}(aq) + 3 e^- \rightarrow Al(s)$	−1.66
$2 H_2O(l) + 2 e^- \rightarrow H_2(g) + 2 OH^-(aq)$	−0.83
$Zn^{2+}(aq) + 2 e^- \rightarrow Zn(s)$	−0.76
$Cr^{3+}(aq) + 3 e^- \rightarrow Cr(s)$	−0.73
$Fe^{2+}(aq) + 2 e^- \rightarrow Fe(s)$	−0.44
$Cd^{2+}(aq) + 2 e^- \rightarrow Cd(s)$	−0.40
$Co^{2+}(aq) + 2 e^- \rightarrow Co(s)$	−0.28
$Ni^{2+}(aq) + 2 e^- \rightarrow Ni(s)$	−0.25
$Sn^{2+}(aq) + 2 e^- \rightarrow Sn(s)$	−0.14
$Pb^{2+}(aq) + 2 e^- \rightarrow Pb(s)$	−0.13
$2 H^+(aq) + 2 e^- \rightarrow H_2(g)$	**0.00**
$Cu^{2+}(aq) + 2 e^- \rightarrow Cu(s)$	+0.34
$O_2(g) + 2 H_2O(l) + 4 e^- \rightarrow 4 OH^-(aq)$	+0.40
$Ag^+(aq) + e^- \rightarrow Ag(s)$	+0.80
$Hg^{2+}(aq) + 2 e^- \rightarrow Hg(l)$	+0.85
$O_2(g) + 4 H^+(aq) + 4 e^- \rightarrow 2 H_2O(l)$	**+1.23**
$Au^{3+}(aq) + 3 e^- \rightarrow Au(s)$	+1.50
$H_2O_2(aq) + 2 H^+(aq) + 2 e^- \rightarrow 2 H_2O(l)$	+1.78
$F_2(g) + 2 e^- \rightarrow 2 F^-(aq)$	+2.87

zinc is oxidized to Zn^{2+} ions. With this combination, iron serves as the oxidizer rather than the reducer.

Standard reduction potential values can also be used to predict the voltage that will be produced by a cell that uses two different metals as electrodes, simply by calculating the difference in reduction potential between the two metals. For example, if you build a cell that uses electrodes of copper (+0.34V) and zinc (−0.76V), that cell will, in theory, provide current at 1.10V, because:

$$-0.76V - (+0.34V) = -1.10V$$

In practice, the actual voltage produced by such cells is always somewhat lower than the theoretical voltage produced by an ideal cell, because real-world physical cells have inefficiencies in ion migration and other issues that reduce the voltage somewhat. Still, using reduction potential values makes it easy to estimate the approximate voltage that will be produced by a cell that uses any two arbitrarily selected metals as electrodes.

But what exactly is a cell? When we drop an iron nail into a solution of copper ions, the iron is oxidized and the copper reduced, but these two half-reactions occur with the reactants in physical contact. Electrons are exchanged, but we have no way to access that electron flow and use it to do useful work. If we separate the reactants physically, the two half-reactions cease—unless, that is, we join them electrically by using a conductor. An arrangement in which the two reactants are physically separated but electrically joined is called a *cell*, and each of the separate reactants is called a *half-cell*.

In a cell, electrons released by the reducing agent in its half-cell travel from its *electrode* as an electric current through a conductor (such as a wire) to the electrode of the half-cell that contains the oxidizing agent, which is reduced by those electrons. That electric current can be used to do useful work, such as lighting a bulb or running a motor. The electrode at which oxidation occurs is called the *anode*, and the electrode at which reduction occurs is called the *cathode*.

In this laboratory session, we'll build such an *electrochemical cell*, also called a *galvanic cell* or *voltaic cell*, using electrodes of various metals. We'll embed those electrodes in a lemon. The semipermeable membranes in the lemon will function as physical barriers to isolate the two half-cells from each other, while the hydronium and citrate ions produced by the citric acid in the lemon juice provide internal electric connectivity between the half-cells. that current to establish that an electrochemical cell exists, and compare our actual measured voltages with the voltage that an ideal cell using those electrodes would produce.

If we stopped at that point, no reactions would occur and no current would be produced, because the semipermeable membranes in the lemon prevent internal migration of metal ions between the electrodes. In other words, no return path for the current exists. But if we provide a return path by connecting the two electrodes directly together with a copper wire, the reaction proceeds and current is produced. We'll measure the voltage of that current to establish that an electrochemical cell exists, and compare our actual measured voltages with the voltage that an ideal cell using those electrodes would produce.

CAUTIONS

Although none of the materials used in this lab session is particularly hazardous, make sure to dispose of the food items after use. **DO NOT EAT ANY FOOD ITEM** used in this session. The food items will be contaminated with metal ions, including toxic heavy-metal ions. Wear splash goggles, gloves, and protective clothing.

REQUIRED EQUIPMENT AND SUPPLIES

- ☐ goggles, gloves, and protective clothing
- ☐ digital multimeter (DMM)
- ☐ sharp knife
- ☐ steel wool or sandpaper
- ☐ lemon (1)
- ☐ Electrodes of as many metals as possible, such as magnesium, aluminum, zinc, iron, nickel, tin, lead, and copper. Ideally, these electrodes should be in the form of metal strips, but if strips are not available, you may use wires or rods. If you cannot obtain electrodes made from all of these metals, get as many as you can. Most laboratory supplies vendors sell inexpensive sets of electrodes that contain most or all of these metals.

SUBSTITUTIONS AND MODIFICATIONS

- You may substitute another citrus fruit, such as an orange or a lime, for the lemon. Dr. Mary Chervenak suggests trying a potato.

PROCEDURE

1. If you have not already done so, put on your splash goggles, gloves, and protective clothing.
2. Use steel wool or sandpaper to polish each of the electrodes to remove surface oxidation, oils, and other contaminants.
3. Use the knife to make two slits in the lemon, spaced as far apart as possible, and large enough to accept the electrodes. It's important that the two electrodes do not touch once they are inserted into the lemon.
4. Insert the copper electrode in one of the slits and insert an electrode made of a different metal in the other slit.
5. Touch or clamp one of the leads from the DMM to each of the two electrodes and observe the reading on the DMM. If necessary, reverse the leads to obtain a positive voltage reading. Record the voltage reading to 0.01V in the appropriate cell of Table 16-4.
6. Leaving the copper electrode in position, remove the second electrode and replace it with another electrode of a different metal.
7. Use the DMM to record the voltage with this second pair of metals, and record that voltage to 0.01V in the appropriate cell of Table 16-4.
8. Repeat steps 6 and 7 until you have tested all combinations of the copper electrode with all of the other electrodes you have available.

FIGURE 16-2: *Measuring the voltage produced by an electrochemical cell with copper and zinc electrodes (note that the actual voltage is lower than the theoretical 1.10V)*

9. Based on the voltage readings you obtained for copper with each of the other metals, choose one of those metals to replace the copper electrode. Test that metal in combination with all of the other metals. (But remember you've already tested it with copper.) Record the voltages you observe to 0.01V in the appropriate cells of Table 16-4.
10. Repeat step 9 until you have tested every metal paired with every other metal.

TABLE 16-4: *Measure electrode potentials—observed data*

Electrode	Mg	Al	Zn	Fe	Ni	Sn	Pb	Cu
Mg		___.___V	___.___V	___.___V	___.___V	___.___V	___.___V	___.___V
Al	___.___V		___.___V	___.___V	___.___V	___.___V	___.___V	___.___V
Zn	___.___V	___.___V		___.___V	___.___V	___.___V	___.___V	___.___V
Fe	___.___V	___.___V	___.___V		___.___V	___.___V	___.___V	___.___V
Ni	___.___V	___.___V	___.___V	___.___V		___.___V	___.___V	___.___V
Sn	___.___V	___.___V	___.___V	___.___V	___.___V		___.___V	___.___V
Pb	___.___V	___.___V	___.___V	___.___V	___.___V	___.___V		___.___V
Cu	___.___V	___.___V	___.___V	___.___V	___.___V	___.___V	___.___V	

DISPOSAL: The lemon and electrodes can be retained for later use, though they must be treated as equally inedible.

REVIEW QUESTIONS

Q1: Of the eight metals shown in Table 16-4, why did we first use copper as the reference electrode for each of the other metals?

_____

_____

_____

_____

Q1: Using copper as the reference electrode, rank the cell voltage for each of the other anodes that you tested.

_____

_____

_____

_____

Q3: Using the standard reduction potentials listed in Table 16-3, calculate the theoretical voltage for each of the cells you tested and recorded in Table 16-4. How do the actual voltages you measured compare to the theoretical voltages you calculated? Propose an explanation for any differences.

_____

_____

_____

_____

_____

_____

_____

_____

_____

_____

_____

LABORATORY 16.4:
OBSERVE ENERGY TRANSFORMATION (VOLTAGE AND CURRENT)

The law of conservation of energy says that energy can neither be created nor destroyed, but can be transformed from one form to another. For example, an electric generator converts mechanical energy into electric energy, a motor converts electric energy into mechanical energy, and an incandescent light bulb converts electric energy into light energy and heat energy. Chemical energy is another form of energy and—as is true of all forms of energy—can be converted to other forms of energy.

In this lab session, we'll transform chemical energy into electric energy and use that electric energy to do work; specifically, to light an LED. Electric energy has two key metrics, which can best be understood by analogy with plumbing. Just as a differential in water pressure forces water to flow through a pipe, a differential in electrical pressure—called *voltage*, denominated in *volts*, and abbreviated *V*—forces electricity to flow through a wire or other conductor. And, just as the amount of water transferred can be specified as a flow rate—denominated in gallons/minute, liters/second, or some other unit of measure—the flow rate of electricity is called *current*, and denominated in *amperes* or *amps* (abbreviated *A*), where one ampere represents a current flow of one coulomb per second. In this lab session, we'll measure voltage and current in a voltaic cell.

REQUIRED EQUIPMENT AND SUPPLIES

☐ goggles, gloves, and protective clothing

☐ digital multimeter (DMM)

☐ patch cables with alligator clips (2)

☐ LED, red (1)

☐ beaker, 250 mL (1)

☐ steel wool or sandpaper

☐ Pieces of magnesium, iron, and copper (one each) to use as electrodes. If they are long and skinny, and flat, that's best.

☐ hydrochloric acid, 2 M (~200 mL)

SUBSTITUTIONS AND MODIFICATIONS

• If you don't have patch cables, you can make your own by using paper clips as field-expedient alligator clips. Wrap each end of a copper wire tightly around a paper clip and crimp it into place with pliers. (If possible, solder the wire to the paper clips.).

• You may substitute another color for the red LED, but different colors of LEDs require different voltages to illuminate. Red LEDs typically require at least 1.8V to 2.1V, which is less than the approximately 2.7V delivered by the magnesium/copper cell used in this session. Orange LEDs typically require 2.2V to 2.3V and yellow LEDs 2.4V to 2.5V, and so will probably illuminate when connected to a magnesium/copper cell. Green LEDs typically require 2.6V or more, and so may not illuminate with a magnesium/copper cell. Blue and white LEDs typically require 3V or more, and so will not illuminate when connected to a magnesium/copper cell. (If you insist on a different color, you can always build two cells and link them in series.)

• If you do not have a bench solution of 2 M hydrochloric acid, you can make it up by transferring about 33 mL of concentrated hydrochloric acid or about 39 mL of 31.45% hardware-store muriatic acid to about 100 mL of water in the beaker and then making it up to about 200 mL. The exact concentration is not critical.

CAUTIONS

Hydrochloric acid is corrosive and emits strong irritating fumes. Wear splash goggles, gloves, and protective clothing.

PROCEDURE

1. If you have not already done so, put on your splash goggles, gloves, and protective clothing.
2. Use steel wool or sandpaper to polish each of the electrodes to remove surface oxidation, oils, and other contaminants.
3. Transfer about 200 mL of 2 M hydrochloric acid to the beaker.
4. Clamp one of the patch cables to the copper electrode and the other patch cable to the magnesium electrode.
5. Clamp the remaining end of each patch cable to one of the wires that extends from the base of the LED.
6. Immerse both electrodes in the beaker of hydrochloric acid. If the LED does not light, reverse the leads. The light emitted by the LED represents work done by the electric current produced by an electrochemical reaction.
7. Leaving the copper electrode in position, remove the magnesium electrode, disconnect it from the patch cable, and rinse it thoroughly with water.
8. Connect the iron electrode to the patch cable, and immerse the iron electrode in the beaker of hydrochloric acid. The LED should not illuminate.
9. Disconnect both leads from the LED and reconnect them to the DMM. Verify that the voltage across the copper/iron electrode pair is approximately 0.78V.
10. Remove the iron electrode from the acid bath and rinse it thoroughly with water.
11. Immerse only 1 cm of the iron electrode in the acid bath, and use the DMM to record the voltage and current (amperage) on the first line of Table 16-5.
12. Increase the distance the iron electrode is immersed in the acid bath centimeter by 1 centimeter, recording the voltage and current at each immersion level in Table 16-5.

TABLE 16-5:
Observe energy transformation—observed data

Level	Voltage	Current
1 cm	__.__V	__.__ mA
2 cm	__.__V	__.__ mA
3 cm	__.__V	__.__ mA
4 cm	__.__V	__.__ mA
5 cm	__.__V	__.__ mA
6 cm	__.__V	__.__ mA
7 cm	__.__V	__.__ mA
8 cm	__.__V	__.__ mA

FIGURE 16-3:
Measuring the current produced by an electrochemical cell

DISPOSAL: The hydrochloric acid waste can be neutralized with sodium carbonate or sodium bicarbonate and flushed down the drain with copious water. The electrodes can be retained for later use.

REVIEW QUESTIONS

Q1: Why did the LED remain unlit with the copper/iron cell?

_____

_____

_____

Q2: What relationship exists between the surface area of the electrodes and the voltage produced by the cell?

_____

_____

_____

Q3: What relationship exists between the surface area of the electrodes and the current produced by the cell?

_____

_____

_____

Q4: Commercial alkaline cells ("batteries") use electrodes made from zinc powder and manganese dioxide powder. Why do these cells use powder rather than solid chemicals?

_____

_____

_____

_____

_____

_____

_____

LABORATORY 16.5:
BUILD A VOLTAIC CELL WITH TWO HALF-CELLS

If you immerse magnesium metal in a solution of copper sulfate, a spontaneous redox reaction occurs. An atom of magnesium releases two electrons and is oxidized to an Mg^{2+} ion. Those two electrons are captured by a Cu^{2+} ion, reducing it to copper metal, which is deposited on the magnesium electrode as a thin plating. This reaction occurs spontaneously, because magnesium ions are less attractive to electrons than are copper ions.

We can write this redox reaction as two half-reactions, with the standard reduction potentials shown in parentheses:

$$Mg(s) \rightarrow Mg^{2+}(aq) + 2e^- \ (+2.37V)$$

$$Cu^{2+}(aq) + 2e^- \rightarrow Cu(s) \ (+0.34V)$$

In the combined reaction, magnesium atoms serve as the reducing agent, which means that they are oxidized to magnesium ions. Conversely, copper ions serve as the oxidizing agent, which means that they are reduced to metallic copper. Adding up the reduction potentials tells us that this cell should be able to provide an electron flow at 2.71V. There's a problem with that, though.

When magnesium metal is in direct contact with a solution of copper ions, electrons are transferred directly from the magnesium metal to the copper ions, so those electrons are not accessible to do work. In effect, both half-reactions are occurring in the same reaction vessel, so there is no externally accessible flow of electrons that can be captured to do work.

But what if we could split the two half-reactions into separate vessels and somehow intercept that electron flow? As it turns out, we can do just that by putting the copper and magnesium parts of the cell in separate containers. That physically isolates the magnesium atoms and ions from the copper atoms and ions, but we still somehow have to provide an external path for the flow of electrons and an electrical connection between the copper and magnesium segments of the cell.

The first part of that circuit is no problem. We can use ordinary copper wire clipped to the electrodes as a way to route electricity outside the cell, where it can do useful work such as lighting a light bulb. But we still need the internal connection

REQUIRED EQUIPMENT AND SUPPLIES

☐ goggles, gloves, and protective clothing

☐ digital multimeter (DMM)

☐ patch cables with alligator clips (2)

☐ beaker, 150 mL (2)

☐ graduated cylinder, 100 mL

☐ Beral pipette or eyedropper

☐ stirring rod

☐ flexible plastic tubing (~15 cm; inside diameter 5 mm or larger)

☐ cotton ball (1)

☐ steel wool or sandpaper

☐ electrodes, magnesium and copper (one each)

☐ sodium chloride, 1 M (a few mL)

☐ magnesium sulfate, 1.0 M (50 mL)

☐ copper sulfate, 1.0 M (50 mL)

between the two half-cells. The answer is a device called a *salt bridge*, which is a tube filled with a conducting solution, with one end immersed in each of the half-cells. The salt bridge conducts electricity while preventing ions from migrating from one half-cell to the other.

In this lab session, we'll build a voltaic cell comprising a magnesium half-cell and a copper half-cell, and measure the voltage and current produced by the cell.

CAUTIONS

Copper sulfate is moderately toxic. Wear splash goggles, gloves, and protective clothing.

SUBSTITUTIONS AND MODIFICATIONS

- You may substitute any containers of similar size for the 150 mL beakers.
- You may substitute a commercially made salt bridge for the plastic tubing and cotton ball.
- You may substitute an aluminum or zinc electrode for the magnesium electrode. If you do that, replace the 1.0 M magnesium sulfate solution with 1.0 M aluminum sulfate or 1.0 M zinc sulfate, respectively.
- If you do not have a bench solution of 1 M sodium chloride, you can make it up by weighing 5.84 g of sodium chloride and dissolving it in sufficient water to make up 100 mL of solution. You may substitute table salt for the sodium chloride. Concentration is not critical.
- If you do not have a bench solution of 1.0 M magnesium sulfate, you can make it up by weighing 6.02 g of anhydrous magnesium sulfate or 12.32 g of magnesium sulfate heptahydrate and dissolving it in sufficient water to make up 50 mL of solution. You may substitute Epsom salts for the magnesium sulfate heptahydrate. Concentration is not critical.
- If you do not have a bench solution of 1.0 M copper sulfate, you can make it up by weighing 7.98 g of anhydrous copper sulfate or 12.49 g of copper sulfate pentahydrate and dissolving it in sufficient water to make up 50 mL of solution. Concentration is not critical.

PROCEDURE

1. If you have not already done so, put on your splash goggles, gloves, and protective clothing.
2. Transfer 50 mL of 1.0 M magnesium sulfate solution to one beaker, and 50 mL of 1.0 M copper sulfate solution to the other beaker.
3. Use steel wool or sandpaper to polish each of the electrodes to remove surface oxidation, oils, and other contaminants.
4. Clip one end of each patch cable to an electrode and the other ends of the patch cables to the DMM terminals.
5. Immerse the magnesium electrode in the beaker of magnesium sulfate solution, and the copper electrode in the beaker of copper sulfate solution.
6. Observe the voltage reading on the DMM. No voltage is detected, because no circuit yet exists.
7. Plug one end of the plastic tubing tightly with cotton. Use the Beral pipette to fill the plastic tubing with 1 M sodium chloride solution, making sure that there are no air bubbles in the tubing. (If the solution runs out of the plugged end as you fill the tubing, the plug isn't tight enough.) When the tubing is full, plug the open end tightly with cotton and add a few more drops of sodium chloride solution to make sure that the plug is saturated with solution.

8. Check again to make sure that there are no air bubbles in the tubing, and then immerse one end of the tubing in the beaker of magnesium sulfate solution and the other end in the beaker of copper sulfate solution.
9. Record the voltage and current readings on the DMM on line A of Table 16-6.
10. Add 50 mL of water to the beaker of magnesium sulfate, stir to mix thoroughly, and record the voltage and current readings on the DMM on line B of Table 16-6.
11. Add 50 mL of water to the beaker of copper sulfate, stir to mix thoroughly, and record the voltage and current readings on the DMM on line C of Table 16-6.

DISPOSAL: The magnesium sulfate solution can be flushed down the drain with copious water. Precipitate the copper ions by adding sodium carbonate solution and filtering the solid copper carbonate precipitate, which may be disposed of with household waste; the supernatant fluid may be flushed down the drain. The electrodes can be retained for later use.

FIGURE 16-4: *A voltaic cell made up of two half-cells*

TABLE 16-6: *Build a voltaic cell—observed data*

Electrolyte concentration	Voltage	Current
A. 1.0 M Cu + 1.0 M Mg	___.___V	___.___ mA
B. 1.0 M Cu + 0.75 M Mg	___.___V	___.___ mA
C. 0.75 M Cu + 0.75 M Mg	___.___V	___.___ mA

REVIEW QUESTIONS

Q1: Which of the electrodes is the anode and which the cathode?

Q2: At which electrode does oxidation occur?

Q3: What is the purpose of the tubing filled with sodium chloride solution?

Q4: If you immerse a magnesium electrode in a solution of copper sulfate, metallic copper spontaneously plates out on the electrode. If you place a copper electrode in a solution of magnesium sulfate, no such spontaneous reaction occurs. Why?

Q5: If you substituted an aluminum electrode for the magnesium electrode, and a silver electrode for the copper electrode, what voltage would you expect the cell to produce? Why?

Q6: Did the voltage and/or current change when you altered the concentration of the electrolytes? Why or why not?

LABORATORY 16.6:

BUILD A BATTERY

In everyday life, the words "cell" and "battery" are often used interchangeably. For example, most people refer to AA cells as "AA batteries." But in chemistry the terms "cell" and "battery" have specific meanings. In the preceding experiments, we've been working with *cells*, which are self-contained or separated containers that use an electrochemical reaction to produce electricity. A *battery* is an interconnected group of two or more cells.

A battery may be constructed by connecting cells in series, which is to say, connecting the anode of one cell to the cathode of the next cell in a daisy chain. In a series battery, the voltages are additive. For example, if you construct three cells, each of which provides 1.2V at a current of 50 mA, and connect those three cells in series, the resulting battery provides 3.6V (1.2V · 3) at a current of 50 mA. The advantage of a series battery is that it provides higher voltage than an individual cell can provide.

Conversely, a battery may be constructed by connecting cells in parallel, which is to say, connecting the anodes of all of the member cells together and the cathodes of all of the member cells together. A battery connected in parallel has the same voltage as any of the individual cells, but the current is additive. For example, if you construct three cells, each of which provides 1.2V at a current of 50 mA, and connect those three cells in parallel, the resulting battery provides 1.2V at a current of 150 mA (50 mA · 3). The advantage of a parallel battery is that it provides higher current than an individual cell can provide.

In theory, it's also possible to construct a hybrid series/parallel battery that contains a group of series batteries connected in parallel (or, another way of looking at it, a group of parallel batteries connected in series). Such hybrid batteries provide both higher voltage and higher current than the individual cells provide. In practice, though, when both higher voltage and higher current are needed, it's more common to construct a series battery (to boost voltage) whose electrodes are physically large with high surface areas and can therefore provide higher current directly. A 12V car battery is an excellent example of such a battery.

Another important aspect of a battery is its internal resistance. Voltage, current, and resistance are interrelated, as stated by Ohm's Law: $E = I \cdot R$

REQUIRED EQUIPMENT AND SUPPLIES

☐ goggles, gloves, and protective clothing

☐ digital multimeter (DMM)

☐ patch cables with alligator clips (see Substitutions and Modifications)

☐ sharp knife

☐ steel wool or sandpaper

☐ lemon (3)

☐ electrodes, copper and magnesium (3 each)

SUBSTITUTIONS AND MODIFICATIONS

- If you do not have patch cables with alligator clips, you may substitute lengths of ordinary copper wire (such as from a discarded lamp or appliance cord or from a discarded electronic device). Use wooden clothespins to clamp the wires to the electrodes, as shown in Figure 16-5.

- You may substitute another citrus fruit, such as oranges or limes, for the lemons.

- You may substitute electrodes of different metals for the copper and/or magnesium electrodes. If you substitute different metals, choose two metals with standard reduction potentials that are as far apart as possible. (Aluminum is a reasonable substitute for magnesium, and much easier to find.)

where E is voltage in volts (V), I is current in amps (A), and R is resistance in ohms (Ω). At a particular voltage, the amount of current in a circuit is determined by the resistance of the circuit, including the internal resistance of the battery. Higher resistance means lower current, and vice versa. For that reason, battery designers take great pains to minimize internal resistance.

In this laboratory session, we'll examine the voltage, resistance, and current of a single cell. We'll then use multiple cells to construct batteries in series and parallel, and examine the characteristics of those batteries.

CAUTIONS

Although none of the materials used in this lab session is particularly hazardous, make sure to dispose of the food items after use. *DO NOT EAT ANY FOOD ITEM* used in this session. The food items will be contaminated with metal ions, including toxic heavy-metal ions. Wear splash goggles, gloves, and protective clothing.

PROCEDURE

1. If you have not already done so, put on your splash goggles, gloves, and protective clothing.
2. Use steel wool or sandpaper to polish each of the electrodes to remove surface oxidation, oils, and other contaminants.
3. Use the knife to make two slits in the one of the lemons, spaced as far apart as possible, and large enough to accept the electrodes. It's important that the two electrodes do not touch once they are inserted into the lemon.
4. Insert the copper electrode in one of the slits and the magnesium electrode in the other slit.
5. Touch or clamp one of the leads from the DMM to each of the two electrodes and observe the voltage reading on the DMM. If necessary, reverse the leads to obtain a positive voltage reading. Record the voltage reading to 0.01V on line A of Table 16-7.
6. Switch the DMM to resistance mode and record the resistance reading on line A of Table 16-7.
7. Switch the DMM to current mode and record the current reading on line A of Table 16-7.
8. Repeat steps 3 and 4 to prepare the other two lemons.
9. Use a patch cable to connect the copper electrode on the first lemon to the magnesium electrode on the second lemon.
10. Use patch cables to connect the magnesium electrode on the first lemon to one of the DMM terminals and the copper electrode on the second lemon to the other DMM terminal. If necessary, reverse the cable connections on the DMM to obtain a positive voltage reading. Record that voltage reading on line B of Table 16-7.
11. Switch the DMM to resistance mode and record the resistance reading on line B of Table 16-7.
12. Switch the DMM to current mode and record the current reading on line B of Table 16-7.
13. Add the third lemon to the circuit by disconnecting the copper electrode on the second lemon from the DMM and reconnecting it to the magnesium electrode on the third lemon. Then connect the copper electrode on the third lemon to the DMM terminal. Record the voltage reading on line C of Table 16-7.

14. Switch the DMM to resistance mode and record the resistance reading on line C of Table 16-7.
15. Switch the DMM to current mode and record the current reading on line C of Table 16-7.
16. Disconnect all of the patch cables.
17. Use a patch cable to connect the copper electrode of the first lemon to the copper electrode of the second lemon.
18. Use a second patch cable to connect the magnesium electrode of the first lemon to the magnesium electrode of the second lemon.
19. Use patch cables to connect the magnesium electrode on the first lemon to one of the DMM terminals and the copper electrode on the first lemon to the other DMM terminal. If necessary, reverse the cable connections on the DMM to obtain a positive voltage reading. Record that voltage reading on line D of Table 16-7.
20. Switch the DMM to resistance mode and record the resistance reading on line D of Table 16-7.
21. Switch the DMM to current mode and record the current reading on line D of Table 16-7.
22. Use a patch cable to connect the copper electrode on the third lemon to the copper electrode on the second lemon, and a second patch cable to connect the magnesium electrode on the third lemon to the magnesium electrode on the second lemon. Record the voltage reading on line E of Table 16-7.
23. Switch the DMM to resistance mode and record the resistance reading on line E of Table 16-7.
24. Switch the DMM to current mode and record the current reading on line E of Table 16-7.

DISPOSAL: Dispose of the lemons with household waste. The electrodes can be retained for later use.

FIGURE 16-5: *A two-cell series lemon-powered battery*

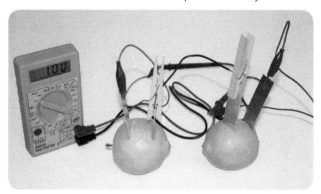

TABLE 16-7: *Build a battery—observed data.*

Configuration	Voltage	Resistance	Current
A. Cell	__.__ V	__.__ Ω	__.__ mA
B. Series battery (2-cell)	__.__ V	__.__ Ω	__.__ mA
C. Series battery (3-cell)	__.__ V	__.__ Ω	__.__ mA
D. Parallel battery (2-cell)	__.__ V	__.__ Ω	__.__ mA
E. Parallel battery (3-cell)	__.__ V	__.__ Ω	__.__ mA

REVIEW QUESTIONS

Q1: From your measurements of voltage, resistance, and current in the cell, series batteries, and parallel batteries, what general conclusions can you draw?

_____

_____

_____

_____

_____

Q2: Which metal is the anode and which is the cathode in these batteries? In which direction do electrons flow?

_____

_____

_____

Q3: A standard 12V lead-acid automobile battery when fully charged has a terminal voltage of 12.6V to 12.8V. How many magnesium/copper cells would be required to reach at least this voltage?

_____

_____

_____

Q4: If starting a particular automobile requires 600 amps (600,000 mA), how many of your magnesium/copper/lemon cells would be needed to construct a battery capable of starting that automobile? Would the internal resistance of the lemon battery be an issue?

_____

_____

_____

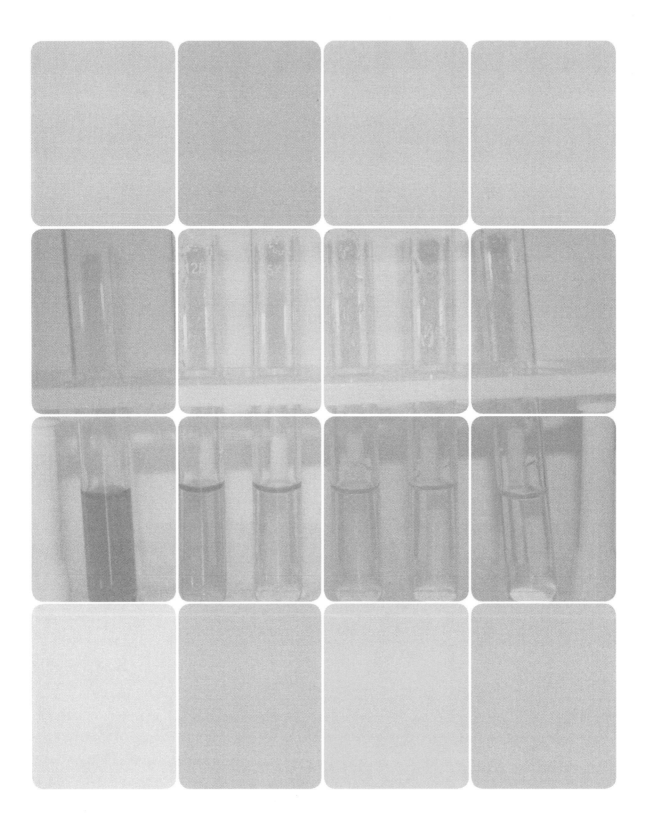

Laboratory: Photochemistry

17

Many chemical reactions occur spontaneously at normal ambient temperatures and pressures, but some chemical reactions require energy from an external source, called *activation energy*, to initiate the reaction. Once the reaction is initiated, it may proceed spontaneously to completion, or it may cease if external energy is not continuously supplied. (Technically, all reactions have an activation energy; however, for a spontaneous reaction at ambient temperature, that amount of energy is present in the system already.)

For example, the gasoline used to fuel an automobile does not react spontaneously with the oxygen in the air. (Well, actually it does, but not quickly enough to be useful.) The spark plug supplies the activation energy required to initiate the reaction. Once the reaction between the gasoline and oxygen begins, it is sufficiently *exothermic* to supply the energy required to sustain the reaction, which proceeds spontaneously to completion. Such reactions are called *self-sustaining*.

Conversely, when you bake a cake, the baking powder or baking soda in the cake mixture reacts to release carbon dioxide, which causes the cake to rise. But that reaction, once initiated, does not proceed spontaneously to completion, because the reaction does not produce enough energy to be self-sustaining. A continuing source of external energy is required to force the reaction to completion, and that energy is supplied by the heat of the oven as the cake bakes. (If the baking powder reaction were self-sustaining, you could bake a cake just by touching a match to it. Yee haw.)

Activation energy can take many forms. In an automobile engine, it's provided by an electrical spark. In many routine laboratory reactions, activation energy is supplied by heating a test tube or flask. In a firearm, activation energy is supplied mechanically when the firing pin strikes the primer. But there is a separate class of reactions in which the activation energy is supplied by visible or ultraviolet light, or some other form of electromagnetic radiation. A reaction of this class is called a *photochemical reaction*. The study of photochemical reactions is called *photochemistry*.

The laboratory session in this chapter demonstrates various aspects of photochemical reactions.

· ·

EVERYDAY PHOTOCHEMISTRY

Photochemical reactions are used frequently in laboratory syntheses, but they are also commonplace in everyday life. Here are just a few examples:

- When you spend the day at the beach, your skin tans as a result of a photochemical reaction. Skin cells called *melanocytes* are stimulated by the ultraviolet light present in sunlight to produce a brown pigment called *melanin*, which diffuses in the skin and protects it by absorbing additional ultraviolet light.

- Smog is produced by a photochemical reaction of the nitrogen oxide (NO) present in automobile exhaust. The raw nitrogen oxide reacts with atmospheric oxygen and unburned hydrocarbons to produce various toxic gases, including nitrogen dioxide (NO_2), ozone (O_3), and peroxyacetyl nitrate (CH_3COONO_2), all of which are components of photochemical smog.

- When you shoot a roll of 35 mm film, you are using a photochemical reaction to record the images. In fact, human vision depends on the photochemical isomerization of retinaldehyde (also commonly called "retinal"), so you are using photochemistry right now to read this page.

- And then there's the most important photochemical reaction of all, photosynthesis, in which the chlorophyll present in green plants uses the energy from sunlight to combine water and carbon dioxide from the atmosphere to produce the carbohydrates that are the foundation of our diets, and without which plant and animal life could not exist.

LABORATORY 17.1:

PHOTOCHEMICAL REACTION OF IODINE AND OXALATE

Ammonium oxalate reacts in solution with elemental iodine to form ammonium iodide and carbon dioxide, but the reaction rate is very low at room temperature. In this laboratory session, we investigate the effects on the reaction rate of elemental iodine with oxalate ions by exposing these reactants to various types and intensities of light for differing periods of time.

The balanced equation shows that one molecule of aqueous ammonium oxalate reacts with one molecule of iodine to form two molecules of ammonium iodide and two molecules of carbon dioxide.

$$(NH_4)_2C_2O_4(aq) + I_2(aq) \rightarrow 2\ NH_4I(aq) + 2\ CO_2(g)$$

Or, looking at the individual atom and ion species,

$$2\ NH_4^+(aq) + C_2O_4^{2-}(aq) + 2\ I^0(aq)$$
$$\rightarrow 2\ NH_4^+(aq) + 2\ I^-(aq)\ + 2\ CO_2(g)$$

Oxalate ions are oxidized to carbon dioxide, and iodine is reduced to iodide ions. Iodine (oxidation state 0) is strongly colored in aqueous solution—an intense orange in the concentration we're using—and iodide ions (oxidation state −1) are colorless. By observing the color change, if any, we can judge

REQUIRED EQUIPMENT AND SUPPLIES

- ☐ goggles, gloves, and protective clothing
- ☐ balance and weighing paper (optional)
- ☐ test tubes (12)
- ☐ stopper (to fit test tubes)
- ☐ test tube rack
- ☐ 150 mL beaker (1)
- ☐ graduated cylinder, 100 mL
- ☐ graduated cylinder (10 mL) or 5 mL measuring pipette
- ☐ dropper or Beral pipette (1)
- ☐ aluminum foil
- ☐ clear household ammonia (25 mL)
- ☐ oxalic acid solution (dissolve 2.5 g or 3/4 tsp. in 25 mL of water)
- ☐ tincture of iodine (small bottle)
- ☐ light sources (incandescent light, fluorescent light, sunlight, open shade)

how far the reaction has proceeded to the right. If the solution remains orange, we know that it contains mostly reactants. If the solution turns colorless, we know that it contains mostly products. If the solution turns an intermediate color, we know that the solution contains a mixture of reactants and products proportionate to the degree of color change.

PROCEDURE

This laboratory session has two parts. In Part I, we prepare standard reference solutions of iodine. In Part II, we test the effects of various types of light on the iodine/oxalate solutions and compare our results against the standard reference solutions we made in Part I. Setting up the experiment and completing the procedures in Parts I and II should take about 30 minutes of actual lab time.

PART I: PREPARE REFERENCE SAMPLES

We expect that some or all of the types of light we use will cause the iodine/oxalate solution to react, partially or completely converting the orange iodine solution to colorless iodide solution. We can use the color of the resulting solutions to judge how far the reaction has proceeded. An intense orange color tells us that the reaction has proceeded little or not at all, and a colorless or

pale yellow solution tells us that the reaction has proceeded to completion, or nearly so.

Just eyeballing each sample gives us imprecise results. We can judge that the reaction has proceeded to completion, mostly to completion, only a bit, or not at all. For quantitative results, we need a set of standard reference samples, with known concentrations of iodine. By comparing the test samples with these reference samples, we can judge the actual concentration of iodine in the test samples (and therefore how far the reaction has proceeded) with some degree of accuracy. (We'll use the same visual colorimetry procedures we used in Laboratory 7.5.)

Follow these steps to prepare a set of standard reference samples for comparison:

1. If you have not already done so, put on your splash goggles, gloves, and protective clothing.
2. Label six test tubes from 1 to 6.
3. Measure 8 mL of water into test tube #1 and add 20 drops (1 mL) of the iodine solution to yield 9 mL of dilute

CAUTIONS

Oxalic acid is a strong organic acid that is very toxic and an irritant. Avoid breathing the dust or allowing it to contact your skin. Ammonia is an irritant. Tincture of iodine is an irritant and stains skin and clothing. (Stains can be removed with a dilute solution of sodium thiosulfate.) Wear splash goggles, gloves, and protective clothing.

SUBSTITUTIONS AND MODIFICATIONS

- Reasonably pure oxalic acid is available from hardware stores as wood bleach powder and from auto parts stores as radiator cleaner crystals. Zud cleanser contains only 5% to 10% oxalic acid by weight, and so should be used only as a last resort. To use Zud, mix about 100 g of the cleanser with about 75 mL of warm water. Strain the resulting goop through a coffee filter, which yields a cloudy solution. Set the solution aside to let the solids settle, which may take overnight, and then carefully decant off 25 mL of clear liquid. This liquid is a nearly saturated solution of oxalic acid.

- Dr. Mary Chervenak adds that Bar Keeper's Friend also contains oxalic acid (as the dihydrate—the product is available at Linens 'n Things, among other places). I couldn't tell the concentration from the MSDS, but I think the powder is quite concentrated.

- Not all drugstores carry tincture of iodine, because much better antiseptics are available. You can substitute a solution of 0.25 g of iodine crystals dissolved in 25 mL of 50% to 70% ethanol. Iodine crystals are sold by camping supply stores for water purification.

- The beaker is used only for mixing. You can substitute any convenient small container.

DR. MARY CHERVENAK COMMENTS:

Some interesting things about oxalic acid: oxalic acid and oxalates are abundantly present in many plants, most notably lamb's quarters, sour grass, and sorrel (including Oxalis). A lot of these plants are weeds that grow right outside your doorstep and have been used, in the not-so-distant past, as salad greens. Common foods that are edible but that still contain significant concentrations of oxalic acid include—in decreasing order—star fruit (carambola), black pepper, parsley, poppy seed, rhubarb stalks, amaranth, spinach, chard, beets, cocoa, chocolate, most nuts, most berries, and beans. Oxalic acid (as oxalate ion) reacts readily with metal ions, including Ca^{+2}, Fe^{+2}, and Mg^{+2}. The gritty "mouth feel" one experiences when drinking milk with a rhubarb dessert is caused by precipitation of calcium oxalate. The primary component of kidney stones is calcium oxalate.

iodine solution. Stopper and shake test tube #1 to mix the contents thoroughly.

4. Use a measuring pipette or Beral pipette and 10 mL graduated cylinder to transfer 4.5 mL (half the contents) from tube #1 to tube #2. Add 4.5 mL of water to tube #2 and mix thoroughly.

5. Transfer 4.5 mL (half the contents) from tube #2 to tube #3. Add 4.5 mL of water to tube #3 and mix thoroughly.

6. Repeat this procedure to create reference comparison samples in tubes #4, #5, and #6. When you finish, you have six test tubes, each of which has half the concentration of iodine that is in the preceding tube. Tube #1 should be an intense orange color, and tube #6 a pale yellow-orange.

Test tube #1 contains the same concentration of iodine that will be the starting point in the reactions in Part II. If we define the concentration of iodine in test tube #1 as 100%, test tubes #2 through #6 contain 50%, 25%, 12.5%, 6.25%, and 3.125%, respectively. Place these test tubes in one of the racks. You'll use them later to estimate how far the reaction has proceeded in each of the test samples.

FIGURE 17-1:
Reference samples with various concentrations of iodine

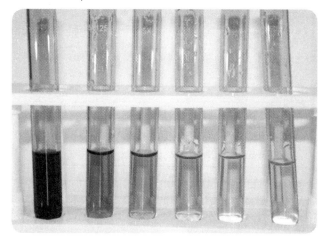

PART II: DETERMINE THE EFFECTS OF DIFFERENT LIGHT SOURCES ON IODINE/OXALATE SOLUTION SAMPLES

1. If you have not already done so, put on your splash goggles, gloves, and protective clothing.

2. Place a weighing paper on the balance pan and tare the balance to read 0.00 g. Add oxalic acid crystals until the balance indicates about 2.5 g. (The exact amount is not critical.) If you do not have a balance, use 3/4 tsp. of oxalic acid crystals.

3. Transfer the oxalic acid crystals to the small beaker and add 25 mL of water, measured with your graduated cylinder. The solubility of pure oxalic acid in water is about 120 g/L at room temperature, so this solution is nearly saturated. You may need to stir the contents of the beaker to dissolve all of the crystals. If some crystals remain undissolved, you can decant the clear solution off to another container or warm the solution slightly to dissolve the few remaining crystals.

4. Once the oxalic acid crystals have dissolved, add 25 mL of clear household ammonia to the beaker and stir to mix the solutions. The oxalic acid and aqueous ammonia react to form a solution of ammonium oxalate.

5. Label six test tubes A through F, and transfer 4 mL of the ammonium oxalate solution to each of the six test tubes. (The exact amount is not critical, but keep the level in each of the test tubes the same.) Using 4 mL per test tube leaves you with about half of the ammonium oxalate solution unused, in case you need to repeat the experiment or collect additional data under different conditions.

6. Wrap test tube A in aluminum foil, leaving an open flap of foil at the top of the test tube, through which you will introduce the iodine solution. The goal is to prevent the solution in this test tube from being exposed to any light at all. The contents of this test tube will serve as the control.

7. Use the dropper or Beral pipette to transfer 10 drops of the iodine solution to test tube A, and immediately close the foil flap to prevent the contents of that test tube from being exposed to light. If necessary, put the test tube in a closed drawer or closet to protect it from exposure to light. (The reaction between ammonium oxalate and iodine produces carbon dioxide gas, so do not seal any of the test tubes with a cork or stopper.)

8. Working as quickly as possible, add 10 drops of the iodine solution to each of the remaining five test tubes. Expose these five test tubes to different light sources, as follows:

   - Expose test tube B to direct sunlight.
   - Expose test tube C to open shade (under an open sky, but not in direct sunlight).
   - Expose test tube D to the ambient light in your working area.
   - Expose test tube E as close as possible to a fluorescent light.
   - Expose test tube F close to a strong incandescent light. (but not close enough to be heated significantly by it).

9. As you expose each test tube to light, record the beginning time. Depending upon the intensity and type of light to which the test tube is exposed, the reaction may take from a few minutes to many hours to proceed to completion.

10. Examine each of the test tubes after 15 minutes, 30 minutes, one hour, two hours, and four hours of exposure.

Compare the tint of the test tube to the comparison samples. Determine the nearest match, interpolating if necessary. Record your results in Table 17-1.

> DISPOSAL: Dilute all solutions with at least three or four times their volume of tap water and then flush them down the drain with plenty of water.

FIGURE 17-2: *Iodine/oxalate solutions (from left) before exposure to light, after extended exposure to a sun lamp, and after extended exposure to strong sunlight*

OPTIONAL ACTIVITIES

If you have time and the required materials, consider performing these optional activities:

- Repeat the experiment using different types of light sources (a plant grow-light, UV light, and so on) Do the results with different types of light most resemble the control test tube, the one exposed to direct sunlight, or some intermediate result? Propose an explanation for your observations.

- Determine whether increasing the intensity or exposure duration for a type of light that in your original experiment did not activate the reaction changes the results. If not, propose an explanation.

- Repeat the experiment using a different concentration of iodine (for example, 5 drops or 20 drops per tube instead of 10). Do your results change for the various types of light you tested?

- Repeat the experiment using four test tubes wrapped completely in red, yellow, green, and blue cellophane and exposed to direct sunlight. Does the color of the cellophane affect the reaction rate? If so, which colors result in the fastest and slowest reaction rates? Propose an explanation.

- Graph the colorimetry data that you recorded in Table 17-1. Is the reaction rate linear, or nearly so? If not, propose an explanation.

TABLE 17-1: *Photochemical reaction of iodine and oxalate—observed data*

	15 minutes	30 minutes	One hour	Two hours	Four hours
Example	Halfway between #1 & #2	Close match to #2	Slightly paler than #6	Colorless	N/A
Test tube A (dark)					
Test tube B (direct sun)					
Test tube C (open shade)					
Test tube D (ambient)					
Test tube E (fluorescent)					
Test tube F (incandescent)					

REVIEW QUESTIONS

Q1: Which of the light sources caused the largest and smallest visible change in the iodine/oxalate solutions? Propose an explanation.

Q2: Examine the example data given in Table 17-1. What type of light source do you think was used? Do the data given for the 30-minutes and 1-hour times match what you would expect from the datum for the 15-minute exposure time? Why or why not?

Q3: If you performed this experiment under a type and intensity of light that caused a very rapid reaction, would you expect to see bubbles of carbon dioxide produced in the solution? If not, why not? (Hint: do the stoichiometry to determine the limiting reagent. How might the excess reagent react?)

Q4: If you adjusted the amounts of reactants to near stoichiometric equivalence, would your answer to Q3 change? If not, why not? (Hint: look up the solubility of carbon dioxide and run the numbers.)

Laboratory:
Colloids and Suspensions

18

A *colloid*, also called a *colloidal dispersion*, is a two-phase heterogeneous mixture that is made up of a *dispersed phase* of tiny particles that are distributed evenly within a *continuous phase*. For example, homogenized milk is a colloid made up of tiny particles of liquid butterfat (the dispersed phase) suspended in water (the continuous phase). In comparison to true solutions, the continuous phase can be thought of as the solvent-like substance and the dispersed phase as the solute-like substance.

Each type of colloid has a name. A *solid sol* is one solid dispersed in another solid, such as colloidal gold particles dispersed in glass to form ruby glass. A *solid emulsion* is a liquid dispersed in a solid, such as butter. A *solid foam* is a gas dispersed in a solid, such as Styrofoam or pumice. A *sol* is a solid dispersed in a liquid, such as asphalt, blood, pigmented inks, and some paints and glues. An *emulsion*, sometimes called a *liquid emulsion*, is a liquid dispersed in another liquid, such as mayonnaise or cold cream. A *foam* is a gas dispersed in a liquid, such as whipped cream or sea foam. A *solid aerosol* is a solid dispersed in a gas, such as smoke and airborne particulates. An *aerosol*, sometimes called a *liquid aerosol*, is a liquid dispersed in a gas, such as fog, which is tiny water droplets suspended in air. All gases are inherently miscible (completely soluble in each other), so by definition there is no such thing as a gas-gas colloid. Some colloidal substances are a mixture of colloid types. For example, smog is a combination of liquid and solid particles dispersed in a gas (air), and latex paint is a combination of liquid latex particles and solid pigment particles dispersed in another liquid. Table 18-1 summarizes the types of colloids and their names.

WHAT ABOUT GELS?

Many reference sources incorrectly list gel as a type of colloid, describing a gel as a liquid dispersed phase in a solid continuous phase, which is properly called a solid emulsion. In fact, a gel is a type of sol in an intermediate physical phase. The density of a gel is similar to the density of the dispersing liquid phase, but a gel is physically closer to solid form than liquid form. Prepared gelatin is a good example of a typical gel. Dr. Mary Chervenak adds, "I think toothpastes are defined as colloidal gels with viscoelastic properties."

	Phase of colloid	Continuous phase	Dispersed phase	Colloid type
TABLE 18-1: *Types of colloids*	solid	solid	solid	solid sol
	solid	solid	liquid	solid emulsion
	solid	solid	gas	solid foam
	liquid	liquid	solid	sol
	liquid	liquid	liquid	emulsion
	liquid	liquid	gas	foam
	gas	gas	solid	solid aerosol
	gas	gas	liquid	aerosol
	gas	gas	gas	n/a

What differentiates a colloid from a solution or a suspension is the size of the dispersed particles. In a solution, the dispersed particles are individual molecules, if the solute is molecular, or ions, if the solute is ionic. Particles in solution are no larger than one nanometer (nm), and usually much smaller. In a colloid, the dispersed particles are much larger, with at least one dimension on the close order of 1 nm to 200 nm (= 0.2 micrometer, µm). In some colloids, the dispersed particles are individual molecules of extremely large size, such as some proteins, or tightly bound aggregates of smaller molecules. In a suspension, the dispersed particles are larger than 100 nm.

These differing particle sizes affect the physical characteristics of solutions, colloids, and suspensions, as follows:

- Solutions and (usually) colloids, do not separate under the influence of gravity; suspensions eventually settle out. In a colloid, the interactions among the tiny particles of the dispersed phase with each other and/or with the continuous phase are sufficient to overcome the force exerted by gravity on the tiny particles of the dispersed phase. In a suspension, the force of gravity on the more massive particles of the dispersed phase is sufficient to cause them to settle out eventually, although it may take a long time for that to occur. (If the particles of the dispersed phase are less dense than those of the continuous phase, as in a mixture of oil dispersed in water, for example, the dispersed phase "settles" out on top of the continuous phase, but the concept is the same.)

- Solutions do not separate when centrifuged; neither do colloids, except those that contain the largest (and most massive) dispersed particles, which may sometimes be separated in an ultracentrifuge.

- The particles in solutions and colloids cannot be separated with filter paper, but suspensions can be separated by filtering.

- Solutions pass unchanged through semipermeable membranes—which are, in effect, filters with extremely tiny pores—while suspensions and all colloids except those with the very smallest particle sizes can be separated by membrane filtration.

- *Flocculants* are chemicals that encourage particulate aggregation by physical means. Adding a flocculant to a solution has no effect on the dispersed particles (unless the flocculent reacts chemically with the solute) but adding a flocculant to a colloid or suspension causes precipitation by encouraging the dispersed particles to aggregate into larger groups and precipitate out.

- The particles in a solution affect the colligative properties of the solution, and the particles in a colloid or suspension have no effect on colligative properties.

- Solutions do not exhibit the *Tyndall Effect*, while colloids and suspensions do. The Tyndall Effect describes the scattering effect of dispersed particles on a beam of light. Particles in solution are too small relative to the wavelength of the light to cause scattering, but the particles in colloids and suspensions are large enough to cause the light beam to scatter, making it visible as it passes through the colloid or suspension.

Figure 18-1 shows the Tyndall Effect in a beaker of water to which a few drops of milk had been added. I used a green laser pointer for this image, because the much dimmer red laser pointer that I used when I actually did the lab session proved impossible to photograph well, even though it was clearly visible to the eye. The bright green line that crosses the beaker is the actual laser beam, reflected by the colloidal dispersion. The green laser pointer is bright enough that the scattered light illuminates the rest of the contents of the beaker as well.

Table 18-2 summarizes the physical characteristics of solutions, colloids, and suspensions. It's important to understand that there are no hard-and-fast boundaries between solutions, colloids, and suspensions. Whether a particular mixture is a colloid or a suspension, for example, depends not just on the particle size, but the nature of the continuous phase and the dispersed phase. For example, note that the particle size of colloids may range from about 1 nm to about 200 nm, and the particle size of suspensions may be anything greater than 100 nm. Furthermore, particle sizes are seldom uniform, and may include a wide range in any particular mixture.

So, is a particular mixture with a mean particle size of 100 nm a colloid or a suspension? It depends on the nature of the particles and the continuous phase. Solutions, colloids, and suspensions are each separated by a large gray area. Near the boundaries between types, it's reasonable to argue that a substance is both a solution and a colloid, or both a colloid and a suspension. As George S. Kaufman said, "One man's Mede is another man's Persian."

FIGURE 18-1: *The Tyndall Effect*

In this chapter, we'll prepare various colloids and suspensions and examine their properties.

TABLE 18-2: *Physical characteristics of solutions, colloids, and suspensions*

Characteristic	Solution	Colloid	Suspension
Type of particle	Individual molecules or ions	Very large individual molecules or aggregates of tens to thousands of smaller molecules	Very large aggregates of molecules
Particle size	< 1 nm	~1 nm to ~200 nm	> 100 nm
Separation by gravity?	No	No (usually; otherwise, very slowly)	Yes
Separation by centrifugation?	No	Yes, for more massive dispersed particles	Yes
Captured by filter paper?	No	No	Yes
Captured by membrane?	No	Yes (usually)	Yes
Precipitatable by flocculation?	No	Yes	Yes
Exhibits Tyndall Effect?	No	Yes	Yes
Affects colligative properties?	Yes	No	No

EVERYDAY COLLOIDS AND SUSPENSIONS

- The protoplasm that makes up our cells is a complex colloid that comprises a dispersed phase of proteins, fats, and other complex molecules in a continuous aqueous phase.

- Detergents are surfactants (surface-active agents) that produce a colloid or suspension of tiny dirt particles in an aqueous continuous phase.

- Photographic film consists of an emulsion of gelatin that serves as a substrate for a suspension of microscopic grains of silver bromide and other light-sensitive silver halide salts.

- Many common foods, including nearly all dairy products, are colloids or suspensions.

- Toothpaste, shaving gel, cosmetic creams and lotions, and similar personal-care products are colloids.

- Water treatment plants use flocculants (chemicals that cause finely suspended or colloidal dirt to clump into larger aggregates and settle out) as the first step in treating drinking water.

CAUTIONS

Be careful with the laser pointer. Although standard 1 mW Class 2 laser pointers are reasonably safe to use, you should never look directly into the beam. (And be cautious about specular reflections, too. A beam accidentally reflected off something shiny can be as hazardous as a direct exposure.) Although none of the chemicals used in this laboratory session are particularly hazardous, it's always good practice to wear splash goggles, heavy-duty gloves, and protective clothing. Discard all food items when you are finished; do not consume them.

LABORATORY 18.1:

OBSERVE SOME PROPERTIES OF COLLOIDS AND SUSPENSIONS

In this laboratory session, we'll use gravitational separation and the Tyndall Effect to test various samples to determine whether they are solutions, colloids, or suspensions.

PROCEDURE

1. If you have not already done so, put on your splash goggles, gloves, and protective clothing.
2. Light the incense or joss stick and blow it out. When it starts to produce smoke, place the 250 mL beaker inverted over the incense and allow the beaker to fill with smoke. Use the watch glass to cover the beaker.
3. Direct the beam from the laser pointer into the beaker and note whether the Tyndall Effect is evident. Allow the beaker to sit undisturbed for at least a minute or two, and then note whether the smoke/air sample separates on standing. Based on your observations, decide whether the smoke/air sample is a solution, colloid, or mixture. Record your observations by circling the appropriate items on line A of Table 18-3.
4. Rinse the beaker thoroughly. Add about a quarter teaspoon of table salt to about 200 mL of water in the beaker and stir until the salt dissolves. Repeat the procedures in step 3 and record your observations on line B of Table 18-3.
5. Rinse the beaker thoroughly. Add about 200 mL of club soda to the beaker. Repeat the procedures in step 3 and record your observations on line C of Table 18-3.
6. Rinse the beaker thoroughly. Add about 200 mL of water to the beaker and then about 20 drops of homogenized milk. Stir until the contents of the beaker are thoroughly mixed. Repeat the procedures in step 3 and record your observations on line D of Table 18-3.
7. Rinse the beaker thoroughly. Add about 200 mL of water to the beaker and then about 20 drops of vegetable oil. Stir until the contents of the beaker are thoroughly mixed. Repeat the procedures in step 3 and record your observations on line E of Table 18-3.
8. Rinse the beaker thoroughly. Add about 200 mL of starch water to the beaker. Repeat the procedures in step 3 and record your observations on Line F of Table 18-3.
9. Rinse the beaker thoroughly. Add about a quarter teaspoon of talcum powder to about 200 mL of water in the beaker and stir until the contents of the beaker are thoroughly mixed. Repeat the procedures in step 3 and record your observations on line G of Table 18-3.

REQUIRED EQUIPMENT AND SUPPLIES

☐ goggles, gloves, and protective clothing

☐ eye dropper or Beral pipette

☐ beaker, 250 mL (1 or more)

☐ watch glass

☐ stirring rod

☐ matches or lighter

☐ laser pointer, red

☐ incense or joss stick (1)

☐ table salt (~1/4 teaspoon)

☐ club soda (~200 mL)

☐ homogenized milk (~20 drops)

☐ vegetable oil (~20 drops)

☐ starch water (~200 mL)

☐ talcum powder (~1/4 teaspoon)

SUBSTITUTIONS AND MODIFICATIONS

- You may substitute any clear glass container of similar size for the 250 mL beaker(s). It saves time to use multiple containers, because you can make up samples and test the Tyndall Effect in one container while waiting for earlier samples to separate by gravity in other containers.

- You may substitute a sheet of thin cardboard or plastic cling-wrap for the watch glass; anything that covers the beaker will do.

- You may substitute any color of laser pointer for the red laser pointer, but be aware that green, blue, and white laser pointers are typically higher power and more hazardous to use.

- You may substitute tobacco (cigar, cigarette, or pipe) smoke for the smoke produced by the incense or joss stick.

(continues ...)

SUBSTITUTIONS AND MODIFICATIONS *(continued)*

- You may substitute any clear, colorless soft drink for the club soda.

- You can produce starch water by boiling a small amount of macaroni or other pasta in 250 mL of water for several minutes. Decant the water into a clean beaker and allow it to cool before use. (Alternatively, you can just save the cooking water the next time you cook pasta; it will keep for at least a day or two if refrigerated.)

- You may substitute any similar powder for the talcum powder, such as foot powder or body powder.

DISPOSAL: All waste solutions from this lab can be flushed down the drain with plenty of water.

TABLE 18-3: *Observe some properties of colloids and suspensions—observed data*

Sample	Tyndall Effect?	Separates on standing?	Classification
A. Smoke in air	yes / no	yes / no	solution / colloid / suspension
B. Salt in water	yes / no	yes / no	solution / colloid / suspension
C. Carbon dioxide in water	yes / no	yes / no	solution / colloid / suspension
D. Milk in water	yes / no	yes / no	solution / colloid / suspension
E. Oil in water	yes / no	yes / no	solution / colloid / suspension
F. Starch in water	yes / no	yes / no	solution / colloid / suspension
G. Talcum powder in water	yes / no	yes / no	solution / colloid / suspension

REVIEW QUESTIONS

Q1: What observable physical characteristic allows you to discriminate a colloid from a suspension?

_____

_____

_____

Q2: What observable physical characteristic allows you to discriminate a solution from a colloid?

_____

_____

_____

Q3: Lunar gravity is about one-sixth Earth's gravity. Might a sample that exhibits the characteristics of a suspension on the moon exhibit the characteristics of a colloid on Earth? Why?

Q4: Some samples are difficult to classify, because their physical properties are intermediate or mixed between the characteristics of solutions, colloids, and suspensions listed in Table 18-2 (for example, they may separate under the force of gravity, but very, very slowly). Why do some samples display such intermediate/mixed properties?

Q5: Consider a mixture of a solid material in water that clearly exhibits the properties of a suspension. If you created a similar mixture, but using a different continuous medium (such as vegetable oil), might that mixture behave as a colloid? Why?

LABORATORY 18.2:
PRODUCE FIREFIGHTING FOAM

A foam is a colloidal gas phase dispersed in a liquid continuous phase. Foams are commonplace in everyday life. The lather produced by shampoo is a foam, as are sea foam, shaving cream, marshmallows, the meringue in a lemon-meringue pie, and the head on a glass of beer.

Foams are an example of an unstable colloidal system. In the ordinary course of things, colloidal gas bubbles dispersed in a liquid quickly coalesce into larger and larger gas bubbles until the gas bubbles are large enough to be displaced by the liquid phase. If it is to persist longer than momentarily, a foam must be stabilized by the addition of a detergent, soap, protein, or other stabilizer to the mixture. Even when stabilized, a foam inevitably collapses into its component liquid and gas, so in that respect a foam can be thought of as a suspension that takes on the characteristics of a colloid for a short time.

Firefighters use foams made up of a carbon dioxide gas phase dispersed in a liquid water phase. Such foams suppress fires in three ways. First, the carbon dioxide dispersed in the foam does not support combustion and is heavier than air. When a foam layer covers a fire, the carbon dioxide covers and smothers the fire as the water cools it. Second, the foam itself presents a physical barrier that prevents air (and oxygen) from reaching the flame. Third, because the foam is elastic and has very low density, it covers and floats upon any burning solid or liquid. These characteristics mean that foam is effective in fighting nearly any type of fire, including burning oils and fats, for which liquid water simply spreads the fire.

In this lab, we'll produce a foam of carbon dioxide gas in water. We'll produce the carbon dioxide by reacting vinegar (acetic acid) and an aqueous solution of sodium hydrogen carbonate (sodium bicarbonate or baking soda), which is represented by the following equation:

$$CH_3COOH(aq) + NaHCO_3(aq)$$
$$\rightarrow H_2O(l) + CH_3COONa(aq) + CO_2(g)$$

The carbon dioxide produced by this reaction constitutes the gas phase of the colloidal foam, and the water the liquid phase. We'll use ordinary liquid dishwashing detergent as the stabilizing agent. And, just to make our foam more attractive, we'll use food coloring for a festive appearance.

REQUIRED EQUIPMENT AND SUPPLIES

- ☐ goggles, gloves, and protective clothing
- ☐ balance and weighing papers
- ☐ beaker, 150 mL
- ☐ beaker, 250 mL
- ☐ graduated cylinder, 100 mL
- ☐ stirring rod
- ☐ vinegar (~100 mL)
- ☐ sodium hydrogen carbonate (~7.5 g)
- ☐ dishwashing liquid (~1 mL)
- ☐ food coloring (a few drops; optional)

SUBSTITUTIONS AND MODIFICATIONS

- You may substitute any containers of similar size for the beakers.
- Sodium hydrogen carbonate is the official name for the substance available in grocery stores as baking soda or sodium bicarbonate.
- The food coloring is optional. Use it if you want the foam you produce to be colorful instead of white. Note that food coloring will stain clothes and surfaces.

PROCEDURE

1. If you have not already done so, put on your splash goggles, gloves, and protective clothing.
2. Transfer about 100 mL of vinegar to the 250 mL beaker.
3. Add about 1 mL (20 drops) of liquid dishwashing detergent to the vinegar and stir gently until it is thoroughly mixed. (You don't want to produce a lather, just to mix the vinegar and detergent.)
4. Transfer about 100 mL of tap water to the 150 mL beaker, and add about 7.5 g of sodium hydrogen carbonate to the water. Stir until the solid dissolves.
5. Dump the contents of the 150 mL beaker into the larger beaker, and watch what happens (preview in Figure 18-2).

CAUTIONS

Although none of the chemicals used in this laboratory session are ordinarily considered hazardous, it's always good practice to wear splash goggles, heavy-duty gloves, and protective clothing.

DISPOSAL: All waste solutions from this lab can be flushed down the drain with plenty of water.

FIGURE 18-2: *Homemade firefighting foam*

REVIEW QUESTIONS

Q1: Why are foams like this particularly effective at putting out fires?

Q2: Would a foam of carbon dioxide dispersed in water be a good choice for putting out burning sodium metal? Why or why not?

Q3: Old-style soda-acid fire extinguishers produce not a stream of foam, but a stream of liquid. Propose an explanation.

LABORATORY 18.3:
PREPARE A GELLED SOL

A *sol* is a solid phase dispersed in a liquid continuous phase. Ordinarily a sol is a liquid, but it can be converted to a semi-solid gel by adding a gelling agent. In some cases, the solid phase itself may also serve as the gelling agent.

In this lab, we'll prepare a flammable gelled sol that comprises ordinary gasoline as the liquid continuous phase with polystyrene plastic serving as both the dispersed solid phase and the gelling agent. This gelled sol is a slightly modified version of the material the military calls Super Napalm B. (Actual Super Napalm B uses low-octane gasoline rather than standard gasoline, and includes a small percentage of benzene, which we'll leave out because it's difficult to obtain.)

PROCEDURE

1. If you have not already done so, put on your splash goggles, gloves, and protective clothing. Verify that there are no open flames or other potential ignition sources nearby, and make sure that you have a fire extinguisher handy.
2. Use the graduated cylinder to measure 50.0 mL of gasoline and transfer it to the 250 mL beaker.
3. Weigh 15.0 g of polystyrene. The density of polystyrene foam (Styrofoam or similar) is so low that you need a large weighing boat to contain a reasonable mass. I used a 1-quart plastic kitchen container that comfortably held 15 g of rigid Styrofoam packing material broken into small chunks.
4. Add a small (thumb-size) chunk of polystyrene to the beaker and observe the reaction. The foam fizzes and appears to dissolve in the gasoline, leaving a small amount of undissolved residue. In fact, what appears to be undissolved residue is the first appearance of the gelled sol.
5. Continue adding the first 15 g of polystyrene in small chunks, using the stirring rod to press the polystyrene down into the liquid. Note that the gelled sol continues to grow in volume. After you've added the first 15 g, the beaker appears to contain mostly gelled sol, but with a significant amount of liquid gasoline remaining.
6. Weigh out another 15 g of polystyrene, and continue adding it in small chunks to the beaker, with stirring. When you've added a total of 30 g of polystyrene, the sol appears to have "soaked up" nearly all of the liquid gasoline, as shown in Figure 18-3.

REQUIRED EQUIPMENT AND SUPPLIES

- ☐ goggles, gloves, and protective clothing
- ☐ balance and weighing boat
- ☐ beaker, 250 mL
- ☐ graduated cylinder, 100 mL
- ☐ stirring rod
- ☐ matches or lighter
- ☐ watch or other timing instrument
- ☐ gasoline (100 mL)
- ☐ rigid polystyrene foam (35 g)

SUBSTITUTIONS AND MODIFICATIONS

- You may substitute any glass container of similar size for the beaker.
- You may use any form of polystyrene, including Styrofoam or the polystyrene foam used in insulated drinking cups and foam packing peanuts.

FIGURE 18-3: *The gelled sol forms*

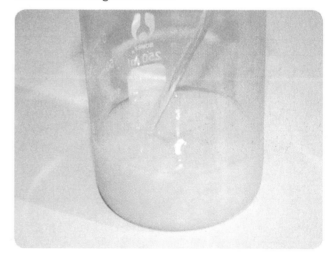

CAUTIONS

To state the obvious, we're making napalm, or at least something very close to napalm. Napalm burns furiously, sticks to anything it touches, and is very difficult to extinguish. Use extreme caution when preparing the sol. Have a fire extinguisher ready, and make absolutely sure that there are no open flames, sparks, or other potential ignition sources nearby. Work outdoors if possible, or at least in an area with excellent ventilation. Do not ignite the product indoors. Wear splash goggles, heavy-duty gloves, and protective clothing.

7. Weigh out a final 5 g of polystyrene foam, and continue adding it in chunks to the beaker, using the stirring rod to make sure that the foam you add is incorporated in the gelled sol. After a total of 35 g of polystyrene foam has been added, the contents of the beaker will appear to be completely gelled, with no liquid visible. At this point, the gelled sol is semi-rigid, enough so that it resists the force of gravity if the beaker is inverted, as shown in Figure 18-4. (Don't invert the beaker indoors, as the gelled sol might fall out unexpectedly.)

FIGURE 18-4:
The gelled sol is rigid enough to resist the force of gravity

8. If you are not already outdoors, take the beaker to a safe outdoor location with a nonflammable surface such as dirt, gravel, or concrete. (Remember that asphalt, made from tar, is flammable.) Invert the beaker, and wait a few moments to see whether the gelled sol separates from

FIGURE 18-5: *Although it is technically a liquid, the gelled sol maintains its form as though it were a solid*

the beaker. If not, tap the beaker gently to release the gelled sol. When the gelled sol separates, note that it retains its form, as shown in Figure 18-5. There should be little or no liquid gasoline remaining. (The stain visible to the right of the gelled sol in Figure 18-5 is a small amount, probably less than 1 mL, of liquid gasoline that was not incorporated in the sol.)

CAUTION

Keep all gasoline containers, including the 50 mL of liquid gasoline (until needed), far away from the ignition site.

9. Make sure that you have a fire extinguisher handy, and verify that there are no children or pets in the vicinity (or indeed, any other life forms who are unaware of what's going on). After you have verified that it is safe to do so, note the time and use the lighter or match to ignite the napalm. While the napalm burns, as shown in Figure 18-6, note your observations, including the appearance and intensity of the flame, whether the flame spreads or stays in one place, how long the flame continues, any unusual odor, and so on.

DR. PAUL JONES COMMENTS:

Be cautious with substances that you haven't used before. We've had good luck taping a match to the end of a metal rod for igniting hydrogen balloons and methane bubbles. That might be a good idea here.

10. After the napalm has finished burning and you have allowed the surface to cool, repeat the burning test using 50 mL of liquid gasoline. (Use extreme care when igniting liquid gasoline, and don't do it in your glass beaker.) Once again, make sure that you have a fire extinguisher handy, and verify that there are no children or pets in the vicinity. After you have verified that it is safe to do so, note the time and use the lighter or match to ignite the gasoline. While the gasoline burns, note your observations, including the appearance and intensity of the flame, whether the flame spreads or stays in one place, how long the flame continues, any unusual odor, and so on.

FIGURE 18-6: *Napalm burning*

DISPOSAL: All of the products used in this lab are burned. Allow the beakers and stirring rod to sit outdoors until any gasoline residue has evaporated, and then wash them with soap and water.

REVIEW QUESTIONS

Q1: What physical characteristic does a gelled sol possess that differentiates it from ordinary liquids?

Q2: What differences did you observe between burning napalm and burning gasoline?

Q3: Napalm was originally developed during World War II as an improved substitute for liquid gasoline for use in flame throwers, fire bombs, and other weapons. Based on your own observations, what advantages does napalm have as a weapon relative to liquid gasoline?

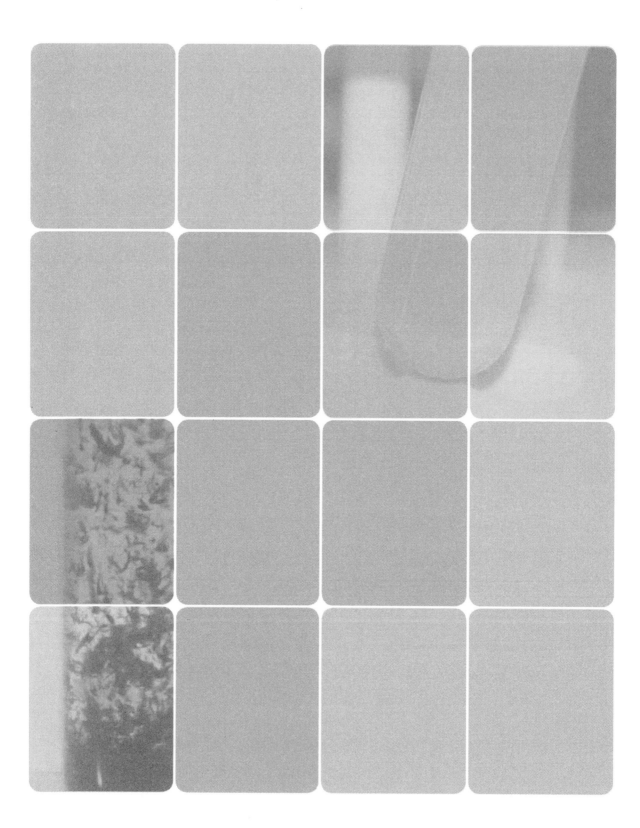

Laboratory:
Qualitative Analysis

19

Qualitative analysis is a structured set of methods used to determine the identities (but not the amounts) of the components that make up a mixture. *Qualitative inorganic analysis* is used to establish the presence or absence of particular elements or inorganic ions or compounds in an unknown sample. For example, an environmental technician might test a water sample to determine whether arsenic, barium, or mercury is present. *Qualitative organic analysis* is used to establish the presence or absence of particular organic compounds or functional groups in an unknown sample. For example, a medical technician might test a urine sample to determine whether glucose is present.

In modern university and corporate laboratories, many qualitative analyses are done instrumentally, using methods such as infrared and mass spectroscopy, nuclear magnetic resonance, neutron activation analysis, x-ray diffraction, spectrophotometry, chromatography, electrophoresis, and others. Before such instruments were introduced, chemists did qualitative analyses using wet-chemistry procedures, reacting the unknown substance with various reagents and observing the results. Such wet-chemistry procedures remain important today, both for field tests or initial screening preliminary to instrumental analysis and as a learning tool.

In this chapter, we'll use various wet-chemistry procedures to do qualitative analyses of inorganic and organic compounds.

EVERYDAY QUALITATIVE ANALYSIS

- Police and drug-enforcement agents use qualitative analysis field tests to do "pass-fail" preliminary analyses of suspect materials.

- Employers use qualitative analysis to test employees and prospective hires for drug use.

- Sports teams and sanctioning committees use qualitative analysis to test players for steroid use.

- Airports and public buildings use instrumental qualitative analysis to "sniff" for explosives.

- Environmental scientists use qualitative analysis to test soil, water, and air samples for the presence of toxic chemicals.

LABORATORY 19.1:

USE FLAME TESTS TO DISCRIMINATE METAL IONS

It has been known since antiquity that the presence of certain metals causes a flame to assume characteristic colors. For example, the presence of sodium causes a flame to assume a bright yellow color, and the presence of copper tints the flame a striking blue-green. This phenomenon occurs because heat excites atoms, boosting the outer electrons of their shells to higher energy states. As those electrons return to their original, lower energy states, the excess energy is released as photons at specific, characteristic wavelengths. The *flame test* uses this phenomenon to determine whether specific elements are present in a sample.

In theory, the flame test can discriminate numerous elements by their flame colors, some of which are listed in Table 19-1. In practice, it's a bit more difficult. The first problem is ambiguity. If a flame test shows a bright red color, for example, does the sample contain lithium or strontium (or both)? Does a pale violet color indicate the presence of potassium in the sample, or is it cesium? Is the flame pale green (antimony or barium), bright green (boron), yellow-green (manganese or molybdenum), or just plain green (copper)? The second problem is masking. Some elements, notably sodium, produce an intense flame coloration if they are present in even tiny amounts. Most elements produce a much subtler coloration, which may be overwhelmed by the more intense color of another element.

So is flame testing useless for real qualitative analytical work? Yes and no. In a home laboratory, flame tests are useful primarily to illustrate the principles involved. There are much more accurate methods available for qualitative analysis of metal ions in a home lab, which we'll explore in later sessions. Conversely, in professional laboratories, flame testing is one of the primary methods used for qualitative inorganic analysis. Professional labs use *flame spectrometers*, expensive instruments that can detect and unambiguously identify elements at the parts-per-million level by charting the emitted spectrum of a sample.

We don't have a flame spectrometer, and you probably don't either, so we'll use the traditional flame test technique in this lab.

REQUIRED EQUIPMENT AND SUPPLIES

☐ goggles, gloves, and protective clothing

☐ test tube

☐ test tube rack

☐ gas burner

☐ inoculating loop (platinum or Nichrome)

☐ cobalt glass square

☐ hydrochloric acid, concentrated (~10 mL)

☐ samples (see Substitutions and Modifications)

SUBSTITUTIONS AND MODIFICATIONS

- An inoculating loop is simply a length of inert wire (usually platinum or Nichrome) with a tiny loop at the end, held in a glass or metal handle. You can make your own by forming a loop at the end of a chrome-plated paper clip and using a wooden clothespin as a handle. If you do not have an inoculating loop, you may substitute strips of paper, toothpicks, or wood splints. Soak the paper or wood in a concentrated solution of each sample and allow it to dry. Hold the dried paper or wood in the burner flame and observe any color imparted to the flame.

- You may substitute a dark blue photographic filter for the cobalt glass square.

- You may substitute hardware-store muriatic acid of at least 30% concentration for the concentrated hydrochloric acid. (But note that hardware store acid may contain impurities that will affect the tests.)

- Obtain samples of as many metal ion species as possible, including barium, boron (use boric acid), calcium, copper(I), copper(II), iron(II), iron(III), lead, lithium, magnesium, manganese, potassium, sodium, strontium, and zinc. You need only a few grains of each sample as a solid or a small amount of the sample in solution. Try to obtain the samples as nitrates or chlorides, which vaporize more readily than carbonates, hydroxides, oxides, sulfates, and salts with other anions.

TABLE 19-1: *Flame test colors by species*

Species	Flame color	Notes
Ammonium	Faint green	Masked by most other species
Antimony	Pale blue-green to blue	Faint and easily masked
Arsenic	Light blue	Faint and easily masked
Barium	Pale green to yellow-green	
Boron	Bright green	
Calcium	Red	Calcium compounds show brick-red to orange to yellow-orange; masked by barium
Cesium	Pale violet	Easily masked
Copper(I)	Blue	
Copper(II)	Green	Copper(II) halides show blue-green
Iron	Yellow	
Lead	Blue to blue-white	
Lithium	Crimson	Masked by barium or sodium
Magnesium	Bright white	
Manganese(II)	Yellow-green	
Molybdenum	Yellow-green	
Phosphates	Blue-green to blue	When moistened with sulfuric acid
Potassium	Pale violet to lilac	Crimson when viewed through cobalt glass; potassium compound pink to lilac to violet; masked by sodium or lithium
Sodium	Bright yellow	
Strontium	Scarlet	Masked by barium
Zinc	Blue-green to green-white	

CAUTIONS

Concentrated hydrochloric acid is corrosive. Wear splash goggles, gloves, and protective clothing.

PROCEDURE

1. If you have not already done so, put on your splash goggles, gloves, and protective clothing.
2. Fill a test tube about half full of concentrated hydrochloric acid and place it in the rack.
3. Light the gas burner and adjust it to produce a small, hot flame.
4. Hold the loop at the tip of the flame and heat it until the loop adds no color to the flame.
5. Test the hydrochloric acid for purity. Dip the loop into the hydrochloric acid, allow it to drain momentarily, and then hold the loop at the tip of the flame. It should add no coloration to the flame. If color appears, particularly the intense yellow color of sodium, the hydrochloric acid is insufficiently pure. Replace it with purer acid.
6. Before you test each sample, verify that the loop is clean by holding it in the flame. If any color appears in the flame, clean the loop by dipping it in the hydrochloric acid and heating it. If necessary, repeat the acid rinse and heat until no color appears. (Hydrochloric acid reacts with any contaminants present on the loop to form chloride salts, which have low boiling points; heating the loop vaporizes the chloride salts, removing them from the loop.)

FIGURE 19-1: *A positive flame test for sodium*

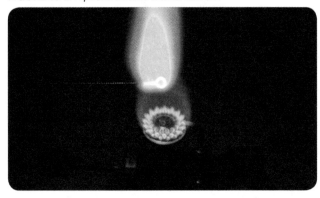

7. Test the first sample by touching the loop to the sample and holding it in the flame. If the sample is solid, dip the clean loop into the hydrochloric acid, allow it to drain momentarily, and then touch the loop to the sample. If the sample is liquid, dip the clean loop into the liquid. Record your observations in Table 19-2. If a sample does not produce a clear coloration, repeat the test using a larger amount of the sample.
8. Repeat step 7 for each of your samples.

TABLE 19-2: *Using flame tests to discriminate metal ions—observed data*

Sample	Observations
Sodium-free salt (example)	**Medium yellow coloration, indicating presence of sodium; pale blue-violet potassium coloration visible through cobalt glass**
A. Barium	
B. Boron	
C. Calcium	
D. Copper(I)	
E. Copper(II)	
F. Copper(II) halide	
G. Iron(II)	
H. Iron(III)	
I. Lead	
J. Lithium	
K. Magnesium	
L. Manganese	
M. Potassium	
N. Sodium	
O. Strontium	
P. Zinc	

DR. PAUL JONES COMMENTS:

You can also dissolve the salt in methanol (in a watch glass) and ignite the methanol. The flame will burn with the color of whatever is in there. This is also a nice way of testing which elements swamp others.

DISPOSAL: Dilute the waste hydrochloric acid with a few hundred mL of water and flush it down the drain.

REVIEW QUESTIONS

Q1: Road safety flares produce a brilliant red light. Which element do you think is used to produce that coloration?

Q2: A pyrotechnic produces brilliant green stars. Salts of what element or elements might be included in the pyrotechnic mixture to produce this effect?

Q3: During the Christmas season, some retailers stock fireplace logs that produce colorful flames. How might you produce such logs yourself from ordinary firewood?

LABORATORY 19.2:
USE BORAX BEAD TESTS TO DISCRIMINATE METAL IONS

The **borax bead test** is a fast, sensitive test for the presence of chromium, copper, cobalt, gold, iron, manganese, nickel, and tungsten. The borax bead test works because very low concentrations of the oxides of these metals in their oxidized and reduced states impart characteristic colors to a transparent borax bead. (Table 19-3 lists the characteristic colors.) Although it is seldom used nowadays for serious qualitative analysis work, the borax bead test was formerly used frequently by geologists and others for field tests of ore-bearing rock. It requires little equipment, and is sufficiently sensitive to use as a screening test.

REQUIRED EQUIPMENT AND SUPPLIES

☐ goggles, gloves, and protective clothing

☐ gas burner

☐ inoculating loop (platinum or Nichrome)

☐ borax (~1 g)

☐ samples (see Substitutions and Modifications)

SUBSTITUTIONS AND MODIFICATIONS

- You may substitute hydrated sodium carbonate (or washing soda) for the borax, although the colors will be different.
- Salts of gold and tungsten are difficult to obtain, but try to obtain samples of salts of each of the other metals listed in Table 19-3. The metal can be present as a cation (for example, cobalt carbonate for cobalt or copper sulfate for copper) or as an anion (for example, potassium permanganate for manganese or sodium dichromate for chromium).

TABLE 19-3: *Borax bead colors by element, flame type, and bead temperature*

Element	Reducing flame (hot bead)	Reducing flame (cold bead)	Oxidizing flame (hot bead)	Oxidizing flame (cold bead)
Chromium	Green	Green	Yellow to yellow-green	Green
Cobalt	Blue	Blue	Blue	Blue
Copper	Colorless	Red to reddish brown	Green to light blue-green	Blue
Gold	Red	Violet	Rose to violet	Rose to violet
Iron	Pale green to green	Green	Brown to yellow	Yellow
Manganese	Colorless	Colorless	Violet	Violet
Nickel	Gray	Gray	Violet to brown	Reddish brown
Tungsten	Green	Blue	Pale yellow	Colorless

CAUTIONS

Concentrated hydrochloric acid is corrosive. Wear splash goggles, gloves, and protective clothing.

DISPOSAL:
Dispose of all waste materials by flushing them down the drain with plenty of water. (The tiny amounts of metals contained in the beads are not hazardous.)

PROCEDURE

1. If you have not already done so, put on your splash goggles, gloves, and protective clothing.
2. Heat the loop portion of a platinum or Nichrome inoculating loop in a gas burner flame until it is red hot.
3. Dip the hot loop into borax powder (sodium tetraborate decahydrate, $Na_2B_4O_7 \cdot 10H_2O$), a small amount of which adheres to the hot loop. Heat the adhering borax in the hottest part of the gas burner flame. At first, the borax swells as it loses its water of crystallization. As you continue to heat the anhydrous sodium tetraborate, it melts and shrinks, forming a colorless, transparent, glassy bead.
4. Allow the bead to cool for a moment, dip it in distilled water to moisten it, and then dip it into the powdered sample, a few grains of which will adhere to the bead.
5. Reheat the bead in the reducing (inner, blue) part of the gas burner flame until the bead remelts. Observe the color of the bead, if any, while it is hot. The bead should remain transparent. If it becomes cloudy or opaque, you've used too much of the sample. Only a grain or two is needed.
6. Allow the bead to cool and again observe the color.
7. Reheat the bead in the oxidizing (outer, colorless) part of the gas burner flame until the bead remelts. Observe the color of the bead, if any, while it is hot.
8. Allow it to cool and again observe the color.
9. Record your observations from steps 5 through 8 in Table 19-4.
10. When you finish the test, reheat the bead and plunge the loop into cold water to remove the bead.

TABLE 19-4: *Using borax bead tests to discriminate metal ions—observed data*

Element	Reducing flame (hot bead)	Reducing flame (cold bead)	Oxidizing flame (hot bead)	Oxidizing flame (cold bead)
A. Chromium				
B. Cobalt				
C. Copper				
D. Iron				
E. Manganese				
F. Nickel				

REVIEW QUESTIONS

Q1: How closely did the bead colors you observed for various metal samples correspond to the colors listed in Table 19-3? Propose an explanation for any significant differences.

_____

_____

_____

_____

_____

_____

_____

_____

Q2: Copper chromate is used as wood preservative for "pressure-treated" wood. What bead colors would you expect if you applied the borax bead test to a sample of copper chromate?

_____

_____

_____

_____

Q3: Copper ferrocyanide is used in water analysis and treatment. What bead colors would you expect if you applied the borax bead test to a sample of copper ferrocyanide?

_____

_____

_____

_____

_____

_____

_____

LABORATORY 19.3:
QUALITATIVE ANALYSIS OF INORGANIC ANIONS

Qualitative inorganic analysis is generally done in two phases. In one phase, the sample is tested to determine which anions (negatively charged ions, usually nonmetals) are present. In the other phase, the sample is tested to determine which cations (positively charged ions, usually metals) are present.

Chemists have developed specific tests for all common anions. These tests depend on chemical reactions that occur between the anion being tested for and one or more reagents. A positive test results in some detectable change—a precipitate, color change, odor, evolution of gas, and so on. Some anion tests yield positive results if any of a group of anions is present. If that test is negative, none of the anions in that group can be present. If that test is positive, a further test is done to determine which anion on anions from that group is present in the sample.

In this lab, we'll use the following tests to identify these eight common anions:

Nitrate (NO_3^-)
The nitrate anion is identified using the ***brown ring test***. Iron(II) (ferrous) ions are added to a portion of the sample in aqueous solution. Concentrated sulfuric acid is then added to the sample by carefully allowing it to run down the side of the test tube. Because concentrated sulfuric acid is very dense, it settles to the bottom of the test tube, under the aqueous layer that contains the sample and ferrous ions. At the phase boundary between the aqueous layer and the sulfuric acid layer, a brownish or purple-brown ring forms if nitrate ion is present in the sample. Testing for nitrate ions must be done on a separate portion of the sample, because the nitrate test introduces sulfate ions into the sample and the other anion tests introduce nitrate ions in the form of nitric acid to the sample.

Sulfite (SO_3^{2-})
Acidifying a solution that contains sulfite ions causes evolution of sulfur dioxide gas, which has a characteristic sharp, choking odor familiar to anyone who has smelled the smoke from a firecracker. The presence of sulfur dioxide can be confirmed by its decolorizing effect on a dilute solution of potassium permanganate.

Carbonate (CO_3^{2-})
Acidifying a solution that contains carbonate ions causes evolution of carbon dioxide gas. Although carbon dioxide

---

> **REQUIRED EQUIPMENT AND SUPPLIES**
>
> ☐ goggles, gloves, and protective clothing
>
> ☐ gas burner
>
> ☐ test tubes (6)
>
> ☐ test tube clamp
>
> ☐ ring stand
>
> ☐ stirring rod (2)
>
> ☐ anion sample solution (see Substitutions and Modifications)
>
> ☐ sulfuric acid, 3 M (a few drops)
>
> ☐ sulfuric acid, concentrated (~2 mL)
>
> ☐ iron(II) sulfate (~2 g)
>
> ☐ nitric acid, 3 M (~15 mL)
>
> ☐ potassium permanganate, 0.02 M (one drop)
>
> ☐ barium hydroxide, saturated (one drop)
>
> ☐ silver nitrate, 0.1 M (a few mL)
>
> ☐ aqueous ammonia, 6 M (a few mL)
>
> ☐ aqueous ammonia, 15 M (a few mL)
>
> ☐ barium nitrate, 0.1 M (a few mL)
>
> ☐ ammonium molybdate, 0.5 M (a few mL)

is odorless, its presence can be confirmed by testing with a solution of barium hydroxide, which turns cloudy in the presence of carbon dioxide.

Chloride (Cl^-), Bromide (Br^-), and Iodide (I^-)
Adding silver nitrate solution to an acidic solution that contains halide ions causes the corresponding silver halide to precipitate. Silver chloride is white, silver bromide cream-colored, and silver iodide pale yellow, but it can be difficult to discriminate by color alone, particularly if the sample contains a mix of halide ions. The particular silver halide or halides present can be confirmed by testing with aqueous ammonia. A 6 M ammonia solution dissolves silver chloride,

SUBSTITUTIONS AND MODIFICATIONS

- Make the anion sample solution by adding a small amount (about 1/8 teaspoon) of each of the following chemicals to a small beaker that contains about 20 mL of distilled water: potassium nitrate, sodium sulfite, sodium carbonate, sodium chloride, potassium bromide, potassium iodide, sodium sulfate, sodium phosphate. Swirl until all of the solids dissolve. If necessary, heat the solution gently or add a bit more distilled water.
- You may substitute salts with the same anions; for example, ammonium nitrate for potassium nitrate or copper sulfate for sodium sulfate.

CAUTIONS

Nitric acid, sulfuric acid, and ammonium hydroxide are corrosive. Potassium permanganate is a strong oxidizer. Silver nitrate is an oxidizer, a corrosive, and stains skin, clothing, and other organic materials. (Stains can removed with a solution of sodium thiosulfate.) Barium hydroxide and barium nitrate are toxic, and barium nitrate is a strong oxidizer. Read the MSDS for each chemical before you use it. Wear splash goggles, gloves, and protective clothing.

but not the bromide or iodide salt. A 15 M ammonia solution dissolves silver bromide, but not silver iodide. (Actually, the halides are not dissolved, but instead form soluble complexes with the ammonium ions.)

Sulfate (SO_4^{2-})

Adding barium nitrate to an acidic solution that contains sulfate ions causes white barium sulfate to precipitate.

Phosphate (PO_4^{3-})

Phosphate ions in solution with nitric acid are precipitated by the addition of ammonium molybdate, which reacts with the phosphate ions to form the insoluble complex salt $(NH_4)_3PO_4 \cdot 12MoO_3$.

There are dozens of other common anions, each of which has a corresponding test. For example, thiocyanate (SCN^-) behaves as a pseudo-halide anion, with characteristic reactions similar to the chloride, bromide, and iodide ions. The presence of thiocyanate ions can be confirmed by adding Fe^{3+} (ferric) ions, which combine with thiocyanate ions to form the blood-red ferrithiocyanate $[Fe(SCN)(H_2O)_5]^{2+}$ complex. Similarly, the presence of the hexacyanoferrate(III) ($[Fe(CN)_6]^{3-}$) anion (usually called ferricyanide) can be confirmed by adding a solution that contains Fe^{2+} (ferrous) ions, which combine with ferricyanide ions to form a characteristic dark-blue precipitate of the pigment Prussian Blue.

PROCEDURE

This lab has two parts. In Part I, we'll test a portion of the anion sample solution for the presence of nitrate ions. In Part II, we'll test a second portion of the anion sample solution for the presence of sulfite, carbonate, chloride, bromide, iodide, sulfate, and phosphate ions. If you have not already done so, put on your splash goggles, gloves, and protective clothing.

PART I: TESTING FOR NITRATE IONS

1. Transfer about 5 mL of the sample to a clean test tube.
2. Acidify the sample by adding two or three drops of 3 M sulfuric acid and swirling or stirring to mix the solution.
3. Add about 2.0 g of anhydrous iron(II) sulfate (or the equivalent mass of hydrated ferrous sulfate) to the test tube and swirl until the solid dissolves. If necessary, heat the solution gently.
4. Hold or clamp the test tube at a 45° angle and slowly add about 2 mL of concentrated sulfuric acid, allowing it to run down the inside of the tube. Do not swirl or stir the liquid. The goal is to allow the concentrated sulfuric acid to form a separate layer at the bottom of the tube.

After a few minutes, examine the phase interface between the sulfuric acid and aqueous layers carefully. If nitrate ion is present in the sample, a hazy brown to brownish-purple ring appears at the interface.

PART II: TESTING FOR OTHER ANIONS

1. If you have not already done so, put on your splash goggles, gloves, and protective clothing.
2. Fill a large clean test tube about one quarter full of the sample solution.
3. Add an equal volume of 6 M nitric acid.
4. Quickly and carefully sniff to detect the odor of sulfur dioxide. Confirm the presence of sulfur dioxide with a drop of 0.02 M potassium permanganate on the tip of a stirring rod. (Don't add it to the test tube; just expose the drop on the tip of the stirring rod to the gas inside the test tube.) Sulfur dioxide reacts with potassium permanganate, turning the violet solution colorless and indicating that sulfite ions are present in the sample.
5. Confirm the presence of carbon dioxide with a drop of saturated barium hydroxide on the tip of a stirring rod. Carbon dioxide reacts with barium hydroxide to form insoluble barium carbonate, which turns the clear

solution cloudy, establishing that carbonate ions are present in the sample.

6. Warm the solution gently to drive off any remaining sulfur dioxide gas and carbon dioxide gas. (Do not boil the solution or heat it strongly; excessive heat can oxidize the iodide ion to elemental iodine.)

7. Add a few drops of 0.1 M silver nitrate solution to the solution in the test tube. A precipitate confirms the presence of one or more of the halide anions.

8. Continue adding 0.1 M silver nitrate until no further precipitation occurs. The precipitate contains all of the halide anions present in the sample. Allow the precipitate to settle completely.

9. Carefully decant the supernatant liquid or use a pipette to draw it off. Get as much of the liquid as possible, while leaving all of the precipitate in the test tube. Transfer the supernatant liquid to another test tube. That liquid contains any sulfate and phosphate ions present in the sample.

10. Wash the precipitate from step 8 with a few mL of distilled water. Allow the precipitate to settle completely. Draw off or decant the wash water, and discard it, keeping the precipitate.

11. Add 5 mL of 6 M aqueous ammonia to the tube containing the precipitate. Stir or swirl the tube. Any silver chloride present in the precipitate dissolves in the 6 M aqueous ammonia. Silver bromide and silver iodide are insoluble in 6 M aqueous ammonia, and so remain as a solid precipitate.

12. Draw off or decant the supernatant fluid and transfer it to another test tube. Add about 5 mL of 6 M nitric acid to the new test tube to neutralize the 6 M aqueous ammonia. Silver chloride forms a white precipitate. Discard the contents of this test tube.

13. Add 4 mL of 15 M (concentrated) aqueous ammonia to the precipitate remaining from step 11. Stir or swirl the tube. Any silver bromide present in the precipitate dissolves in the 15 M ammonia. Silver iodide is insoluble in 15 M ammonia and so remains a solid light yellow precipitate, confirming the presence of the iodide anion.

14. Draw off or decant the supernatant fluid and transfer it to another test tube. Add about 10 mL of 6 M nitric acid to that test tube to neutralize the 15 M aqueous ammonia. Silver bromide forms a cream-colored precipitate. Discard the contents of this test tube.

DISPOSAL: Dispose of all waste materials by neutralizing the acids with sodium bicarbonate and then flushing the neutralized solutions down the drain with plenty of water.

15. The test tube containing the supernatant fluid from step 9 includes the sulfate and phosphate ions present in the sample. Add a few drops of 0.1 M barium nitrate solution to the supernatant fluid. A white precipitate confirms the presence of sulfate ions. Continue adding barium nitrate until no further precipitation occurs.

FIGURE 19-2: *A white precipitate upon addition of barium nitrate confirms the presence of sulfate ions*

16. Draw off or decant about 4 mL of the supernatant liquid and transfer it to another test tube. (You may discard the remaining supernatant fluid and the precipitate from step 15.) This liquid must contain only phosphate ions, because all of the other anions have been removed by earlier tests.

17. Add about 2 mL of 6 M nitric acid to the supernatant liquid; stir to mix the solution. Add about 2 mL of 0.5 M ammonium molybdate solution to the test tube. An insoluble precipitate will form.

FIGURE 19-3: *A yellowish precipitate upon addition of ammonium molybdate confirms the presence of phosphate ions*

OPTIONAL ACTIVITIES

If you have time and the required materials, consider performing these optional activities:

- Have a friend or lab partner make up an unknown anion sample solution for you. Test that solution to determine which anions are present.

- Use the ammonium molybdate reagent to test a sample of laundry detergent labeled "phosphate-free" or "low phosphates" for the presence of phosphate ions. Compare your observations with those you make using the same reagent to test a sample of "ordinary" laundry detergent.

- Use the ammonium molybdate reagent to test for the presence of phosphate ions in any solid or liquid fertilizers or plant foods that you have access to. After you run the tests, compare your results to the contents listed on the product label. You can also test soil samples from around the yard (a flower bed is a particularly likely candidate).

- Nutrient pollution is a major source of fouling in ponds and inlets. Try testing water samples from such sources for phosphates and nitrates. Until recently, a significant percentage of phosphate pollution in standing water originated from phosphates in laundry detergents. The ban on phosphates in detergents means that most phosphate pollution in standing natural waters originates from phosphates in fertilizers.

REVIEW QUESTIONS

Q1: The precipitates produced by the anion tests take some time to settle. What alternative procedures might you use to minimize the time necessary?

Q2: Why did we wash the precipitate in step 10?

Q3: If silver nitrate were unavailable to you, what alternative reagent or reagents might you use to detect and discriminate among the halide anions? (Hint: look up the solubility of the various halide salts to find other insoluble halides.)

LABORATORY 19.4:

QUALITATIVE ANALYSIS OF INORGANIC CATIONS

Qualitative analysis of inorganic cations refers to a structured method of analyzing an unknown solid or solution to determine whether specific cations are present. Such an analysis is an iterative process of separating cations into groups based on reactions characteristic to each group and then further analyzing each separated group to determine which specific cations in that group are present in the sample. Cations are conventionally grouped as follows:

Group I cations (Ag^+, Pb^{2+}, Hg^+)

Group I cations are the only cations that produce insoluble chlorides. Group I cations can be precipitated by adding dilute hydrochloric acid, leaving all other cations in solution.

Group II cations (Cu^{2+}, Bi^{3+}, Cd^{2+}, Hg^{2+}, As^{3+}, Sb^{3+}, Sn^{4+})

Group II cations produce extremely insoluble sulfides ($K_{sp} < 10^{-30}$). Group II cations can be precipitated by adding sulfide ions at low concentration, which is most conveniently achieved by adding thioacetamide to the acidic solution. In aqueous solution, thioacetamide produces hydrogen sulfide gas. Because hydrogen sulfide is a very weak acid, the equilibrium in acidic solution stays far to the left, with most of the hydrogen sulfide present in molecular form rather than dissociated to form sulfide ions.

Group III cations (Al^{3+}, Cr^{3+}, Fe^{3+}, Zn^{2+}, Ni^{2+}, Co^{2+}, Mn^{2+})

Group III cations produce slightly soluble sulfides ($K_{sp} > 10^{-20}$). Group III cations can be precipitated by adding sulfide ions at high concentration, which is most conveniently achieved by adding aqueous ammonia to the acidic thioacetamide solution used to precipitate Group II cations until the solution is strongly basic. The addition of ammonia drives the equilibrium to the right, forcing more of the aqueous hydrogen sulfide gas produced by the thioacetamide to dissociate, forming sulfide ions.

Group IV cations (Mg^{2+}, Ca^{2+}, Sr^{2+}, Ba^{2+})

Group IV cations produce insoluble carbonates. Group IV cations can be precipitated by adding carbonate ions.

Group V cations (Na^+, K^+, NH_4^+)

Group V cations are not precipitated by chloride, sulfide, or carbonate ions, and so are the only cations remaining in solution after an unknown solution has been treated with those reagents.

REQUIRED EQUIPMENT AND SUPPLIES

☐ goggles, gloves, and protective clothing

☐ 2-liter soft drink bottle labeled "hazardous waste"

☐ test tubes (7)

☐ test tube rack

☐ eye dropper or Beral pipette (5)

☐ graduated cylinder, 10 mL

☐ stirring rod (1 or more)

☐ litmus paper (red and blue)

☐ cation sample solutions (see Substitutions and Modifications)

☐ primary and secondary reagents (see Substitutions and Modifications)

In a standard qualitative cation analysis, the first step is to treat the unknown with chloride ions, which causes all Group I cations to precipitate out. The precipitate and supernatant liquid are then separated by filtration or centrifugation. The liquid at this point contains only Group II, III, IV, and V cations. The precipitate, which may contain any or all of the Group I cations, is subsequently treated with hot water. Lead chloride, which is insoluble in cold water, is reasonably soluble in hot water, and so can be separated from any silver chloride and/or mercury(I) chloride present in the precipitate by removing the supernatant fluid from the precipitate by filtration or centrifugation. The supernatant fluid is tested with potassium chromate solution. If lead(II) ions are present, they combine with the chromate ions to produce insoluble lead chromate, which precipitates as a solid with a characteristic yellow color, confirming the presence of lead(II) ions in the sample.

The precipitate from the first step, which may contain silver chloride and/or mercury(I) chloride, is then treated with aqueous ammonia, with which silver chloride forms a soluble complex. Any solid precipitate remaining after treatment with aqueous ammonia must be mercury(I) chloride, and so confirms the presence of mercury(I) ions in the sample. To confirm the presence of silver, the aqueous ammonia is neutralized with nitric acid. If a precipitate occurs, the presence of silver ions in the sample is confirmed.

SUBSTITUTIONS AND MODIFICATIONS

- You may substitute a reaction plate for the test tubes and rack. We recommend using the test tubes, despite the larger quantities of solutions required, because the reactions are much easier to observe if you work with larger quantities of the solutions.

- You may use bench solutions of any of the reagents or analytes rather than making them up specifically for this session.

- Make up approximately 0.1 M cation sample solutions using the nitrate salts of as many of the following metals as possible: aluminum, barium, calcium, chromium, cobalt, copper, iron(II) (ferrous sulfate), iron(III) (ferric), lead, manganese (as sulfate), nickel, silver, strontium, and zinc.

- Make up primary reagent solutions, including 3 M sulfuric acid and 6 M solutions of hydrochloric acid, sodium hydroxide, aqueous ammonia, and nitric acid. (Do not store the 6 M sodium hydroxide in a glass bottle; it dissolves glass, literally. Use a Nalgene or similar plastic bottle.)

- Make up secondary reagents: 0.25 M solutions of potassium ferricyanide, potassium ferrocyanide, and potassium thiocyanate.

- How much of the reagents and cation sample solutions you need depends on whether you observe the reactions with a reaction plate or with test tubes. Using a reaction plate, you'll need 5 mL to 10 mL of each reagent (3 M sulfuric acid and 6 M solutions of hydrochloric acid, sodium hydroxide, ammonia, and nitric acid) and of each cation sample solution. Using test tubes, you'll need 25 mL to 50 mL of each reagent and of each cation sample solution.

The next step is to separate Group II ions, which form extremely insoluble sulfide salts. The supernatant liquid from the first step is made strongly acid, and thioacetamide is added to the solution. In acidic solution, thioacetamide produces a low concentration of sulfide ions, which causes the Group II ions to precipitate. Although we don't do so in this book, that precipitate can be further analyzed to isolate and identify the specific Group II ion or ions present in the sample. The supernatant liquid can contain only Group III, Group IV, and Group V cations.

That solution is treated with aqueous ammonia until it is strongly basic, which causes the thioacetamide to dissociate further, providing a higher concentration of sulfide ions. The higher sulfide concentration causes Group III cations—whose sulfides are more soluble than Group II sulfides—to precipitate, leaving only Group IV and Group V cations in solution. Again, the Group III precipitate can be further analyzed to isolate and identify the specific Group III ion or ions present in the sample, though that is not covered in this book.

Finally, the supernatant liquid, which can contain only Group IV and Group V cations, is treated with carbonate ion, which causes Group IV cations to precipitate and leaves only Group V cations in solution. The Group IV precipitate is further analyzed to isolate and identify the specific Group IV ion or ions present in the sample. The solution, which can now contain only Group V cations, can be further analyzed to isolate and identify the specific Group V ion or ions present in the sample.

Order is critical when you separate cations. For example, the carbonate ions used to precipitate Group IV cations also precipitate all of the Group I, Group II, and Group III cations. If you treat an unknown solution with carbonate ions first, you precipitate all Group I through Group IV ions without achieving any separation.

In a real qualitative inorganic analysis, any or all of these cations might be present in the unknown, and the separation and analysis would be done using microscale or semi-micro procedures. For simplicity and to avoid using expensive reagents, in this lab we'll instead observe the reactions of individual cations with various reagents and build a matrix of the observable changes that occur with various combinations of cation and reagent.

We'll use only five primary reagents (six, counting distilled water): 3 M sulfuric acid, and 6 M solutions of hydrochloric acid, sodium hydroxide, aqueous ammonia, and nitric acid. We'll also use three secondary reagents to confirm the presence of some cations: 0.25 M solutions of potassium ferricyanide, potassium ferrocyanide, and potassium thiocyanate.

PROCEDURE

1. If you have not already done so, put on your splash goggles, gloves, and protective clothing.
2. Use the graduated cylinder to transfer about 5 mL of 0.1 M aluminum nitrate solution to each of seven test tubes in a rack.
3. Add a few drops of 3 M sulfuric acid to the first test tube and swirl or stir the tube gently. If no observable change occurs, note that fact in Table 19-5 and continue to the next step. If adding sulfuric acid produces a visible change such as a precipitate, color change, bubbling, or some other action, continue adding sulfuric acid until no further change occurs. If a precipitate occurs, add aqueous ammonia to neutralize the sulfuric acid (until a drop of the solution in the test tube turns litmus paper

CAUTIONS

Hydrochloric, nitric, and sulfuric acids and concentrated aqueous ammonia are corrosive. Some of the metal salts used in this lab session are poisons, oxidizers, corrosives, or otherwise hazardous. Some are known or suspected carcinogens. Read the MSDS for each chemical before you use it. Wear splash goggles, gloves, and protective clothing. Wear a disposable N100 respirator mask if you handle any of the hazardous chemicals in solid form. Dispose of all chemical wastes properly, in accordance with hazardous material disposal laws and regulations.

turns blue) and then add the same volume of ammonia again to make the ammonia in large excess. Determine whether the precipitate dissolves in excess ammonia. Note your observations in Table 19-5.

4. Add a few drops of 6 M hydrochloric acid to the second test tube and swirl or stir the tube gently. If no observable change occurs, note that fact in Table 19-5 and continue to the next step. If adding hydrochloric acid produces a visible change such as a precipitate, color change, bubbling, or some other action, continue adding hydrochloric acid until no further change occurs. If a precipitate occurs, continue adding hydrochloric acid with stirring until you have added twice the amount of acid required to reach the first endpoint and see if the precipitate redissolves. If the precipitate does not redissolve, neutralize the hydrochloric acid to litmus paper with aqueous ammonia and continue until the ammonia is in large excess to determine whether the precipitate dissolves in excess ammonia. Note your observations in Table 19-5.

5. Add a few drops of 6 M sodium hydroxide to the third test tube and swirl or stir the tube gently. If no observable change occurs, note that fact in Table 19-5 and continue to the next step. If adding sodium hydroxide produces a visible change such as a precipitate, color change, bubbling, or some other action, continue adding sodium hydroxide until no further change occurs. If a precipitate occurs, continue adding sodium hydroxide with stirring until you have added twice the amount of sodium hydroxide required to reach the first endpoint and see whether the precipitate redissolves. If the precipitate does not redissolve with excess sodium hydroxide, test it with excess aqueous ammonia and then with excess nitric acid. Note your observations in Table 19-5.

6. Add a few drops of 6 M aqueous ammonia to the fourth test tube and swirl or stir the tube gently. If no observable

change occurs, note that fact in Table 19-5 and continue to the next step. If adding ammonia produces a visible change such as a precipitate, color change, bubbling, or some other action, continue adding ammonia until no further change occurs. If a precipitate occurs, continue adding ammonia with stirring until you have added twice the amount of ammonia required to reach the first endpoint and see whether the precipitate redissolves. If the precipitate does not redissolve with excess ammonia, test it with excess nitric acid. Note your observations in Table 19-5.

7. Add a few drops of 0.25 M potassium ferricyanide to the fifth test tube and swirl or stir the tube gently. If no observable change occurs, note that fact in Table 19-5 and continue to the next step. If adding potassium ferricyanide produces a visible change such as a precipitate or color change, continue adding potassium ferricyanide until no further change occurs. If no visible change occurs, add one or two drops of nitric acid to the test tube and note any visible change. If acidifying the solution causes no change, add three or four drops of sodium hydroxide and note any visible changes. Note your observations in Table 19-5. See Figure 19-4.

8. Add a few drops of 0.25 M potassium ferrocyanide to the sixth test tube and swirl or stir the tube gently. If no observable change occurs, note that fact in Table 19-5 and continue to the next step. If adding potassium ferrocyanide produces a visible change such as a precipitate or color change, continue adding potassium ferrocyanide until no further change occurs. If no visible change occurs, add one or two drops of nitric acid to the test tube and note any visible change. If acidifying the solution causes no change, add three or four drops of sodium hydroxide and note any visible changes. Note your observations in Table 19-5.

9. Add a few drops of 0.25 M potassium thiocyanate to the seventh test tube and swirl or stir the tube gently. If no observable change occurs, note that fact in Table 19-5 and continue to the next step. If adding potassium thiocyanate produces a visible change such as a precipitate or color change, continue adding potassium thiocyanate until no further change occurs. If no visible change occurs, add one or two drops of nitric acid to the test tube and note any visible change. If acidifying the solution causes no change, add three or four drops of sodium hydroxide and note any visible changes. Note your observations in Table 19-5.

10. Discard all solutions and precipitates in the hazardous waste container, and wash all of the test tubes thoroughly.

11. Repeat steps 1 through 10 for each of the other cation sample solutions.

TABLE 19-5: *Qualitative analysis of inorganic cations—observed data*

Cation	H_2SO_4 (3 M)	HCl	NaOH	NH_3
Example	White ppt; not soluble in excess ammonia	White ppt; not soluble in excess HCl; soluble in excess ammonia; re-ppt'd by excess nitric acid	Black ppt; not soluble in excess NaOH; soluble in excess ammonia; re-ppt'd by excess nitric acid	Black ppt; soluble in excess ammonia; re-ppt'd by excess nitric acid
Al3+				
Ag+				
Ba2+				
Ca2+				
Co2+				
Cr3+				
Cu2+				
Fe2+				
Fe3+				
Mn2+				
Ni3+				
Pb2+				
Sr2+				
Zn2+				

K$_3$[Fe(CN)$_6$]	K$_4$[Fe(CN)$_6$]	KSCN

CAUTIONS: CYANIDE GAS!

Depending on the reaction conditions, mixing strong acids with ferricyanide, ferrocyanide, or thiocyanate salts may cause deadly hydrogen cyanide gas to be evolved. The amounts we use are much too small and the solutions too dilute to present any hazard, but be very careful when handling these salts in solid form around strong acids.

DISPOSAL: Dispose of all waste materials by pouring them into the hazardous waste 2-liter soft drink bottle. When you complete this lab session, make up a saturated solution of sodium carbonate (washing soda). Add the sodium carbonate solution to the soft drink bottle until no further precipitation occurs. Decant off the supernatant liquid and flush it down the drain with plenty of water. The precipitate contains small amounts of barium, chromium, and lead. Depending on your local environmental laws and regulations, you may be permitted to dispose of this precipitate with your ordinary solid household waste, or you may need to take it to your local hazardous waste disposal facility.

OPTIONAL ACTIVITIES

If you have time and the required materials, consider performing this optional activity:

• Have a friend or lab partner make up an unknown cation sample solution for you. Test that solution to determine which cations are present.

FIGURE 19-4:
A blue precipitate upon addition of potassium ferricyanide confirms the presence of ferrous ions

REVIEW QUESTIONS

Q1: Examine the matrix that you filled in for Table 19-5. Which, if any, of the cations cannot be unambiguously identified using only the five primary reagents?

Q2: If any of the cations cannot be unambiguously identified using only the five primary reagents, can they be identified using one or more of the secondary reagents? If so, which secondary reagents are useful for identifying which cations?

Q3: If you have a sample that is known to contain only iron(II) and/or iron(III) and/or copper(II) cations, what is the minimum number and identity of the reagents needed to identify unambiguously the cations present in the unknown? What characteristic reactions would serve to discriminate these three cations?

LABORATORY 19.5:
QUALITATIVE ANALYSIS OF BONE

Animal bone is a complex structure that incorporates both organic and inorganic components. Most of the inorganic component, which makes up about 70% of bone mass, is **hydroxylapatite**, which is produced by the body but also occurs naturally as a mineral. Hydroxylapatite has the empirical formula $Ca_5(PO_4)_3(OH)$, but the formula is usually written as $Ca_{10}(PO_4)_6(OH)_2$, because the crystalline structure of hydroxylapatite is bimolecular. In bone, the hydroxyl ion is sometimes replaced by chloride, carbonate, or other anions. Other inorganic and organic components of bone include smaller amounts of other anions and cations.

In this laboratory, we'll apply what we've learned in the preceding lab sessions to analyze a sample of bone for several anions and cations.

SUBSTITUTIONS AND MODIFICATIONS

- You need a 1.0 g sample of bone. Any type of bone is acceptable, including a pork, chicken, or beef bone from kitchen scrap. You may also use a sample of bone meal.

- Although all common sodium, potassium, and ammonium salts are readily soluble in water, there is one exception. Sodium cobaltinitrite is soluble, but the potassium and ammonium cobaltinitrite salts are insoluble. We'll use sodium cobaltinitrite in this laboratory session to precipitate any potassium ions present in the original sample. You can purchase sodium cobaltinitrite, but it's just as easy and less expensive to synthesize it yourself. To do so, dissolve 25.0 g of sodium nitrite in 75 mL of water. Add 2.0 mL of concentrated (glacial) acetic acid. Add 2.5 g of cobalt nitrate hexahydrate, and stir until the cobalt salt dissolves. Allow the solution to stand for a few days, filter out any solid precipitate, and dilute the solution to 100.0 mL.

REQUIRED EQUIPMENT AND SUPPLIES

- ☐ goggles, gloves, and protective clothing
- ☐ balance and weighing papers
- ☐ beaker, 150 mL (2)
- ☐ graduated cylinder, 10 mL
- ☐ eye dropper or Beral pipette (5)
- ☐ test tubes (6)
- ☐ test tube rack
- ☐ test tube clamp
- ☐ gas burner
- ☐ ring stand
- ☐ support ring
- ☐ wire gauze
- ☐ inoculating loop
- ☐ stirring rod
- ☐ funnel and filter paper
- ☐ litmus paper (red and blue)
- ☐ bone sample (see Substitutions and Modifications)
- ☐ nitric acid, 6 M (~30 mL)
- ☐ silver nitrate, 0.1 M (a few mL)
- ☐ barium chloride, 0.1 M (1 mL)
- ☐ ammonium molybdate, 0.1 M (5 mL)
- ☐ ammonium oxalate, 0.2 M (15 mL)
- ☐ potassium thiocyanate, 0.1 M (a few drops)
- ☐ hydrochloric acid, concentrated (~1 mL)
- ☐ sodium hydroxide, 6 M (5 mL)
- ☐ sodium cobaltinitrite reagent (see Substitutions and Modifications)

CAUTIONS

Hydrochloric acid, nitric acid, and sodium hydroxide are corrosive. Some of the salts used in this lab session are poisons, oxidizers, corrosives, or otherwise hazardous. Read the MSDS for each chemical before you use it. Wear splash goggles, gloves, and protective clothing.

PROCEDURE

This laboratory has two parts. In Part I, we'll prepare the bone sample for analysis. In Part II, we'll determine whether eight specific ions are present in the sample: chloride, sulfate, phosphate, calcium, iron(III), sodium, ammonium, and potassium.

PART I: SAMPLE PREPARATION

In Part I, we prepare the bone sample for analysis by digesting it in dilute nitric acid. This procedure should be done outdoors, under an exhaust hood, or in an otherwise well-ventilated area. The reaction of nitric acid with the bone sample evolves toxic and irritating fumes of nitrogen oxides. Do not breathe these fumes or allow them to contact your eyes or skin.

1. If you have not already done so, put on your splash goggles, gloves, and protective clothing.
2. Weigh a 1.0 g sample of bone. If possible, use small pieces rather than one large chunk.
3. Add 25 mL of 6 M nitric acid to 25 mL of distilled or deionized water in a 150 mL beaker. Place the beaker on a support ring and wire gauze over the gas burner.
4. Add the bone sample to the beaker and heat the beaker gently, with stirring, to dissolve the bone sample. Do not boil the solution. As the bone sample dissolves, toxic fumes are evolved.
5. Continue heating and stirring the solution in the beaker for 10 or 15 minutes. As the reaction progresses, the quantity of fumes evolved rapidly diminishes. Some solid matter remains undissolved.
6. Remove the heat and allow the beaker to cool to room temperature.
7. Set up a funnel with filter paper and the second 150 mL beaker as the receiving vessel.
8. Filter the solution into the second beaker and discard the filter paper and solid filtrand. The filtrate is your sample.

PART II: ANALYSIS OF IONS

In Part II, we analyze the sample to determine whether chloride, sulfate, phosphate, calcium, iron(III), sodium, ammonium, and/or potassium ions are present.

1. If you have not already done so, put on your splash goggles, gloves, and protective clothing.
2. Transfer 5 mL of the sample to the first test tube. Add several drops of 0.1 M silver nitrate solution. A precipitate confirms the presence of chloride ions. (Actually, a precipitate confirms the presence of chloride, bromide, or iodide ions, but the bromide and iodide ions are present in extremely low concentrations in animal bone, so we can assume that any precipitate is caused by chloride ions.) Note your observations on line A of Table 19-6.
3. Transfer 5 mL of the sample to the second test tube. Add 10 drops of 0.1 M barium chloride to the test tube, stir or swirl to mix the solution, and look for a white precipitate that confirms the presence of sulfate ions. Record your observations on line B of Table 19-6.
4. Transfer 5 mL of the sample to the third test tube. Add about 4 mL of 6 M nitric acid and 6 mL of ammonium molybdate to the test tube and mix the solution well. A yellow precipitate confirms the presence of phosphate ion. If no precipitate occurs immediately, continue observing the test tube for a minute or two. If there is still no precipitate, heat the test tube very gently for a minute or two. (Do not heat the solution strongly. Doing so may cause a precipitate of white molybdenum trioxide, which is *not* a positive test for phosphate ions.) Record your observations on line C of Table 19-6.
5. Transfer about 3 mL of the sample to the fourth test tube. Add about 15 mL of 0.2 M ammonium oxalate and mix the solutions thoroughly. A white precipitate or white cloudiness confirms the presence of the calcium ion. (Under the acidic conditions of this test, calcium oxalate is more soluble than it is in neutral solution. Examine the test tube carefully; even a slight white cloudiness is a positive result.) Record your observations on line D of Table 19-6.
6. Transfer 5 mL of the sample to the fifth test tube. Add 5 mL of 6 M nitric acid and mix thoroughly. Add 5 drops of 0.1 M potassium thiocyanate, mix thoroughly, and look for a color change. A blood red color (Figure 19-5) indicates the presence of Fe(III) ion in relatively high concentration. At lower concentrations of Fe(III) ion, the solution assumes anything from a light red color to a very pale straw yellow color. Record your observations on line E of Table 19-6.

FIGURE 19-5:
Ferric ions react with potassium thiocyanate to produce a blood red complex

7. Dip the end of the platinum or Nichrome inoculating loop into concentrated hydrochloric acid and then hold it in the gas burner flame until the loop adds no color to the gas flame. Dip the clean loop into the sample, and then hold it in the gas flame. A bright yellow flame coloration confirms the presence of sodium ions in the sample. Record your observations on line F of Table 19-6.

8. Transfer 5 mL of the sample to the sixth test tube. Add 3 mL of 6 M sodium hydroxide solution to the test tube and mix thoroughly. Carefully sniff the test tube by using your hand to waft any vapors from the mouth of the test tube toward your nose. An odor of ammonia confirms the presence of ammonium ion in the sample. If no odor is obvious, gently warm the test tube for a minute or so, periodically testing for the scent of ammonia. Record your observations on line G of Table 19-6.

9. Continue heating the test tube until it comes to a very gentle boil. Continue the gentle boil for a minute or two to drive off all of the ammonium ions as gaseous ammonia. Allow the test tube to cool and then add about 3 mL of the sodium cobaltinitrite reagent. A precipitate of insoluble potassium cobaltinitrite confirms the presence of potassium ions in the sample. Record your observations on line H of Table 19-6.

> DISPOSAL: Dispose of all waste materials by flushing them down the drain with plenty of water.

TABLE 19-6: *Qualitative analysis of bone—observed data*

Ion	Reagent	Observations
A. Chloride(Cl^-)	Silver nitrate	
B. Sulfate (SO_4^{2-})	Barium chloride	
C. Phosphate (PO_4^{3-})	Ammonium molybdate	
D. Calcium (Ca^{2+})	Ammonium oxalate	
E. Iron(III) (Fe^{3+})	Potassium thiocyanate	
F. Sodium (Na^+)	Flame test	
G. Ammonium (NH_4^+)	Sodium hydroxide	
H. Potassium (K^+)	Sodium cobaltinitrite	

REVIEW QUESTIONS

Q1: Considering the results of your tests, name and give formulae for the compounds that you believe may be present in the bone sample.

_____

_____

_____

_____

_____

_____

_____

_____

_____

_____

_____

Q2: If a sample produces a precipitate when treated with silver nitrate, how would you proceed to establish unambiguously that the precipitate is silver chloride rather than silver bromide or silver iodide?

_____

_____

_____

_____

_____

_____

_____

_____

_____

_____

_____

Q3: Is it possible to test the dissolved bone sample to determine whether carbonate ion was present in the original solid sample? If not, explain why and devise a test procedure to determine whether

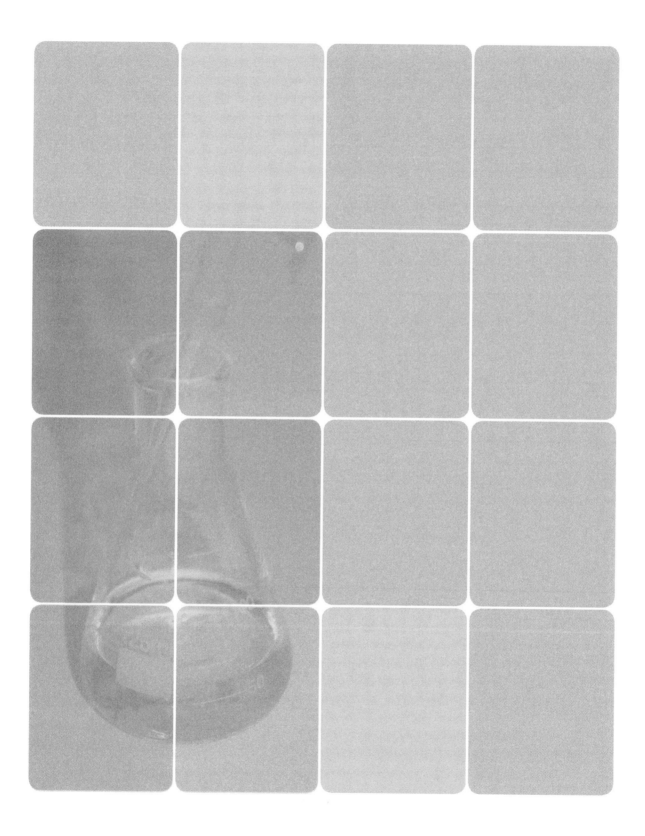

Laboratory: Quantitative Analysis

20

In the last chapter, we looked at qualitative analysis, which is used to identify the chemical species present in an unknown. After you identify which species are present, the obvious next step is to determine the relative amounts of each species that are present. The methods used to make that determination are collectively called *quantitative analysis*.

The key difference between qualitative analysis and quantitative analysis is that the former is not concerned with quantities. If you do a qualitative analysis on an unknown for the presence of potassium, for example, you care only *whether* potassium is present or absent. If you do a quantitative analysis on that sample, your goal is to determine *how much* potassium is present.

If you've completed all of the laboratory sessions in the preceding chapters, you've already done one quantitative analysis without even being aware of it. In Chapter 11, we standardized a solution of hydrochloric acid. In other words, we determined how much hydrochloric acid was present in a sample. In doing that, we did a quantitative analysis.

Because the goal of quantitative analysis is to determine amounts, it's critical to measure masses and volumes as accurately as possible. In the preceding chapter on qualitative analysis, we often gave instructions like "transfer about 5 mL of the sample," because the exact amount didn't really matter. In this chapter, the instructions are considerably more precise—such as "transfer 5.00 mL of the sample"—because for quantitative analysis, the amount is critical. The lab sessions in this chapter reflect that careful attention to quantities.

. . . .

EVERYDAY QUANTITATIVE ANALYSIS

The importance of quantitative analysis isn't limited to chemistry labs. Quantitative analysis is important in everyday life, and is sometimes literally a matter of life and death. Here are just a few examples.

Pharmaceuticals must be of known strength and purity. Quantitative analytical techniques are used to ensure both. For example, drug tablets and capsules are assayed using titration and other quantitative analytical techniques to make sure that each contains the nominal dosage, within allowable limits. Conversely, it's just as important to ensure that contaminants are present only at or below acceptable levels. Again, quantitative analytical tests are used to verify that the drug is sufficiently pure to be safe for human consumption.

Quantitative analysis is used to determine the content and purity of food. For example, wine is analyzed to determine the percentage of alcohol that it contains and to ensure that contaminants such as sulfites are at acceptably low levels. When you read the nutrition facts panel on a can of food or a bottle of soda and learn that one serving contains 3 g of fat or 8 g of protein or 5 g of carbohydrate, you're reading information that was obtained by quantitative analysis.

When you turn on your faucet, the water that comes out is safe to drink because quantitative analytical procedures are used to ensure that safety. Chemists at the water department regularly test the water to determine levels of arsenic, lead, mercury, and other dangerous heavy metals, as well as pesticides and other poisons. In poor countries, thousands of people die every week from drinking water that contains high levels of such contaminants.

The gasoline you pump into your car is actually a complex blend of raw gasoline with numerous additives that improve performance, increase mileage, and reduce air pollution. Because gasoline is blended in a continuous process (when fuel delivery trucks fill up at the terminal) rather than in batch mode, gasoline is tested regularly using quantitative analysis to ensure that the concentration of the additives is within acceptable limits.

Police officers use instrumental quantitative analysis to test the blood alcohol content of suspected drunk drivers.

LABORATORY 20.1:

QUANTITATIVE ANALYSIS OF VITAMIN C BY ACID-BASE TITRATION

Vitamin C is a vital part of the human diet. A deficiency of vitamin C causes a wasting disease called scurvy, which until the eighteenth century plagued sailors and others whose regular diets provided few fresh fruits and vegetables. In fact, the chemical name for vitamin C, ascorbic acid, is derived from scorbutus, the Latin name for scurvy.

In 1753, British Royal Navy surgeon James Lind published a book, *A Treatise on the Scurvy*, which concluded that scurvy could be prevented by including citrus fruits in the diet, and that active cases of scurvy could be treated almost miraculously simply by having the patient consume citrus fruits or juices. The Royal Navy began adding lime juice to the regular tots of rum provided to all sailors, accidentally creating the world's first rum punch. Scurvy disappeared, and British sailors became known as Limeys.

REQUIRED EQUIPMENT AND SUPPLIES

- ☐ goggles, gloves, and protective clothing
- ☐ balance and weighing papers
- ☐ hotplate
- ☐ Erlenmeyer flask, 125 mL
- ☐ graduated cylinder, 100 mL
- ☐ burette, 50 mL
- ☐ ring stand
- ☐ burette clamp
- ☐ vitamin C tablet, 500 mg
- ☐ sodium hydroxide, 0.1000 M (~50 mL)
- ☐ phenolphthalein indicator (a few drops)

Scurvy is now almost unheard of in the developed world, because our regular diets include more than sufficient vitamin C to prevent scurvy. But many people take vitamin supplements that provide much higher dosages of vitamin C than the recommended daily minimum.

In this lab session, we'll use acid-base titration with standardized 0.1000 M sodium hydroxide titrant to determine the actual amount of ascorbic acid ($C_6H_8O_6$) present in a nominally 500 mg vitamin C tablet. One mole of ascorbic acid reacts with one mole of sodium hydroxide to yield one mole of sodium ascorbate and one mole of water, according to the following balanced equation:

$$C_6H_8O_6 + NaOH \rightarrow C_6H_7O_6Na + H_2O$$

Using phenolphthalein indicator provides a sharp endpoint for the reaction. The solution remains colorless while ascorbic acid is in excess. When sodium hydroxide is even slightly in excess, the pH of the solution rapidly rises to a point above the color change range of phenolphthalein, and the solution quickly turns bright pink.

CAUTIONS

Sodium hydroxide is corrosive. Wear splash goggles, gloves, and protective clothing.

PROCEDURE

Before we begin the analysis, we need to decide what sample sizes are appropriate for our quantitative tests.

We know that a 500 mg vitamin C tablet nominally contains 500 mg of ascorbic acid, and we expect the stated value to be reasonably close to the actual value. The gram molecular weight of ascorbic acid is 176.14 g/mol, so at nominal value one 500 mg tablet should contain about 0.0028+ moles of ascorbic acid. One mole of ascorbic acid reacts stoichiometrically with one mole of sodium hydroxide, so we'll require about 0.0028+ moles of sodium hydroxide to reach the equivalence point. Our 0.1000 M solution of sodium hydroxide contains 0.1 mol/L or 0.0001 mol/mL, so we should need about 28+ mL of titrant to neutralize the ascorbic acid present in one 500 mg tablet.

1. If you have not already done so, put on your splash goggles, gloves, and protective clothing.
2. Weigh a clean, dry 125 mL Erlenmeyer flask and record its mass to 0.01 g on line A of Table 20-1.
3. Add one 500 mg vitamin C tablet to the flask, reweigh the flask, and record the combined mass of the flask and tablet to 0.01 g on line B of Table 20-1. Subtract the empty mass of the flask from the mass with the vitamin C table, and enter the mass of the vitamin C table to 0.01 g on line C of Table 20-1.
4. Use the graduated cylinder to transfer about 50 mL of distilled water to the 125 mL Erlenmeyer flask and swirl the flask gently to break up the tablet.
5. Place the 125 mL flask on the hotplate and heat it gently (do not boil) for 5 to 10 minutes with occasional swirling to dissolve the vitamin C from the sample. It's normal for a small amount of binder and other components to remain undissolved. Set the flask aside and allow it to cool to room temperature.
6. Rinse the 50 mL burette with a few mL of 0.1 M sodium hydroxide, and allow it to drain into a waste container.
7. Transfer at least 40 mL of 0.1 M sodium hydroxide to the burette. Drain a couple mL into the waste container and make sure that there are no bubbles in the burette, including the tip. Record the initial volume as accurately as possible, interpolating between the graduation marks on the burette, and record that volume on line D of Table 20-1.
8. Add a few drops of phenolphthalein indicator solution to the flask and swirl the contents until they are thoroughly mixed.
9. We calculated that we should need about 28 mL of sodium hydroxide titrant, so begin by running 25 mL or so of titrant into the receiving flask with constant swirling. As you near the endpoint, a pink color will appear in the flask where the titrant is being added, but will disappear quickly as you swirl the flask. When you reach this point, slow the addition rate to a fast drip, and continue swirling.
10. When you reach the point where the pink color persists for a few seconds, begin adding the titrant dropwise, and swirl the flask to mix the contents thoroughly after each drop. You've reached the endpoint when the pink color persists for at least 30 seconds. Record the final volume on the burette on line E of Table 20-1. Subtract the initial volume of titrant from the final volume, and record the volume of titrant used on line F of Table 20-1.
11. Using the actual molarity of your nominal 0.1 M sodium hydroxide titrant, calculate the number of moles of sodium hydroxide required to reach the endpoint and enter that value on line G of Table 20-1.
12. Sodium hydroxide reacts stoichiometrically with ascorbic acid at a 1:1 mole ratio, so the number of moles of ascorbic acid present in the sample is identical to the number of moles of sodium hydroxide needed to reach the endpoint. Calculate the mass of ascorbic acid present in the sample by multiplying the number of moles of sodium hydroxide required by the gram molecular mass of ascorbic acid (176.14 g/mol) and enter that value on line H of Table 20-1.

SUBSTITUTIONS AND MODIFICATIONS

- You may substitute any suitable container of similar size for the 125 mL Erlenmeyer flask.

- The 0.1000 M sodium hydroxide solution should be standardized, if possible. (You can standardize it against the hydrochloric acid you standardized in chapter 11. If you do not have standardized sodium hydroxide available, you can make do with a nonstandardized solution of 4.00 g of sodium hydroxide dissolved in water and made up to 100.0 mL in a volumetric flask.

DR. MARY CHERVENAK COMMENTS:

An introductory biology class that I took had students consume a known amount of Vitamin C and then collect urine for the 24-hour period following consumption. The urine was then analyzed for Vitamin C. It really was astonishing how little Vitamin C was actually absorbed. Easily one of my favorite experiments.

TABLE 20-1: *Quantitative analysis of vitamin C by acid-base titration—observed and calculated data*

Item	Value
A. Mass of empty 125 mL Erlenmeyer flask	_____.__ g
B. Mass of empty 125 mL flask + vitamin C table	_____.__ g
C. Mass of vitamin C table (B – A)	_____.__ g
D. Initial volume of sodium hydroxide titrant	_____.___ mL
E. Final volume of sodium hydroxide titrant	_____.___ mL
F. Volume of sodium hydroxide titrant used (C – B)	_____.___ mL
G. Moles of sodium hydroxide required to reach endpoint	__._____ mol
H. Mass of ascorbic acid present in the sample	_____.__ g

OPTIONAL ACTIVITIES

If you have time and the required materials, consider performing these optional activities:

- Repeat the experiment, using the same unknown and titrant, and compare the result of your second titration with that of the first. Repeat again as necessary until you have at least two titrations whose results are closely matched.

- Repeat the experiment, using a 10.00 mL aliquot of lemon juice concentrate (assume that all acid present in the sample is ascorbic acid).

- Repeat the experiment, using a 100.0 mL aliquot of orange juice.

- Reproduce the experiment that Dr. Mary Chervenak describes in the introduction to this lab session.

> **DISPOSAL:**
> The waste solutions from this lab can be flushed down the drain with plenty of water.

REVIEW QUESTIONS

Q1: How closely did the mass of vitamin C that you determined experimentally correspond to the nominal mass of vitamin C listed by the manufacturer?

Q2: What percentage of the tablet's mass is vitamin C and what percentage of its mass is binder and other inert ingredients?

Q3: Ascorbic acid and atmospheric oxygen can react spontaneously by the following redox reaction:

$$2\ C_6H_8O_6(s) + O_2(g) \leftrightarrow C_6H_6O_6(s) + 2\ H_2O(l)$$

If this reaction occurred before you completed the titration, how might it affect your results?

Q4: When chemists use titration for quantitative analysis, they ordinarily repeat the titration several times. Why?

LABORATORY 20.2:
QUANTITATIVE ANALYSIS OF CHLORINE BLEACH BY REDOX TITRATION

Chlorine laundry bleach is an aqueous solution of sodium hypochlorite (NaOCl). Chlorine bleach is produced by running an electric current through a solution of sodium chloride. The electric current oxidizes Cl^- ions to Cl^0_2 gas and reduces H^+ ions to H^0_2 gas, increasing the concentration of OH^- ions. In effect, this electrolysis can be thought of as producing an aqueous solution of sodium hydroxide and chlorine, which immediately react to form sodium hypochlorite.

In solution, sodium hypochlorite dissociates into sodium ions and hypochlorite (OCl^-) ions. The bleaching action of chlorine bleach is entirely due to the presence of hypochlorite ion, which is a strong oxidizer. The hypochlorite ion reacts with the colored unsaturated organic compounds that make up stains, oxidizing them to colorless saturated organic compounds.

Inexpensive no-name chlorine bleaches usually contain nothing but 5.25% sodium hypochlorite by mass. (That is, 100.00 g of bleach solution contains 5.25 g of sodium hypochlorite and 94.75 g of water.) Name-brand bleaches are more expensive and often contain scents and brighteners. Name-brand bleaches labeled "extra strength," "ultra," or something similar contain slightly higher concentrations of sodium hypochlorite, typically about 6%. Because the hypochlorite ion is the only active ingredient in any chlorine bleach, the sodium hypochlorite content determines bleaching effectiveness.

In this lab session, we'll use redox titration to quantitatively determine hypochlorite in a chlorine bleach sample. We use a common quantitative analysis procedure called **back titration**, in which the concentration of an unknown sample is determined by reacting it with an excess of one reagent to produce a product. That product is then back titrated with a second reagent of known concentration until the endpoint is reached. Back titration is used for many reasons, including for samples that are insoluble in water or that contain impurities that interfere with forward titration. But back titration is most commonly used when the endpoint is more apparent by titrating back than by titrating forward.

Rather than attempt to determine hypochlorite concentration by forward titration, we'll first react the hypochlorite sample in an acetic acid solution that contains excess iodide ions, which are colorless in aqueous solution. The hypochlorite ions oxidize the

REQUIRED EQUIPMENT AND SUPPLIES

- ☐ goggles, gloves, and protective clothing
- ☐ balance and weighing papers
- ☐ Erlenmeyer flask, 125 mL
- ☐ volumetric flask, 100 mL
- ☐ graduated cylinder, 10 mL
- ☐ graduated cylinder, 100 mL
- ☐ pipette, Mohr or serological, 2.00 mL
- ☐ burette, 50 mL
- ☐ dropper or Beral pipette
- ☐ funnel
- ☐ ring stand
- ☐ burette clamp
- ☐ chlorine bleach (100 mL)
- ☐ potassium iodide (~2 g per titration)
- ☐ acetic acid, glacial (~5 mL per titration)
- ☐ sodium thiosulfate, 0.1 M (~60 mL per titration)
- ☐ starch indicator (see Substitutions and Modifications)

iodide ions, forming elemental iodine, which is brown in aqueous solution. We'll then back-titrate with a standardized solution of sodium thiosulfate ($Na_2S_2O_3$), which reduces elemental iodine to iodide ions. Here are the balanced redox equations for these two reactions:

$$OCl^-(aq) + 2\ CH_3COOH^0 + 2\ I^-(aq)$$
$$\rightarrow Cl^-(aq) + 2\ CH_3COO^- + I^0_2(aq) + H_2O^0(l)$$

$$2\ S_2O_3{}^{2-}(aq) + I^0_2(aq) \rightarrow 2\ I^-(aq) + S_4O_6{}^{2-}(aq)$$

As we add sodium thiosulfate titrant to the solution of iodine, the brown color of aqueous iodine becomes fainter and fainter. Because there is no abrupt color change, it is very difficult to judge the endpoint of this titration directly. Fortunately, even at

SUBSTITUTIONS AND MODIFICATIONS

- You may substitute any suitable container of similar size for the 125 mL Erlenmeyer flask.

- You may substitute a 1.00 mL or 10.0 mL pipette for the 2.0 mL pipette at the expense of some loss in accuracy.

- Ideally, the 0.1 M sodium thiosulfate solution should be standardized, but doing so is time-consuming and requires the relatively expensive chemical potassium iodate. You can make do with a nonstandardized sodium thiosulfate solution. Make a 0.1 M solution of sodium thiosulfate by adding 1.58 g of anhydrous sodium thiosulfate or 2.48 g of sodium thiosulfate pentahydrate to your 100 mL volumetric flask and making up the solution to 100.0 mL.

- Starch indicator solution does not keep for more than few days even if refrigerated, so you should make it up fresh for each laboratory session. To so so, mix about 1 g of cornstarch or other starch with about 25 mL of cold water until it forms a paste. Add this paste to about one liter of boiling water and stir until it forms a clear suspension. Allow the liquid to cool and use it as is. This indicator is used in such small quantity that it's fine to make it up with tap water rather than distilled water. You'll need only a couple mL for each titration. Alternatively, you can use water in which potatoes, macaroni, or other pasta has been boiled.

ALTERNATIVE PROCEDURE

If you don't have a 100 mL volumetric flask, or if the mass of the filled flask would exceed the capacity of your balance, you can substitute a 25 mL or 50 mL volumetric flask. Alternatively, tare a small beaker or a foam cup on your balance and use a 10 mL or larger pipette to transfer a measured volume of bleach to the beaker. The value you obtain for density will be less accurate using the smaller volume, but is sufficiently accurate for this experiment.

DR. MARY CHERVENAK COMMENTS:

Back titration is often used when an appropriate indicator isn't available. For example, aqueous silver ions react strongly with halides, but indicators don't respond to the presence of silver ions. Back-titrating the excess silver ions with thiocyanate will enable the determination of the halide ion concentration. Ferric chloride acts as an indicator for the back titration—the ferric ion forms a colorless complex in water, but the ferric thiocyanate complex is bright red.

CAUTIONS

Chlorine bleach is corrosive and a strong oxidizer. Glacial acetic acid is corrosive. Wear splash goggles, gloves, and protective clothing.

very low concentrations, aqueous iodine reacts with a starch solution to produce an intense blue color. By using starch as an indicator, we can judge the endpoint very precisely. We'll begin the titration without the indicator. When we've added enough sodium thiosulfate titrant to reduce the brown coloration to a faint tint, we know the titration is near the endpoint. We'll then add starch indicator, which turns the solution bright blue, and continue adding sodium thiosulfate titrant dropwise until the blue color disappears.

PROCEDURE

This lab session is in two parts. In Part I, we'll determine the density of chlorine bleach solution, which we'll need later to calculate the mass percentage of sodium hypochlorite in the sample. In Part II, we'll react the chlorine bleach sample with a solution of potassium iodide in acetic acid, and titrate the iodine formed by that reaction with sodium thiosulfate titrant.

PART I: DETERMINE THE DENSITY OF CHLORINE BLEACH

In Part I, we'll determine the mass of a 100.00 mL sample of chlorine bleach, which allows us to calculate its density.

1. If you have not already done so, put on your splash goggles, gloves, and protective clothing.
2. Weigh a clean, dry, empty 100 mL volumetric flask and record its mass to 0.01 g on line A of Table 20-2.
3. Use the funnel to fill the volumetric flask with the chlorine bleach sample nearly to the index line. Use the dropper or Beral pipette to add the last mL or two of chlorine bleach, bringing the level of the solution up to the index mark on the flask.
4. Reweigh the filled flask and record its mass to 0.01 g on line B of Table 20-2.
5. Subtract the empty mass of the flask from the combined mass of the flask and chlorine bleach, and enter the mass of the chlorine bleach on line C of Table 20-2.
6. Calculate the density of chlorine bleach in g/mL and enter that value on line D of Table 20-2.

PART II: DETERMINE HYPOCHLORITE CONCENTRATION

Before we begin the analysis, we need to decide what sample size is appropriate for our quantitative test. We know that 1 g of chlorine bleach nominally contains 0.0525 g of sodium hypochlorite. The gram molecular mass of sodium hypochlorite is 74.44 g/mol, so 1 g of chlorine bleach contains about 0.0007 moles of sodium hypochlorite. In the first redox equation, one mole of sodium hypochlorite reacts with two moles of iodide ion to form one mole of molecular iodine (I_2), so the stoichiometric ratio of hypochlorite:iodine is 1:1. In the second redox equation, one mole of molecular iodine reacts with two moles of sodium thiosulfate to produce two moles of iodide ions, so the stoichiometric ratio of iodine:thiosulfate is 1:2. It follows, then, that the stoichiometric ratio for hypochlorite:thiosulfate is 1:2, which means that two moles of thiosulfate ion react stoichiometrically with one mole of molecular iodine.

Our sodium thiosulfate titrant contains 0.1 mol/L or 0.0001 mol/mL. Because two moles of thiosulfate are required per mole of hypochlorite and the chlorine bleach contains about 0.0007 moles of hypochlorite, about 14 mL of titrant should be required to neutralize 1 g of chlorine bleach. As we determined in Part I, the density of chlorine bleach is slightly more than 1 g/mL (my observed value was 1.0841 g/mL), so just over 15 mL of titrant should be required per mL of chlorine bleach. Using a 3 mL sample of bleach would require 45+ mL of titrant, uncomfortably close to the 50 mL capacity of our burette. We decided to use a 2.0 mL sample, which should require just over 30 mL of titrant, well within the capacity of our burette.

1. If you have not already done so, put on your splash goggles, gloves, and protective clothing.
2. Weigh about 2.0 g of potassium iodide and add it to about 25 mL of distilled water in a 125 mL Erlenmeyer flask. (Alternatively, you can use 25 mL of a 0.5 M bench solution of potassium iodide.)
3. Swirl the contents of the flask until the potassium iodide dissolves and then, with swirling, add about 5.0 mL of glacial (concentrated) acetic acid to the flask.
4. Use the pipette to transfer 2.00 mL of the chlorine bleach sample to the flask. The solution should immediately turn dark brown as the chlorine bleach oxidizes the iodide ions to free iodine. Record the volume of the sample to 0.01 mL on line E of Table 20-2.
5. Rinse the 50 mL burette with a few mL of 0.1 M sodium thiosulfate, and allow it to drain into a waste container.
6. Transfer 50 mL or slightly more of 0.1 M sodium thiosulfate to the burette. Drain a couple mL into the waste container, until the level of the solution is below the zero index mark on the burette, and make sure that there are no bubbles in the burette, including the tip. Record the initial volume as accurately as possible, interpolating between the graduation marks on the burette, and record that volume on line F of Table 20-2.
7. We calculated that we should need about 30 mL of sodium thiosulfate titrant, so begin by running 25 mL or so of titrant into the receiving flask with constant swirling. As you near the endpoint, the intense brown color of the

solution in the flask will begin to fade noticeably. When you reach that point, slow the addition rate to a fast drip, and continue swirling.
8. When the solution in the flask reaches a pale brown color, stop adding titrant.
9. Add about 2 mL of the starch indicator to the flask and swirl the flask to mix the contents, which immediately take on an intense blue-black color due to the formation of a starch-iodine complex.
10. Continue adding sodium thiosulfate titrant dropwise, with swirling. When the color of the solution fades to a slight bluish tint, continue swirling the flask for 30 seconds. (See Figure 20-1.) If the blue tint persists, add one more drop of titrant and swirl the solution. The endpoint occurs when the blue tint disappears.
11. Record the final volume as accurately as possible, interpolating between the graduation marks on the burette, and record that volume on line G of Table 20-2.
12. Determine the volume of titrant used by subtracting the initial volume reading from the final volume reading. Record that value on line H of Table 20-2.
13. Using the actual molarity of your nominal 0.1 M sodium thiosulfate titrant, calculate the number of moles of sodium thiosulfate required to reach the endpoint and enter that value on line I of Table 20-2.
14. Sodium thiosulfate reacts stoichiometrically with iodine at a 1:2 mole ratio, so the number of moles of iodine (and therefore hypochlorite) present in the sample is half the number of moles of sodium thiosulfate needed to reach the endpoint. Calculate the number of moles of sodium hypochlorite present in the sample by halving the number of moles of thiosulfate required, and enter that value on line J of Table 20-2.
15. Determine the mass of sodium hypochlorite present in the sample by multiplying the number of moles present by the gram molecular weight of sodium hypochlorite (74.4422 g/mol). Enter your calculated value on line K of Table 20-2.
16. Using your values for the density of chlorine bleach (line D) and the volume of the chlorine bleach sample (line E), calculate the mass of the chlorine bleach sample and enter the result on line L of Table 20-2.
17. Calculate the mass percentage of sodium hypochlorite in the chlorine bleach sample, and enter the result on line M of Table 20-2.

DISPOSAL:
Return the unused chlorine bleach from Part I to the original bottle. The waste solutions from this lab can be flushed down the drain with plenty of water.

FIGURE 20-1: *Nearing the endpoint of the titration*

TABLE 20-2: *Quantitative analysis of chlorine bleach by redox titration—observed and calculated data*

Item	Value
A. Mass of empty 100 mL volumetric flask	_____ . ____ g
B. Mass of volumetric flask + 100.0 mL of chlorine bleach	_____ . ____ g
C. Mass of chlorine bleach (B − A)	_____ . ____ g
D. Density of chlorine bleach (C/100)	_____ . ____ g/mL
E. Volume of chlorine bleach sample	_____ . ____ mL
F. Initial volume of sodium thiosulfate titrant	_____ . ____ mL
G. Final volume of sodium thiosulfate titrant	_____ . ____ mL
H. Volume of sodium thiosulfate titrant used (F − G)	_____ . ____ mL
I. Moles of sodium thiosulfate required to reach endpoint	___ . _____ mol
J. Moles of sodium hypochlorite present in the sample (G/2)	___ . _____ mol
K. Mass of sodium hypochlorite present in the sample	_____ . _____ g
L. Mass of sample	_____ . _____ g
M. Mass percentage of sodium hypochlorite	_____ . _____ %

REVIEW QUESTIONS

Q1: How closely does your experimental value for the mass percentage of sodium hypochlorite in the chlorine bleach sample correspond to the mass percentage on the label? Give at least three possible sources of error that might explain any discrepancy.

Q2: How would you modify the procedure in Part II to determine the sample size gravimetrically rather than volumetrically? What advantage or advantages would substituting a gravimetric procedure confer?

Q3: Why did we use back titration rather than forward titration for this quantitative analysis?

LABORATORY 20.3:
QUANTITATIVE ANION ANALYSIS OF SEAWATER

Seawater is a complex solution of dissolved solid salts. Although salinity varies with location, depth, time of year, and other factors, one kilogram of standard seawater as defined for analytical purposes contains about 964.83 g of water and about 35.17 g of dissolved salts. That 35.17 g is made up of about 19.35 g (55.0%) chloride ions, 10.78 g (30.7%) sodium ions, 2.71 g (7.71%) sulfate ions, 1.28 g (3.64%) magnesium ions, 0.41 g (1.17%) calcium ions, 0.40 g (1.13%) potassium ions, and 0.24 g (0.68%) other ions, including strontium, bromide, fluoride, bicarbonate, carbonate, and others.

In this lab, we do a quantitative analysis for chloride and sulfate ions, the only two anions that are present in seawater in sufficient quantity to be realistic targets for quantitative analysis in a home chemistry lab. For illustrative purposes, we'll use different methods to analyze chloride ions and sulfate ions. We'll determine chloride ion content by titrimetric (volumetric) analysis, and sulfate ion content by gravimetric (mass) analysis.

As we learned in the preceding chapter, chloride ions can be precipitated by silver ions in the form of insoluble silver chloride. For a qualitative analysis, the simple fact that adding silver nitrate to a solution of chloride ions causes a precipitation is sufficient to establish the presence of those two ions. But a quantitative analysis requires that we add just sufficient silver ions to precipitate all of the chloride ions, and no more.

Unfortunately, it's almost impossible to determine that endpoint directly. Silver chloride precipitates as a fluffy whitish solid that forms clumps and takes a long time to settle. Visually, it's impossible to tell if you've added insufficient silver nitrate, just the right amount, or too much. Fortunately, there's an elegant solution to this problem, based on the differential molar solubilities of silver chloride and silver chromate. (See Lab 13.3 for more details.)

We'll add a small amount of lemon-yellow potassium chromate solution to our seawater sample as an indicator. Although silver chromate is considered insoluble, it is slightly more soluble than silver chloride. As we titrate the seawater sample with standard silver nitrate solution, the less-soluble silver chloride precipitates

> **REQUIRED EQUIPMENT AND SUPPLIES**
>
> ☐ goggles, gloves, and protective clothing
>
> ☐ balance and weighing papers
>
> ☐ hotplate
>
> ☐ beaker, 150 mL
>
> ☐ Erlenmeyer flask, 125 mL
>
> ☐ volumetric flask, 100 mL
>
> ☐ graduated cylinder, 100 mL
>
> ☐ evaporating dish
>
> ☐ burette, 50 mL
>
> ☐ serological or Mohr pipette, 5 mL
>
> ☐ stirring rod
>
> ☐ funnel
>
> ☐ ring stand
>
> ☐ support ring
>
> ☐ wire gauze
>
> ☐ burette clamp
>
> ☐ filter paper
>
> ☐ seawater (200 mL)
>
> ☐ silver nitrate, 0.1000 M (~50 mL)
>
> ☐ potassium chromate, 0.1 M (a few mL)
>
> ☐ barium nitrate, 0.1000 M (~40 mL)

first. (That precipitate appears yellow rather than white because of the chromate ion present in the solution.) Only after all of the silver chloride has precipitated does the silver nitrate titrant begin to react with the chromate ion, forming insoluble, brick-red silver chromate. The end point is readily visible as an abrupt color change from lemon yellow to a distinct orange that resembles orange juice.

The sulfate ion is present in seawater at a much lower concentration than the chloride ion, but at a level high enough

SUBSTITUTIONS AND MODIFICATIONS

- You may substitute any suitable containers of similar size for the 150 mL beaker and/or the 125 mL Erlenmeyer flask.

- You may substitute a 100 mL graduated cylinder for the 100 mL volumetric flask, with some loss of accuracy.

- You may substitute a 150 mL beaker for the evaporating dish.

- You may substitute a 10 mL pipette for the 5 mL pipette.

- If you use actual seawater, allow it to settle completely or filter it to remove any algae or other suspended solids.

- If you do not have access to seawater, you can make 100 mL of artificial seawater by adding about 3.5 g of sea salt (sold in grocery stores) to a 100 mL volumetric flask and making up the solution to 100 mL. Record the actual mass of the salt to 0.01 g, because you'll need that mass later to determine the mass percentages of chloride and sulfate ions. Dr. Mary Chervenak adds that there's also a product called "Instant Ocean" for salt water fish tanks available in most pet stores that works well.

to be easily analyzed quantitatively in a home lab. We'll analyze the sulfate ion by precipitating it with a slight excess of barium nitrate solution. Barium ions react with sulfate ions to produce a precipitate of insoluble barium sulfate, which we'll separate by filtration. After determining the mass of the barium sulfate, we can easily calculate the concentration of sulfate ions in the seawater sample.

CAUTIONS

Silver nitrate is toxic, corrosive, and stains skin, clothing, and other organic materials. Potassium chromate is extremely toxic and a known carcinogen. Soluble barium salts are toxic. Wear splash goggles, gloves, and protective clothing. Wear a disposable respirator mask if you work with solid potassium chromate or barium nitrate. (An N95 or N100 disposable mask from the hardware store or home center is fine.)

PROCEDURE

Before we begin the analysis, we need to decide what sample sizes are appropriate for our quantitative tests.

We know that a seawater sample should contain roughly 20 g of chloride ions per liter. The atomic mass of chlorine is about 35.45 g/mol, so the molarity of seawater with respect to chloride ion is about 20/35.45 = 0.56+ M. If we use a 5 mL aliquot, that sample should contain about 0.0028 moles of chloride ion. One mole of chloride ion reacts stoichiometrically with one mole of silver nitrate, so we'd need about 0.0028 moles of silver nitrate to react with that 5 mL aliquot of seawater. Our 0.1 M silver nitrate titrant contains 0.1 mol/L or 0.0001 mol/mL, so we should need about 28 mL of titrant, well within the range of our 50 mL burette.

We know that a seawater sample should contain roughly 3 g of sulfate ions per liter. The formula weight of the sulfate ion is about 96.06 g/mol, so the molarity of seawater with respect to sulfate ion is about 3/96.06 = 0.03+ M. If we use another 5 mL aliquot of seawater, that sample should contain about 0.00016 moles of sulfate ion. One mole of sulfate ion reacts stoichiometrically with one mole of barium nitrate, so about 0.00016 moles of barium nitrate would react with that 5 mL aliquot of seawater to produce about 0.00016 moles of barium sulfate. The formula weight of barium sulfate is 233.43 g/mol, so this reaction would produce only about 0.04 g of barium sulfate.

Such a small amount of product is likely to introduce significant errors, so we'll use a much larger sample. It's convenient to use the 100 mL volumetric flask to obtain that sample. If we start with 100.00 mL of seawater in the volumetric flask and remove a 5.00 mL aliquot for chloride ion testing, that leaves 95.00 mL of seawater in the flask. If a 5 mL aliquot would yield about 0.04 g of barium sulfate, a 95 mL aliquot will yield about 0.76 g of barium sulfate, an amount that is much easier to weigh accurately. (Of course, if you have plenty of seawater available, you can simply use a 100.0 mL aliquot. Use the 95 mL aliquot only if you've made up an artificial seawater sample.)

We can also calculate about how much 0.1 M barium nitrate titrant should be needed. If about 0.00016 moles of barium nitrate are needed to react completely with a 5 mL aliquot of seawater, the amount needed to react with a 95 mL aliquot of seawater is (95 mL/5 mL) · 0.00016 mol = 0.0030 mol. Our 0.1 M barium nitrate titrant contains 0.1 mol/L or 0.0001 mol/mL, so we'll need about 30 mL to react completely with the 95 mL aliquot of seawater. To ensure that all of the sulfate ions are precipitated, we'll add a slight excess of barium nitrate.

PART I: DETERMINE SEAWATER DENSITY AND TOTAL DISSOLVED SOLIDS

In Part I, we'll determine the mass of a 100.00 mL sample of seawater, which allows us to calculate its density. We'll then evaporate the water to determine the mass of the dissolved solids. (If you make your own artificial seawater from purchased sea salt, you can skip the evaporation step and simply enter the mass of sea salt you dissolved to make your artificial seawater on line G of Table 20-3.)

1. If you have not already done so, put on your splash goggles, gloves, and protective clothing.
2. Weigh the clean, dry, empty 100 mL volumetric flask and record its mass to 0.01 g on line A of Table 20-3.
3. Fill the 100 mL volumetric flask to the index line with seawater. Weigh the filled flask, and record its mass to 0.01 g on line B of Table 20-3.
4. Subtract the mass of the empty flask from the mass of the filled flask, and record the mass of 100.0 mL of seawater to 0.01 g on line C of Table 20-3.
5. Calculate the density of your seawater sample in g/mL, and enter that value on line D of Table 20-3.
6. Weigh a clean, dry 150 mL beaker and record its mass to 0.01 g on line E of Table 20-3.
7. Transfer the contents of the 100 mL volumetric flask to the weighed beaker. To make sure you've done a quantitative transfer, rinse the volumetric flask several times with a few mL of distilled water, and add the rinse water to the beaker.
8. Place the beaker on the hotplate and boil the solution gently to evaporate the water. Keep a close eye on the beaker. When the liquid is nearly boiled off, use the stirring rod to break up any solid masses, reduce the heat and continue heating until the contents of the beaker appear to be completely dry.

DR. MARY CHERVENAK COMMENTS:

If you have access to an oven, use it to dry the sample. Although the temperature inside the oven is lower than the temperature of the hotplate, extended drying even at a lower temperature is the most effective way to eliminate all moisture from the sample because the heat is uniform across the sample, instead of being concentrated on the bottom.

---

9. Allow the beaker to cool completely and reweigh it. Jot down the mass to 0.01 g and return the beaker to the hotplate. Heat the beaker strongly for several minutes, allow it to cool completely, and reweigh it. If the mass is the same as the mass you jotted down earlier, you can assume the sample is completely dry. If the second reweighing yields a smaller mass, the sample is not yet completely dry. Heat the beaker again and reweigh it until two sequential weighings yield the same mass. Record that mass to 0.01 g on line F of Table 20-3.
10. Subtract the mass of the empty beaker from the combined mass of the beaker and solid residue to determine the mass of the total dissolved solids. Record that mass to 0.01 g on line G of Table 20-3. Calculate the mass percentage of total dissolved solids and enter that value on line H of Table 20-3.

PART II: DETERMINE CHLORIDE ION QUANTITATIVELY

In Part II, we'll quantitatively analyze the chloride ion content of a seawater sample by using silver nitrate as a titrant to precipitate chloride as silver chloride. Because the endpoint of this titration is difficult or impossible to determine directly, we'll use potassium chromate as a titration indicator. A change in color from bright yellow to orange indicates that all of the chloride ion has been precipitated and that silver chromate is beginning to form. At its endpoint, the titrated solution resembles orange juice.

1. If you have not already done so, put on your splash goggles, gloves, and protective clothing.
2. Use the pipette to transfer 5.00 mL of seawater to the 125 mL Erlenmeyer flask. Note the volume of the aliquot as accurately as possible, interpolating between the graduation marks on the pipette. Record the volume on line I of Table 20-3.
3. Use the graduated cylinder to transfer about 45 mL of distilled water to the 125 mL Erlenmeyer flask and swirl the flask to mix the contents. Place a stirring rod in the flask, and leave it there until the titration is complete.
4. Add sufficient 0.1 M potassium chromate solution to the flask to give the solution a bright lemon-yellow color. (The exact amount is not critical. The yellow chromate ion serves as a titration indicator, so if in doubt, use more rather than less.)
5. Rinse the 50 mL burette with a few mL of 0.1 M silver nitrate solution, allow it to drain into a waste container.
6. Transfer at least 40 mL of 0.1 M silver nitrate solution to the burette. Drain a couple mL into the waste container and make sure there are no bubbles in the burette, including the tip. Record the initial volume as accurately as possible, interpolating between the graduation marks on the burette, and record that volume on line J of Table 20-3.
7. We calculated that we should need about 28 mL of silver nitrate titrant, so begin by running 20 mL or so of titrant into the receiving flask with constant swirling. As the silver nitrate is added, it reacts with the chloride ions in the seawater, forming a milky precipitate of silver chloride, which is tinted bright yellow by the chromate ions present in the indicator.
8. Continue adding titrant more slowly, and with constant swirling and stirring. (The stirring helps to break up the curds of silver chloride that otherwise prevent complete mixing of the solutions.) As you approach the endpoint of the titration, you'll see a orange tint begin to appear in the yellow liquid at the point where the titrant comes into contact with it (Figure 20-2). That signals that the endpoint of the titration is near. Continue adding titrant dropwise with constant swirling and stirring until a noticeable color change occurs, from yellow to a distinct orange.

9. Continue stirring vigorously for at least 30 seconds to make sure that no unreacted chloride ions remain trapped within the silver chloride curd. As you stir, the contents of the flask will probably return to the original bright yellow color as additional chloride ions are released from the silver chloride curd.

10. Continue adding titrant dropwise with vigorous stirring until the orange coloration remains for at least 30 seconds before fading. At that point, you're about one drop short of the end point. Add that drop of titrant with stirring and verify that the orange tint persists for at least a full minute. Record the final volume of silver nitrate on line K of Table 20-3. Subtract the initial volume reading from the final volume reading, and enter the volume of titrant required on line L of Table 20-3.

11. Using the actual molarity of the nominally 0.1 M silver nitrate titrant and the actual volume used, calculate the number of moles of silver ions required to reach the endpoint. Enter that value on line M of Table 20-3.

12. One mole of silver ions react with one mole of chloride ions, so the number of moles of chloride ions in the 5.00 mL aliquot is identical to the number of moles of silver ions in the volume of titrant required to reach the endpoint. On that basis, calculate the mass of the chloride ions present in the aliquot, and enter that value on line N of Table 20-3.

13. Calculate the mass of chloride ions per liter of seawater, and enter that value on line O of Table 20-3.

14. Dispose of the contents of the flask as noted in the Disposal section. Wash and dry all of the glassware you used. The silver chloride precipitate can be quite tenacious, covering the inside of the flask with a milky white film even after you've washed it. To remove this film, rinse the inside of the flask with a 6 M or higher solution of aqueous ammonia solution.

PART III: DETERMINE SULFATE ION QUANTITATIVELY

In Part III, we'll quantitatively analyze the sulfate ion content of a seawater sample by using barium nitrate as a titrant to precipitate sulfate as barium sulfate. After we add sufficient barium nitrate solution to the aliquot of seawater to ensure that all of sulfate ions present in the sample have been precipitated, we'll filter that precipitate, rinse it thoroughly with water to remove any traces of soluble salts, dry the precipitate, and determine its mass. With the mass of barium sulfate known, it requires only a simple calculation to determine the mass of sulfate ions present in the seawater sample.

1. If you have not already done so, put on your splash goggles, gloves, and protective clothing.

2. Transfer a 100.0 mL aliquot (or a 95.0 mL aliquot; see the introduction to this section) of seawater from the

FIGURE 20-2: *The first appearance of silver chromate tints the solution orange*

100 mL volumetric flask to the 250 mL beaker. Record the volume of the aliquot on line P of Table 20-3.

3. Rinse the flask two or three times with a few mL of distilled water to make sure you've transferred all of the sample to the beaker.

4. Measure about 40 mL of 0.1 M barium nitrate in the 100 mL graduated cylinder, and pour it slowly, with stirring, into the 250 mL beaker that contains the seawater sample. The solution immediately assumes a white, milky appearance as insoluble barium sulfate is formed by the reaction of the barium nitrate with the sulfate ions present in the seawater sample.

5. Continue stirring the solution for 30 seconds or so to make sure that the solutions are completely mixed and that the reaction is complete. When you remove the stirring rod from the beaker, use your wash bottle to rinse any solids on the stirring rod into the beaker.

6. Fold a piece of filter paper to prepare it for use, weigh it, and record its mass to 0.01 g on line Q of Table 20-3.

7. Insert the folded filter paper in the funnel, place a receiving vessel under the funnel, and pour the contents of the beaker into the funnel. Rinse the beaker thoroughly several times with a few mL of distilled water and transfer the rinse to the filter funnel to make sure that all of the solids have been transferred to the filter paper.

FILTERING BARIUM SULFATE

Barium sulfate can be obnoxious to filter. If your filter paper is too fast (porous), some of the finely divided barium sulfate passes through the filter paper. If your filter paper is too slow, the barium sulfate may plug the filter paper, causing the filtration to proceed at a snail's pace (or stop entirely).

After the liquid drains through the filter paper, examine the filtrate in the receiving container. It should appear clear. Any milkiness indicates that some of the solid barium sulfate has passed through the filter paper. If that occurs, the only real option is to repeat the procedure, using more retentive filter paper. It's pointless to evaporate the liquid in the receiving container, because that liquid contains a substantial mass of soluble salts in addition to the barium sulfate that passed through the filter paper. Alternatively, if the amount of barium sulfate that passed through the filter paper appears to be small, you can simply note the fact that some of the barium sulfate was lost during filtration and use the value you obtain in the following steps for the mass of barium sulfate obtained.

8. After all of the liquid has passed through the filter paper, rinse the filtrand two or three times with several mL of distilled water to ensure that any soluble salts are eliminated from the filtrand.

9. Remove the filter paper from the funnel and place it in a drying oven or under a heat lamp to evaporate any remaining liquid. After the filter paper and filtrand are completely dry, reweigh the filter paper and product and record the mass to 0.01 g on line R of Table 20-3. Subtract the initial mass of the filter paper from the combined mass of filter paper and filtrand to determine the mass of barium sulfate, and record that value to 0.01 g on line S of Table 20-3.

10. The gram molecular mass of barium sulfate, $BaSO_4$, is 233.43 g/mol, of which barium constitutes 137.33 g. Use these values to calculate the mass of sulfate ions present in the aliquot, and enter that value on line T of Table 20-3. Calculate the mass of sulfate ions present per liter of seawater, and enter this value on line U of Table 20-3.

OPTIONAL ACTIVITIES

If you have time and the required materials, consider performing these optional activities:

• Repeat the experiment, using seawater collected from a different location or at a different time of year. In near shore waters, you can find the line where fresh water meets the salt. At that point, especially on a calm day and an incoming tide, there is likely to be some degree of stratification, with the top water being fresher than the bottom.

• If you used actual seawater, repeat the experiment using sea salt from the grocery store and compare your results.

• Repeat the experiment using "Instant Ocean" from the pet store.

• Repeat the experiment using water from a lake or pond.

DISPOSAL:

The waste solutions from this lab contain small amounts of silver and barium, both of which are toxic heavy metals. Any excess silver or barium ions present in solutions can be precipitated with excess sulfate or carbonate ions (for example, from a solution of sodium sulfate or sodium carbonate) and filtered. Dispose of the solids as hazardous waste. Any remaining solutions can be flushed down the drain with plenty of water.

TABLE 20-3: *Quantitative anion analysis of seawater—observed and calculated data*

Item	Value
A. Mass of empty 100 mL volumetric flask	_____ . ___ g
B. Mass of flask + 100.00 mL seawater	_____ . ___ g
C. Mass of 100.00 mL seawater (B − A)	_____ . ___ g
D. Density of seawater (C/100)	_____ . ___ g/mL
E. Mass of 150 mL beaker	_____ . ___ g
F. Mass of 150 mL beaker plus residue after evaporation	_____ . ___ g
G. Mass of total dissolved solids	_____ . ___ g
H. Mass percentage of dissolved solids ([G/C] · 100)	_____ . ___ %
I. Volume of seawater aliquot for chloride analysis	_____ . _____ mL
J. Initial volume of silver nitrate titrant	_____ . _____ mL
K. Final volume of silver nitrate titrant	_____ . _____ mL
L. Volume of silver nitrate titrant used	_____ . _____ mL
M. Moles of silver ion required to reach endpoint	___ . _____ mol
N. Mass of chloride ions in 5.00 mL sample	_____ . ___ g
O. Mass of chloride ions per liter of seawater	_____ . _____ g/L
P. Volume of seawater aliquot for sulfate analysis	_____ . _____ mL
Q. Mass of filter paper	_____ . ___ g
R. Mass of dry filter paper and filtrand	_____ . ___ g
S. Mass of barium sulfate (R − Q)	_____ . ___ g
T. Mass of sulfate ions present in the aliquot	_____ . ___ g
U. Mass of sulfate ions per liter of seawater	_____ . _____ g/L

REVIEW QUESTIONS

Q1: How do the values for chloride and sulfate ion concentrations you determined experimentally compare with the standard values given in the introduction? If the values differ significantly, propose at least one explanation. (Reasonable explanations may differ depending on whether you're using natural or synthetic seawater.)

Q2: List at least five possible sources of experimental error.

Q3: Other than sodium, the two cations that are present in the highest concentrations in seawater are magnesium (1.17%) and calcium (1.13%). Look up the solubility product constants for magnesium and calcium in combination with various anions and use that information to choose a common reagent anion that could be used to selectively precipitate magnesium while leaving calcium in solution. Choose another common reagent anion that could be used to selectively precipitate calcium ions while leaving magnesium ions in solution.

Laboratory:
Synthesis of Useful Compounds

21

Chemical reactions are often themselves interesting, but most chemical reactions have the practical goal of synthesizing some useful compound. With the exception of purely natural products, nearly every material you come into contact with in your daily life is the product of a chemical synthesis. Those products were first synthesized on a small scale by chemists in laboratories, and later produced on an industrial scale using processes designed by chemists and chemical engineers.

In choosing the sessions for this chapter, I was faced with an embarrassment of riches. I could have chosen to synthesize any of literally thousands of common, everyday compounds, from soaps to plastics, dyes and pigments, common drugs such as aspirin, and . . . well, you get the idea. From that huge list of possibilities, I chose two representative syntheses.

In the first lab session, we synthesize methyl salicylate from aspirin and methanol. Methyl salicylate, better known as synthetic oil of wintergreen, is an interesting compound. Although it is toxic if a large quantity is ingested, it is commonly used to provide wintergreen flavoring in candies and other foods. It's also the primary active ingredient in many muscle rubs and balms, giving them their minty odor.

In the second lab session, we begin by synthesizing the copper-ammonia coordination compound tetraamminecopper(II) hydroxide, $[Cu(NH_3)_4](OH)_2$, better known as *Schweizer's Reagent*. That compound has the interesting property of being a solvent for cellulose. We'll use the Schweizer's Reagent we synthesize to dissolve some paper and then use that cellulose solution to produce rayon, which was the first semisynthetic fabric.

DR. MARY CHERVENAK COMMENTS:

When a pathogen infects one part of the plant, the tissue under attack notifies surrounding healthy plants of impending danger. The molecular messenger is methyl salicylate.

EVERYDAY SYNTHESES

With the exception of natural materials such as wood, metals, and raw food products, nearly every material with which you come into daily contact exists because a chemist synthesized it. As a thought experiment, I began looking around my office as I wrote this section, counting objects made from or that contain synthetic materials. I stopped when I reached 100, and I was still counting objects on my desk. (Yes, I have a very cluttered desk.)

All of us are utterly dependent on artificial materials synthesized by chemists. Life without synthetic materials is difficult to imagine.

LABORATORY 21.1:

SYNTHESIZE METHYL SALICYLATE FROM ASPIRIN

Esters are a class of organic compounds. An ester comprises an organic or inorganic acid in which one or more hydroxy (OH) groups has been replaced by an alkoxy (O-alkyl) group. For example, the simplest ester, methyl formate (CHO-OCH$_3$), is made up of formic acid (CHO-OH) in which the hydroxy group has been replaced by a methoxy group (-OCH$_3$). Similarly, ethyl acetate (CH$_3$CO-OCH$_2$CH$_3$) is made up of acetic acid (CH$_3$CO-OH) in which the hydroxy group has been replaced by an ethoxy group (-OCH$_2$CH$_3$).

Although esters can be produced by many mechanisms, the most commonly used method is called *esterification*, which is a condensation reaction between an alcohol and an acid, typically in the presence of a strong acid catalyst, such as sulfuric acid. For example, ethyl acetate can be produced by reacting ethanol (ethyl alcohol) with acetic acid and isopropyl butyrate by reacting isopropanol (isopropyl alcohol) with butyric acid.

Esters were traditionally named by combining the name of the alcohol with the root name of the acid and adding "ate"

as a postfix. Traditional names are still widely used by most chemists, particularly for the simpler esters. The IUPAC naming system uses the systematic names for the alcohol and root name of the acid, followed by "oate." For example, the traditional name n-amyl acetate (n-amyl alcohol with acetic acid) is represented in IUPAC nomenclature as 1-pentyl ethanoate (1-pentyl alcohol, the systematic name for n-amyl alcohol, with ethanoic acid, the systematic name for acetic acid).

Esters typically have strong, often pleasant, scents and tastes, so many esters are used as flavoring and perfume agents, either individually or in combination. For example, the scent and taste of strawberries is due to the presence of (among others) methyl cinnamate, ethyl formate, ethyl butyrate, ethyl caproate, isobutyl acetate, and benzyl acetate.

In this lab session, we'll synthesize methyl salicylate, whose common name is oil of wintergreen. You might expect that we'd synthesize this compound by reacting methyl alcohol (methanol) with salicylic acid, and that is indeed one possible method. Instead, however, we'll synthesize methyl salicylate by reacting methanol with aspirin, which is much easier to come by than salicylic acid. Aspirin is actually a substituted salicylic acid, called acetylsalicylic acid, which is itself both an acid and an ester.

FIGURE 21-1: *Conversion of aspirin to salicylic acid to methyl salicylate (image courtesy Dr. Paul B. Jones)*

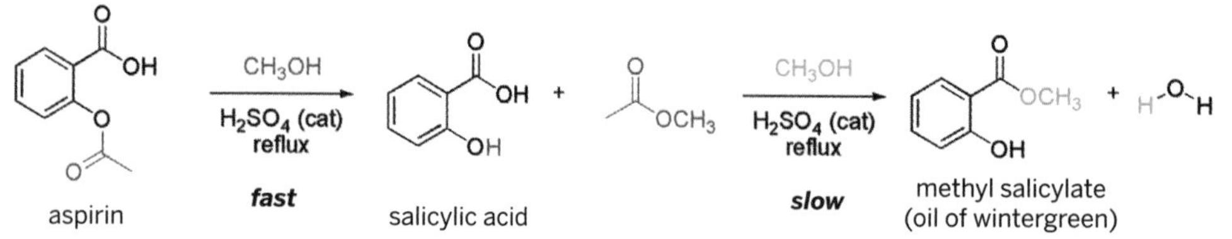

REQUIRED EQUIPMENT AND SUPPLIES

- ☐ goggles, gloves, and protective clothing
- ☐ balance
- ☐ hot water bath (see Substitutions and Modifications)
- ☐ ice bath (see Substitutions and Modifications)
- ☐ test tube
- ☐ beaker, 50 mL
- ☐ beaker, 250 mL
- ☐ beaker, 600 mL
- ☐ separatory funnel (optional)
- ☐ Erlenmeyer flask, 125 mL
- ☐ graduated cylinder, 10 mL
- ☐ graduated cylinder, 100 mL
- ☐ pipette, Mohr or serological, 10.00 mL
- ☐ ring stand
- ☐ utility clamp (for flask)
- ☐ stirring rod
- ☐ eye dropper or Beral pipette
- ☐ thermometer
- ☐ aspirin (13.0 g)
- ☐ methanol (~100 mL)
- ☐ sulfuric acid, concentrated (~10 mL)
- ☐ saturated sodium bicarbonate solution (50 mL; 3.9 g/50 mL)

SUBSTITUTIONS AND MODIFICATIONS

- The reaction requires that the contents of the flask be kept at about 60°C for 60 to 90 minutes by partially immersing the reaction flask in a hot water bath. You can make a hot water bath by placing a large beaker half-filled with water on a hotplate, adding water as necessary, and keeping an eye on the temperature. Alternatively, you can simply partially fill a large beaker with water at about 60°C and add small amounts of boiling water periodically to keep the temperature near 60°C. Pure methanol boils at 64.7°C, and the boiling point of the aspirin solution in methanol is slightly higher. Our goal is to keep the methanol solution near the boiling point without actually bringing it to a boil.

- You can make an ice bath by filling a medium or large beaker about half full of chipped ice, adding a layer about 1 cm thick of rock salt or table salt to the top of the ice, and then stirring to mix.

- If you do not have a separatory funnel, you may use the separation procedure described in Chapter 5.

- You may substitute any container of similar size for the 50 mL beaker, which is used as a weighing boat when determining the density of the product.

- Use ordinary (unbuffered) aspirin tablets. Generic aspirin is fine. You may assume that the aspirin content of aspirin tablets is accurately labeled. For the 13.0 g aspirin sample, you may use 40 standard (325 mg) tablets or 26 extra-strength (500 mg) tablets.

- This reaction is acid-catalyzed, which means that you may substitute another strong acid, such as concentrated hydrochloric acid, for the sulfuric acid. The presence of any water in the reaction mixture reduces yield. Concentrated sulfuric acid contains almost no water, and concentrated hydrochloric acid contains a significant amount of water.

CAUTIONS

Methanol is toxic (particularly to the optic nerve; eye protection is essential), extremely flammable, and has a *flash point* of only 11°C. (The flash point is the lowest temperature at which a flammable liquid can form an ignitable mixture in air.) Have a fire extinguisher handy and avoid all open flames. Concentrated sulfuric acid is extremely corrosive. Methyl salicylate is toxic, irritating, and penetrates the skin. Do not taste the methyl salicylate produced in this lab, and be cautious when you test its odor. Wear splash goggles, gloves, and protective clothing.

PROCEDURE

This lab is broken into three parts. In Part I, we'll synthesize methyl salicylate. In Part II, we'll isolate and purify the product. In Part III, we'll determine the density and freezing point of our product.

Before you begin this lab (or, indeed, any organic synthesis), you should calculate the theoretical yield for the reaction. We'll react 13.0 g of aspirin with excess methanol to form methyl salicylate. Enter the actual mass of aspirin you use to 0.01 g on line A of Table 21-1. The formula weight of aspirin is 180.160 g/mol. Based on the actual mass of your aspirin sample, calculate the number of moles of aspirin to be reacted and enter that value on line B of Table 21-1. In the presence of an acid catalyst, one mole of salicylic acid reacts with one mole of methanol to produce one mole of methyl salicylate. (One mole of aspirin reacts with two moles of methanol to produce one mole of methyl acetate in addition to one mole of methyl salicylate.) The formula weight of methyl salicylate is 152.1494 g/mol. Calculate the theoretical yield of methyl salicylate in grams, and enter that value on line C of Table 21-1. The density of methyl salicylate is 1.1825 g/mL. Using this density and the theoretical yield in grams, calculate the theoretical yield of methyl salicylate in mL and enter that value on line D of Table 21-1.

PART I: SYNTHESIZE METHYL SALICYLATE

Perform this part of the lab under an exhaust hood or in a well-ventilated area. The reaction produces a strong odor of methyl salicylate and methanol vapor. Make certain that there are no open flames or other ignition sources nearby.

1. If you have not already done so, put on your splash goggles, gloves, and protective clothing.
2. Transfer about 60 mL of methanol to the 125 mL Erlenmeyer flask. Add 13.00 g of aspirin to the flask and swirl or stir the contents until the aspirin tablets dissolve. (Aspirin tablets contain binders and other inactive ingredients that may not dissolve in methanol, which is no cause for concern.)
3. Add about 10.0 mL of concentrated sulfuric acid to the reaction vessel and swirl to mix the solutions.
4. Clamp the flask to a ring stand and partially immerse it in a hot water bath at about 60°C.
5. Allow the reaction to proceed, stirring the reaction mixture occasionally, for 60 minutes. As the methyl salicylate forms, you'll notice its distinct wintergreen odor. Keep the reaction mixture at about 60°C, adding water to the bath if necessary to keep its level up. The level of the liquid in the reaction vessel will decrease as methanol evaporates from the flask. Add more methanol as needed to keep the liquid in the reaction vessel near its original volume.
6. After 60 minutes, stop adding methanol to the flask. Increase heat slightly to bring the liquid in the reaction

flask to a gentle boil. Boil the solution long enough to vaporize most of the remaining methanol.

7. When enough methanol has boiled off to reduce the volume of liquid in the reaction vessel to about half its original volume, remove the flask from the water bath and set it aside to cool.

PART II: ISOLATE AND PURIFY THE PRODUCT

The brown liquid in the reaction vessel is a complex solution that contains methanol, crude methyl salicylate, sulfuric acid, unreacted aspirin, and other impurities. Methyl salicylate is freely soluble in methanol, but only very slightly soluble in water. We'll take advantage of that differential solubility to extract most of the water-soluble impurities from the crude product.

1. If you have not already done so, put on your splash goggles, gloves, and protective clothing.
2. Pour the contents of the reaction flask into the sep funnel.
3. Add about 50 mL of ice-cold tap water to the sep funnel, cap the funnel, and agitate the contents vigorously for 30 seconds.

> **DR. MARY CHERVENAK COMMENTS:**
>
> Generally, sep funnels need to be vented after vigorous shaking. Even if gas isn't generated, it's good practice. The sodium bicarbonate wash in step 6 will definitely have to be vented after shaking. Enough gas can be generated to blow the stopper off the back of the sep funnel.

4. Allow the contents of the sep funnel to separate into two layers. The aqueous layer contains nearly all of the sulfuric acid and most of any other water-soluble impurities. (Make sure you know which layer is which.) Separate the two layers, and transfer the aqueous layer to the 250 mL beaker.
5. Do a second washing by repeating steps 3 and 4.
6. Add about 50 mL of sodium bicarbonate solution to the sep funnel, cap the funnel, and agitate the contents vigorously for 30 seconds. (See previous note on venting.)
7. Allow the contents of the sep funnel to separate into two layers. The aqueous layer contains an excess of sodium bicarbonate and a small amount of sodium sulfate produced by the neutralization of any sulfuric acid that remained in the organic layer. Separate the two layers, and transfer the aqueous layer to the 250 mL beaker.
8. Neutralize the sulfuric acid solution in the 250 mL beaker with sodium carbonate or sodium bicarbonate and flush the neutralized solution down the drain with plenty of water. Retain the organic layer in the sep funnel, which contains the crude methyl salicylate.

PART III: DETERMINE DENSITY AND FREEZING POINT OF THE PRODUCT

Our product is a few mL of semi-pure methyl salicylate. If we had the equipment needed for microscale distillations, we could further purify our raw product by distillation. We don't have that equipment, so we'll test our product as-is to determine its density and freezing point.

We know that the density of pure methyl salicylate is 1.1825 g/mL, so the density of our product will give us some idea of its purity. That's not sufficient, however. Even if the density of our product is very close to 1.1825 g/mL, for all we know our product might be a small amount of methyl salicylate mixed with a large amount of some impurity that has the same density. The freezing point of our product will give us a better idea of its purity. Pure methyl salicylate freezes (or melts) at −8.3°C. Furthermore, the sharpness of the freezing/melting point gives a good indication of a product's purity. Pure substances tend to freeze/melt very sharply, at one particular temperature. Impure substances freeze/melt more gradually, over a range of a few degrees.

1. If you have not already done so, put on your splash goggles, gloves, and protective clothing.
2. Weigh a clean, dry, empty 50 mL beaker and record its mass to 0.01 g on line E of Table 21-1.
3. Draw up as much as possible of your product into the 10.0 mL Mohr or serological pipette. Record the volume of product (if possible, to 0.01 mL) on line F of Table 21-1. (If you somehow produced more than 10 mL of product, use as close as possible to a 10.00 mL sample.)

4. Transfer the measured product to the 50 mL beaker. Reweigh the beaker and record the combined mass of the beaker and methyl salicylate to 0.01 g on line G of Table 21-1. Subtract the mass of the empty beaker from the combined mass of the beaker and sample, and enter that value to 0.01 g on line H of Table 21-1. Divide the mass of the sample (line H) by the volume of the sample (line F) to determine the density of the sample. Enter that value on line I of Table 21-1.
5. Transfer as much as possible of the sample to a test tube. Immerse the test tube in the ice bath, and use a thermometer (carefully) to stir the contents of the test tube. When the temperature of the sample falls to 0°C, check the contents of the tube frequently to see whether solid crystals of methyl salicylate have begun to form. (You'll need to remove the tube from the ice bath and examine it against a strong light to detect the first formation of crystals.) Record the temperature at which crystals just begin to form on line J of Table 21-1.
6. Continue cooling the contents of the tube until the methyl salicylate freezes solid. (Leave the thermometer embedded in the sample.) Record that temperature on line K of Table 21-1.
7. Remove the test tube from the ice bath and allow it to begin warming. Record the temperature at which the frozen sample first begins to liquefy on line L of Table 21-1.
8. Allow the tube to continue warming. Stir gently with the thermometer, and record the temperature at which the sample again becomes completely liquid on line M of Table 21-1.

FIGURE 21-2: *Determining the freezing point of the semi-pure methyl salicylate*

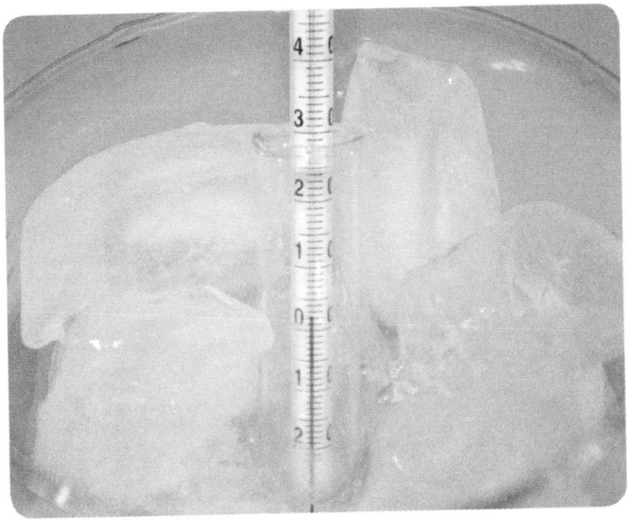

DISPOSAL:
The methyl salicylate produced in this lab is not safe for human consumption under any circumstances. It contains numerous impurities, some of which may be toxic. (Or perhaps I should say even more toxic than methyl salicylate itself.) Do not taste it or allow it to contact your skin. Discard it by flushing it and all other waste solutions from this lab down the drain with plenty of water. Alternatively, you can use your crude methyl salicylate in a scent candle by mixing it with paraffin.

TABLE 21-1: *Synthesize methyl salicylate from aspirin—observed and calculated data*

Item	Value
A. Mass of aspirin	_____.____ g
B. Moles of aspirin (A/[180.160 g/mol])	__._____ mol
C. Theoretical yield of methyl salicylate (B · [152.1494 g/mol])	_____.____ g
D. Theoretical yield of methyl salicylate (C/[1.1825 g/mL])	_____.____ mL
E. Mass of empty 50 mL beaker	_____.____ g
F. Volume of methyl salicylate sample	_____.____ mL
G. Mass of beaker + methyl salicylate sample	_____.____ g
H. Mass of methyl salicylate sample (G – E)	_____.____ g
I. Density of methyl salicylate sample (H/F)	_____.____ g/mL
J. Temperature at which crystals begin to form	_____.___ °C
K. Temperature at which sample freezes solid	_____.___ °C
L. Temperature at which sample begins to melt	_____.___ °C
M. Temperature at which sample is fully melted	_____.___ °C

REVIEW QUESTIONS

Q1: How did your actual yield compare to the theoretical yield? Propose at least five possible explanations for the actual yield being smaller than theoretical.

Q2: Why did we use ice-cold water to extract impurities from the product?

Q3: If we had used salicylic acid, would you expect the percent yield to be higher, lower, or the same as the percent yield using aspirin? Why?

Q4: Aspirin that is very old or has been stored improperly often has a strong odor of vinegar. On that basis, what ester other than methyl salicylate would you expect to be present in your crude product?

Q5: Did your tests show a sharp freezing/melting point? If not, what does this suggest about the purity of your product?

LABORATORY 21.2:
PRODUCE RAYON FIBER

In this lab session we produce rayon, the first practical artificial fiber. We use the word "artificial," because rayon is neither a natural fiber nor a synthetic fiber. What other option is there? Half and half. Rayon, first produced commercially in 1899, is actually a semi-synthetic or reconstituted form of the natural polymer cellulose.

Rayon was first produced commercially using the *cuprammonium process*, a process that is still in limited use today but has been largely supplemented by alternative processes that are more environmentally friendly. In the cuprammonium process, which we use in this lab session, solid cellulose (such as paper or wood chips) is dissolved in *Schweizer's Reagent*, a coordination compound of copper(II) hydroxide in aqueous ammonia. The cellulose is then reconstituted by reacting the tetraamminecopper(II) hydroxide solution with an acid, to neutralize the ammonia and destroy the coordination compound that makes cellulose soluble. The cellulose precipitates as a solid.

CAUTIONS

Concentrated aqueous ammonia is corrosive and produces strong irritating fumes. Sulfuric acid and sodium hydroxide are corrosive. Copper sulfate is moderately toxic. Wear splash goggles, gloves, and protective clothing.

PROCEDURE

Perform this lab under an exhaust hood or in a well-ventilated area to dissipate the fumes of concentrated aqueous ammonia.

1. If you have not already done so, put on your splash goggles, gloves, and protective clothing.
2. Weigh out about 25 g of copper sulfate pentahydrate and transfer it to the 250 mL beaker. Add about 100 mL of water and swirl or stir until the copper sulfate dissolves completely. (This is a nearly saturated solution; you may warm the solution to dissolve the copper sulfate faster.)
3. Transfer about 50 mL of water to the 250 mL Erlenmeyer flask and sufficient crushed or chipped ice to bring the total volume to about 125 mL. Weigh out about 8 g of sodium hydroxide and transfer it in small portions, with

- ☐ goggles, gloves, and protective clothing
- ☐ balance and weighing papers
- ☐ beaker, 250 mL
- ☐ beaker, 600 mL
- ☐ Erlenmeyer flask, 250 mL
- ☐ graduated cylinder, 100 mL
- ☐ filtering funnel and support
- ☐ stirring rod
- ☐ syringe, plastic, 10 mL to 50 mL
- ☐ filter paper (5 pieces)
- ☐ stopper, solid (to fit flask)
- ☐ copper sulfate pentahydrate (25 g)
- ☐ sodium hydroxide (8 g)
- ☐ aqueous ammonia, concentrated (70 mL)
- ☐ sulfuric acid, ~1.5 M (~300 mL)
- ☐ crushed or chipped ice (~75 mL)

SUBSTITUTIONS AND MODIFICATIONS

- You may substitute any containers of similar size for the beakers.
- You may substitute a 10 mL Mohr or serological pipette or a disposable plastic pipette for the syringe.
- You may substitute about 11 mL of concentrated (18 m) aqueous ammonia for the 8 g of sodium hydroxide. If you use aqueous ammonia, add only a sufficient amount to precipitate all of the copper present as the powder-blue copper salt. As you add aqueous ammonia to the copper sulfate solution, you'll notice a very dark blue coloration appear where the ammonia contacts the copper sulfate solution, but that dark blue disappears as the solution is mixed. Add ammonia only until the dark blue color first begins to persist in the solution.

swirling, to the flask. (Caution: this process is extremely exothermic. Add the sodium hydroxide slowly.)

4. With both solutions at about room temperature, pour the sodium hydroxide solution into the copper sulfate solution, with stirring. The solution immediately turns a milky powder-blue color as the two solutions react to form a precipitate of insoluble copper(II) hydroxide.

5. Rinse the 250 mL Erlenmeyer flask immediately and set it aside to drain.

6. Set up your filter funnel, using the 600 mL beaker as the receiving container.

7. Pour the blue mixture from the beaker into the filter funnel. The blue precipitate is voluminous, so you may have to use several pieces of filter paper to get all of it. (I used 11 cm filter paper, and it took me three filtrations to capture all of the precipitate. If you use 15 cm filter paper, you may be able to get all of the precipitate in one filtering pass.)

8. When the filter paper fills with precipitate, allow the remaining filtrate to drain into the receiving vessel and then wash the filtrand with about 10 mL of water and allow the water to drain.

9. Transfer the filtrand, filter paper and all, to the 250 mL Erlenmeyer flask. If you weren't able to capture all of the precipitate on one piece of filter paper, repeat steps 7 and 8 until you have transferred all of the precipitate to the flask.

10. Measure about 70 mL of concentrated aqueous ammonia, and add it to the 250 mL Erlenmeyer flask with swirling. (The copper(II) hydroxide in the flask reacts with the aqueous ammonia to form tetraamminecopper(II) hydroxide, also known as Schweizer's Reagent.)

11. Including the pieces you used to do the filtration, add pieces of filter paper to the flask to a total of 5 pieces of 11 cm filter paper or 3 pieces of 15 cm filter paper. (The filter paper provides the cellulose that will be dissolved by the tetraamminecopper(II) hydroxide.)

12. Stopper the flask and set it aside to allow the tetraamminecopper(II) hydroxide to dissolve the filter paper. Swirl the flask periodically, and allow the reaction to proceed for at least 24 to 48 hours.

13. Prepare about 300 mL of about 1.5 M sulfuric acid solution in the 600 mL beaker. (Caution: this process is extremely exothermic. Add the acid slowly to the water with stirring.)

14. Pour the cellulose solution from the 250 mL Erlenmeyer flask into the 250 mL beaker. If there is undissolved filter paper in the flask, discard it.

15. Fill the syringe with the cellulose solution, submerge the tip of the syringe in the sulfuric acid bath, and slowly expel a stream of cellulose solution into the sulfuric acid. Refill the syringe and repeat until you have transferred all of the cellulose solution to the sulfuric acid bath. The threads and clumps of solid material that appear are rayon.

16. Carefully pour off the liquid into a waste container, retaining the raw rayon. Rinse the rayon several times with tap water, discard the rinse water, and allow the rayon to dry.

DISPOSAL:
The waste solution produced in this lab session is a solution of copper sulfate in sulfuric acid. Neutralize this solution by adding sodium carbonate or sodium bicarbonate until all of the copper precipitates as blue copper carbonate. Discard the supernatant solution of sodium sulfate by flushing it down the drain with plenty of water. You can retain the crude copper carbonate for later use or dispose of it with household solid waste.

FIGURE 21-3: *Dissolving cellulose (paper) in a solution of Schweizer's Reagent*

REVIEW QUESTIONS

Q1: What compound forms the powder-blue precipitate when you add sodium hydroxide solution to the copper sulfate solution?

Q2: What compound produces the intense blue color when the powder-blue precipitate is dissolved in excess aqueous ammonia?

Q3: Would you expect the solid rayon you produced to be soluble or insoluble in a solution of tetraamminecopper(II) hydroxide? Why?

Q4: Is copper sulfate consumed in this reaction, or could it be recycled from the spent cellulose solution? If the latter, propose a general method for reclaiming and reusing the copper.

Q5: Structurally and chemically, rayon resembles the natural fiber cotton, and in fact rayon can be dyed, Mercerized, and otherwise treated using many of the same finishing processes that are used for cotton. Based on your examination of your product, why, then, would anyone go to the trouble of producing rayon rather than just using natural cotton fiber?

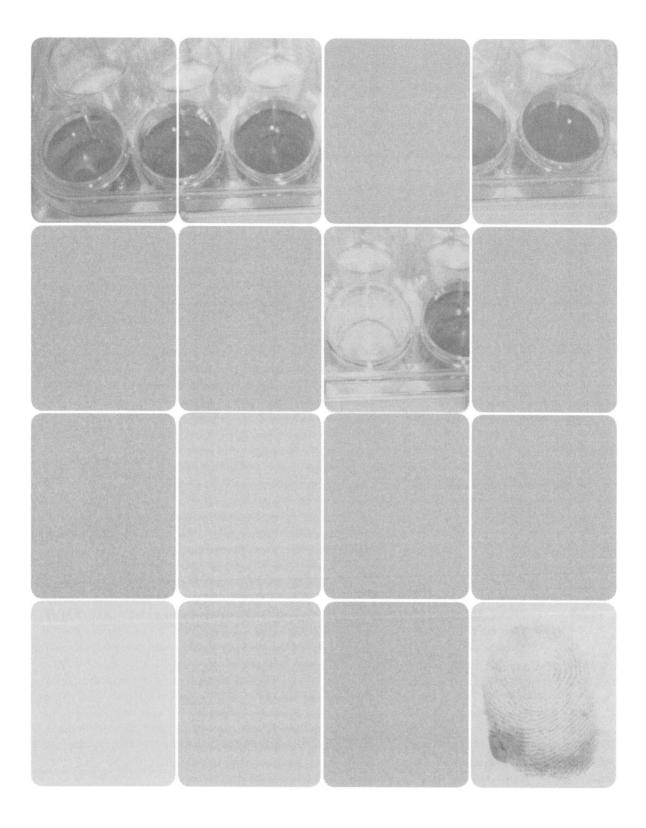

Laboratory:
Forensic Chemistry

22

Chemistry and forensics have been inextricably linked since 1836, when Scottish chemist James Marsh announced the eponymous named Marsh Test, the first reliable forensic test for arsenic and antimony. In the early nineteenth century, arsenic compounds were widely used for many purposes, and were readily available at low cost. Would-be murderers then considered arsenic an ideal poison, because arsenic compounds were lethal in small doses, readily soluble, odorless, tasteless, and impossible to detect. Because the symptoms of arsenic poisoning are very similar to the symptoms of gastroenteritis, an untold number of murder victims were instead believed to have died of natural causes.

But even if murder by arsenic was suspected, there was no easy way to prove it. Marsh developed his test in frustration after watching just such a murderer go free. In 1832, Marsh testified as an expert witness at the trial of John Bodle, who was accused of murdering his grandfather by putting arsenic in his coffee. Marsh used the then-standard test for arsenic, mixing a sample of the suspect material with hydrochloric acid and hydrogen sulfide to precipitate the arsenic as insoluble arsenic trisulfide.

Unfortunately, by the time Marsh testified the sample had degraded and the jury was not convinced by Marsh's scientific testimony. (This problem persisted for many years; until the late nineteenth century, many juries and even judges gave little weight to the testimony of forensics experts.) Bodle was acquitted, and Marsh watched a guilty man walk free. Determined not to let that happen again, Marsh devoted much of his free time over the next several years to developing a reliable test for arsenic. The test Marsh finally devised was revolutionary then, and is still used today by forensics labs in jurisdictions that cannot afford mass spectrometers and all of the other expensive equipment typically found in modern forensics labs.

By the late nineteenth century, forensic toxicology was well developed. Hundreds of specific forensic tests had been devised. A forensic chemist still had to know what he was looking for, but if that substance was present, the tests would identify it. Reliable chemical tests had been devised for most common poisons, and would-be poisoners could no longer assume that poisonings would go undetected. In the early twentieth century, the world-renowned forensic scientist Bernard Spilsbury was testifying for the prosecution at the trial of an accused murderer, when he was asked by the prosecutor if common poisons could be detected reliably in the lab. Spilsbury testified that all common vegetable and mineral poisons could easily be detected, with only one exception. At that point, the judge ordered Spilsbury to stop speaking, concerned that he might reveal the name of this "undetectable" poison. (That poison was aconite, derived from the plant monkshood, which is lethal in doses too small to be detected by any means at that time, and remains difficult to detect today.)

Although early forensic chemists devoted most of their efforts to detecting and identifying poisons, forensic chemistry soon expanded into other areas, notably the unambiguous detection of blood and revealing latent (hidden) fingerprints. Modern forensics laboratories now substitute instrumental analysis for much of the "wet chemistry" that was used formerly, but wet chemistry remains important, particularly for field work. In fact, the majority of forensics tests done today are wet chemistry field tests. When police officers or drug enforcement agents confiscate a substance they believe may be an illegal drug, they use **presumptive tests** on a sample of the suspect material. Presumptive tests are not perfect—they can produce false positives and false negatives—but they're a fast, inexpensive, and reasonably reliable way to confirm or deny the presence of a particular compound (such as cocaine) or class of compound (such as opioids).

The laboratory sessions in this chapter explore various aspects of forensic chemistry.

LABORATORY 22.1:

USE THE SHERLOCK HOLMES TEST TO DETECT BLOOD

Until the early twentieth century, forensic scientists had no reliable test for blood. The best method available was to examine the suspected blood stain with a microscope to detect the presence of red blood cells. Unfortunately, this microscopic test was considered unreliable unless the blood stains were very fresh. Many probable murderers escaped justice because their attorneys claimed that apparent blood stains could be rust or even red paint, claims that forensic scientists at that time were unable to refute definitively.

In 1901, Dr. Joseph H. Kastle of Kentucky University introduced a simple, cheap, reliable presumptive test for blood. This test, with modifications suggested by another forensic scientist, became known as the **Kastle-Meyer test**, and is still used today. But I think this test may actually have been devised in 1888 by a forensic scientist whose name is known to everyone:

He seized me by the coat-sleeve in his eagerness, and drew me over to the table at which he had been working. "Let us have some fresh blood," he said, digging a long bodkin into his finger, and drawing off the resulting drop of blood in a chemical pipette. "Now, I add this small quantity of blood to a

REQUIRED EQUIPMENT AND SUPPLIES

☐ goggles, gloves, and protective clothing

☐ dropper or Beral pipette (3)

☐ sample of dried blood

☐ sample(s) of rust and/or other materials that look like blood stains

☐ cotton swabs to collect samples

☐ distilled or deionized water

☐ ethanol, 70% (a few drops per test)

☐ phenolphthalein solution (a few drops per test)

☐ hydrogen peroxide solution, 3% (a few drops per test)

SUBSTITUTIONS AND MODIFICATIONS

• This test reacts to human or animal blood, so you may use fresh meat as a source of blood.

• You may substitute small pieces of filter paper for the cotton swabs.

litre of water. You perceive that the resulting mixture has the appearance of pure water. The proportion of blood cannot be more than one in a million. I have no doubt, however, that we shall be able to obtain the characteristic reaction." As he spoke, he threw into the vessel a few white crystals, and then added some drops of a transparent fluid. In an instant the contents assumed a dull mahogany colour, and a brownish dust was precipitated to the bottom of the glass jar.

Sherlock Holmes, A Study in Scarlet

Although this description doesn't correspond exactly to the test we're about to run, there's little doubt that Sherlock Holmes was the true inventor of this test. Arthur Conan Doyle, who was at best an unreliable chronicler, probably didn't take notes as he watched Holmes perform the test, or perhaps Doyle obfuscated the details for reasons of his own. (Holmes was very precise in speech and action; the numbers given for the dilution are an indication of Doyle's sloppiness. One drop of blood—about 0.05 mL—in one litre of water is a dilution of about 1:20,000, not 1:1,000,000. Coincidentally, or perhaps not, 1:20,000 is about the threshold of detectability for this test, so Holmes was speaking literally.)

The Sherlock Holmes test, usually called the Kastle-Meyer test, is a presumptive test for blood, or, more precisely, for the hemoglobin present in blood. The test is fast and inexpensive, requiring only the sample and solutions of ethanol, hydrogen peroxide, and phenolphthalein ($C_{20}H_{14}O_4$). The test is non-destructive, so the sample can be preserved for later testing.

The ethanol functions as a solvent, but does not take part in the reaction. The peroxidase enzyme present in the hemoglobin component of human or animal blood catalyzes the reaction, during which colorless phenolphthalein molecules are converted to pink phenolphthalein ions. The peroxidase enzyme in hemoglobin is extremely efficient catalyst. One peroxidase molecule can catalyze the conversion of hundreds of thousands of peroxide molecules per second, so the presence of even a tiny amount of blood yields a distinct color change in the phenolphthalein. The high sensitivity of this test means that, as Sherlock Holmes stated, one drop of blood can be detected even if it is diluted with a liter of water. In addition to being very sensitive, this test is also very specific for blood.

CAUTIONS

The chemicals used in this experiment are reasonably safe. Ethanol and the phenolphthalein solution are flammable, so avoid open flame. Hydrogen peroxide is an oxidizer and irritant, and may bleach clothing and other materials with which it comes into contact. Wear splash goggles, gloves, and protective clothing.

PROCEDURE

1. If you have not already done so, put on your splash goggles, gloves, and protective clothing.
2. Moisten a cotton swab with distilled or deionized water, and touch it to the sample of dried blood. You needn't rub vigorously, but try to get at least a small amount of the sample transferred to the swab.
3. Put a drop or two of ethanol on the sample. The goal is to moisten the swab with ethanol, not to drench it. The ethanol does not take part in the reaction, but merely serves to clean up the sample and expose the hemoglobin contained in the sample.
4. Allow the ethanol to work for a few seconds, and then put a drop or two of phenolphthalein solution on the sample. (The sample should remain colorless. If the phenolphthalein turns pink at this point, it is an indication that the phenolphthalein is old or oxidized. Rerun the test using fresh phenolphthalein.)
5. After a few seconds, put a drop or two of hydrogen peroxide on the sample. The sample should turn bright pink almost immediately, indicating the presence of blood.
6. Repeat steps 2 through 5, but using a sample or rust or another material that looks like dried blood but does not contain blood. When you apply the hydrogen peroxide, the sample should remain colorless. (Even in the absence of blood, a pink color will develop after 30 seconds or so as the hydrogen peroxide converts the colorless phenolphthalein molecules to pink phenolphthalein ions, but anything other than a nearly immediate color change does not indicate the presence of blood.) Figure 22-1 shows swabs with negative and positive test results.

DISPOSAL: Used swabs can be discarded as ordinary household waste.

FIGURE 22-1: *Swabs showing negative (left) and positive results for the presence of blood*

REVIEW QUESTIONS

Q1: You perform this test on a control sample that is known not to contain blood, but the test results are positive. What might the sample contain that produces the false positive results?

_____

_____

_____

Q2: As a forensic specialist, you are provided with a sample and asked to use this test to determine whether blood is present. You know that you will have to testify in court about the results of your test. Other than testing the sample itself, what other actions might you take to ensure that your testimony is accepted as valid?

_____

_____

_____

_____

_____

_____

_____

_____

_____

_____

_____

LABORATORY 22.2:
PERFORM A PRESUMPTIVE TEST FOR ILLICIT DRUGS

In making arrests, police officers and drug enforcement agents are often faced with the problem of determining whether an unknown material is a controlled substance. Agents may strongly suspect, for example, that a confiscated vial of white powder contains cocaine, heroin, or another illegal drug, but that's not sufficient reason to make an arrest. After all, it's conceivable that the guy was carrying around a vial of baking soda. Not likely, but conceivable.

In an ideal world, the police would deliver the suspect substance by courier to a forensics lab, which would immediately test the substance and report the results. In the real world, forensics labs are understaffed and overburdened, and it sometimes takes days or weeks for them to determine definitively whether the confiscated substance is an illegal drug.

As a workaround for this problem, forensic chemists developed *presumptive drug tests*, which can be performed by a police officer in the field. These tests are fast, easy, and inexpensive, and are sufficiently sensitive and selective that courts consider a positive test as sufficient cause for making an arrest. Most presumptive drug tests are available as packaged kits that include sample containers, small dropper bottles of various reagents, Munsell Color Charts for comparison, and so on. Sales of these kits are generally restricted to law enforcement personnel, presumably to prevent them being used by buyers of illegal drugs. As is often true of such restrictions, they're easy enough to get around, because the test procedures and the required reagents are readily available to anyone.

False positives are a limitation of most presumptive drug tests. For example, a presumptive test for cocaine may provide a positive result even if no cocaine is present because the test is "triggered" by a different, completely legal substance. Police eliminate this uncertainty by running two or more different presumptive drug tests against the suspect sample. If the results are negative for one or more of the tests, the sample is unlikely to contain the suspected substance. If the results are positive for all of the tests, the sample is likely to contain the suspected substance, because it's very unlikely that an innocuous substance triggered a false positive for all of the tests.

In "Color Test Reagents/Kits for Preliminary Identification of Drugs of Abuse" (*http://www.ncjrs.gov/pdffiles1/nij/183258.pdf*),

---

REQUIRED EQUIPMENT AND SUPPLIES

- ☐ goggles, gloves, and protective clothing
- ☐ balance and weighing paper
- ☐ reaction plate
- ☐ Barnes bottles or dropping bottles (as required)
- ☐ 10 mL graduated cylinder (1)
- ☐ timer (optional)
- ☐ samples to be tested (see note)
- ☐ chloroform (a few mL)
- ☐ methanol (a few mL)
- ☐ sulfuric acid, concentrated (a few mL)
- ☐ formaldehyde, 40% (a few drops)
- ☐ nitric acid, concentrated (a few mL)
- ☐ ferric chloride, anhydrous (0.10 g)
- ☐ distilled or deionized water

DR. MARY CHERVENAK COMMENTS:

The vial could also contain artificial sweetener. Paul carries this around. I, of course, carry vials of baking soda.

---

which I call 183258 for short, the National Institute of Justice (NIJ) specifies a dozen different reagents/kits that can be used to perform presumptive drug tests. Many of these reagents/kits use chemicals that are difficult to obtain or extremely hazardous. For the purposes of this lab, we'll use three of those presumptive tests, all of which use chemicals that are relatively easy to obtain (although two of them use concentrated acids that must be handled with extreme caution).

Marquis reagent
Marquis reagent, which the NIJ designates A.5, is one of the most frequently used presumptive drug tests. The Marquis test is very sensitive and detects a very wide range of illicit drugs, including amphetamines, mescaline, and opioids like morphine and codeine. Marquis reagent uses concentrated sulfuric acid

and 40% formaldehyde. Samples are tested in powder form and with chloroform as a solvent.

Nitric acid reagent

Nitric acid reagent, which the NIJ designates A.6, is another frequently used presumptive drug test. The nitric acid test is very sensitive, particularly for mescaline hydrochloride, and detects a wide range of illicit drugs, including amphetamines, mescaline, and opioids like morphine and codeine. The nitric acid test uses, as you might expect, concentrated nitric acid. Samples are tested in powder form and with chloroform as a solvent.

Ferric chloride reagent

Ferric chloride reagent, which the NIJ designates A.8, is much less frequently used than Marquis reagent and much less sensitive. We'll use ferric chloride reagent for this lab, because it uses a readily available chemical that reacts interestingly with many common household substances. Ferric chloride reagent uses, as you might have guessed, ferric chloride. Samples are tested in powder form and with methanol as a solvent.

Table 22-1 shows the color changes expected when the analyte shown in the left column is treated with Marquis reagent, concentrated nitric acid, or ferric chloride solution. An asterisk in a color change column indicates that this is a preferred test for this substance.

DR. MARY CHERVENAK COMMENTS:

Ferric chloride is also used in wastewater treatment, but good luck finding a local wastewater treatment store.

SUBSTITUTIONS AND MODIFICATIONS

- You may substitute test tubes for the spot plate, but doing so increases the quantities of test reagents and the sample sizes needed.

- The Barnes bottles or dropping bottles are used to store the reagents. You may substitute any small bottle or stoppered test tube for storage and use Beral pipettes or droppers to dispense the reagents.

- You may substitute 0.17 g of ferric chloride hexahydrate for the 0.10 g of anhydrous ferric chloride. Ferric chloride solution is used as a circuit board etchant, and is sold by some professional electronics stores and suppliers.

CAUTIONS

Perform this experiment only in a well-ventilated area. Chloroform is toxic and carcinogenic. Methanol is flammable. Sulfuric acid is corrosive and toxic. Formaldehyde is toxic, corrosive, and a lachrymator. Nitric acid is toxic, corrosive, a strong oxidizer, and reacts vigorously with many organic and inorganic substances. Ferric chloride is an oxidizer and corrosive. Wear splash goggles, gloves, and protective clothing.

TEST SAMPLES

Obviously, I don't recommend you attempt to buy samples of heroin, cocaine, methamphetamine, and other illegal drugs. No matter how convincing you are, the judge is likely to take a dim view of your claims that you bought them only in the spirit of scientific inquiry. Fortunately, there are alternatives that are just as good for the purposes of this experiment.

In addition to common household substances, you may be able to get legal samples of some controlled substances. For example, some prescription cough medicines contain codeine, as does Tylenol with Codeine (Tylenol 1, 2, 3, or 4, in different dosages). In some jurisdictions, some OTC drugs include codeine in very small amounts, which are nonetheless more than sufficient for these tests. Also, if anyone in your household uses a prescription pain reliever, the smallest scraping from a tablet or even the dust from an empty bottle is sufficient. If you have a pet, try testing tiny samples of any medication that your veterinarian has prescribed.

You can also test samples of whatever prescription and OTC medications you have available. In particular, try testing pain relievers (including aspirin and acetaminophen, naproxen sodium, ibuprofen, etc.) and allergy medications (including those based on pseudoephedrine hydrochloride, chlorpheniramine maleate, brompheniramine maleate, and so on). The kitchen spice rack is also a fertile source of herbs, spices, and other possible test samples (poppy seeds!), as is the garden shed, with its selection of insecticides, fertilizers, and so on. Use your imagination, and test as many substances as you have time for.

TABLE 22-1: *Color changes with Marquis reagent, nitric acid reagent, or ferric chloride reagent (data from NIJ 183258). An asterisk in a color change column indicates that this is a preferred test for this substance.*

Analyte	Marquis reagent color change	Nitric acid color change	Ferric chloride color change
Acetaminophen	N/A	Brilliant orange yellow	Dark greenish yellow
Aspirin (powder)	Deep red	N/A	N/A
Baking soda (powder)	N/A	N/A	Deep orange
Benzphetamine HCl	Deep reddish brown*	N/A	N/A
Chlorpromazine HCl	Deep purplish red*	N/A	Very orange
Codeine	Very dark purple*	Light greenish yellow*	N/A
d-Amphetamine HCl	Strong reddish orange to dark reddish brown*	N/A	N/A
d-Methamphetamine HCl	Deep reddish orange to dark reddish brown*	N/A	N/A
Diacetylmorphine HCl (heroin)	Deep purplish red*	Pale yellow*	N/A
Dimethoxy-meth HCl	Moderate olive	Very yellow	N/A
Doxepin HCl	Blackish red	Brilliant yellow	N/A
Dristan (powder)	Dark grayish red	Deep orange	Moderate purplish blue
Excedrin (powder)	Dark red	Brilliant orange yellow	Moderate purplish blue
LSD	Olive black	Strong brown	N/A
Mace (crystals)	Moderate yellow	Moderate greenish yellow	N/A
MDA HCl	Black*	Light greenish yellow	N/A
Meperidine HCl	Deep brown	N/A	N/A
Mescaline HCl	Strong orange*	Dark red*	N/A
Methadone HCl	Light yellowish pink	N/A	N/A
Methylphenidate HCl	Moderate orange yellow	N/A	N/A
Morphine monohydrate	Very deep reddish purple*	Brilliant orange yellow*	Dark green*
Opium (powder)	Dark grayish reddish brown*	Dark orange yellow*	N/A
Oxycodone HCL	Pale violet*	Brilliant yellow	N/A
Propoxyphene HCl	Blackish purple	N/A	N/A
Sugar (crystals)	Dark brown	N/A	N/A

PROCEDURE

This lab has two parts: preparation of the reagents and running the tests.

PART I: PREPARE REAGENTS

Before running the tests, we need to prepare the reagents. The amount of each reagent you need depends on the number of tests you plan to run and whether you run the tests on a spot plate or in test tubes. We decided to prepare about 5 mL of each reagent, which is sufficient for many tests using a spot plate. If you are using test tubes, we recommend preparing at least 10 mL of each reagent, depending on how many tests you plan to run.

1. If you have not already done so, put on your splash goggles, gloves, and protective clothing.
2. To prepare the Marquis reagent, add five drops (0.25 mL) of 40% formaldehyde to 5.0 mL of concentrated sulfuric acid. Stir thoroughly to mix, and store the Marquis reagent in a labeled Barnes bottle or other airtight container.
3. The nitric acid reagent is simply concentrated (~70%) nitric acid. For convenience, you may wish to transfer 5.0 mL of concentrated nitric acid to a dropper bottle or whatever other container you're using for reagents. Note that concentrated nitric acid may react with rubber or plastic stoppers, so a dropper bottle with a ground glass seal is the best choice.
4. To prepare the ferric chloride reagent, dissolve 0.10 g of anhydrous ferric chloride (or 0.17 g of ferric chloride hexahydrate) in 5.0 mL of water.

Use freshly prepared reagents to ensure that your results are reproducible.

PART II: RUN THE TESTS

Although you can run several tests simultaneously if time is short, it's better to make a separate test run for each substance. We suggest that you test each substance with each reagent both in solid form and dissolved in the solvent recommended for that reagent (chloroform for Marquis reagent and the nitric acid reagent, and methanol for the ferric chloride reagent.) For each substance you are testing, take the following steps:

1. If you have not already done so, put on your splash goggles, gloves, and protective clothing.
2. Add a few granules of the substance (if solid) or a few drops of the substance (if liquid) to a well in the test plate. If you are testing a solid substance in solution, add a few drops of the appropriate solvent and allow the solid to dissolve.
3. Add a few drops of the reagent and start the timer. Record any obvious changes in color in Table 22-2, noting the time required for the change to occur. After the change appears to be complete, continue observing for 15 to 30 seconds to see if any further change occurs. For example, a sample may immediately turn a bright color that fades or changes color over a short period. Figure 22-2 shows our spot plate with several positive and negative tests. (We won't tell you which reagents or which substances are shown; run your own experiment and see for yourself.)
4. Repeat steps 2 and 3 for all of your samples, recording the names or descriptions of samples that are not labeled in the table.

OPTIONAL ACTIVITIES

If you have time and the required materials, consider performing these optional activities:

- If you have access to a Munsell color chart, match the color of the positive test results you obtain as closely as possible to the standard colors, and compare your results to published color standards for the substances you tested.

- Test various household substances using one or more of the other presumptive test reagents described in the 183258 document referenced earlier. (Before you consider doing this, look up each of the chemicals used in each reagent and decide whether it is worth using that reagent, based on cost and safety considerations.)

FIGURE 22-2: *A spot plate showing positive and negative results for various substances*

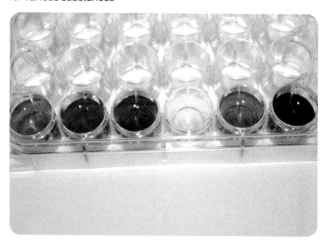

TABLE 22-2: *Observed data*

Analyte	Marquis reagent color change	Nitric acid color change	Ferric chloride color change
Acetaminophen			
Aspirin (powder)			
Baking soda (powder)			
Cough medicine (prescription)			
Dristan (powder)			
Excedrin (powder)			
Loperamide HCl (Imodium)			
Mace (spice)			
Sugar (crystals)			
a.			
b.			
c.			
d.			
e.			
f.			
g.			
h.			
i.			
j.			
k.			
l.			
m.			
n.			
o.			
p.			

DISPOSAL:
Dilute all solutions with at least three or four times their volume of
tap water and then flush them down the drain with plenty of water.

REVIEW QUESTIONS

Q1: You have used Marquis reagent to test a sample that is suspected to contain codeine. The reagent yields a purple color, which apparently confirms that codeine is present, but the purple color is less intense than you expected. Using the reagents you have on hand, what might you do to verify or disprove the positive Marquis test?

Q2: You have a sample that you suspect contains morphine or opium, but not both. Which reagent would you use to discriminate between the two? If the sample tested negative for morphine, would you run additional presumptive tests? If so, why, and which test or tests would you run?

LABORATORY 22.3:
REVEAL LATENT FINGERPRINTS

Developing methods to reveal latent fingerprints was one of the early focuses of forensic chemistry. Latent fingerprints on glass, polished metal, and other glossy surfaces can be revealed by dusting them with a fine powder whose color contrasts with the background surface. But latent fingerprints on paper and similar materials are difficult or impossible to reveal by dusting.

Iodine fuming was the first method developed for revealing latent fingerprints on such surfaces. The paper or other material to be treated is placed in a chamber that also contains a few crystals of iodine. As the iodine is heated, it sublimates as a violet vapor, which is deposited on the material being treated. The iodine condenses to a solid and selectively adheres to the oils present in the fingerprints, revealing the fingerprints as faint orange marks. Because it depends on the physical binding of a marking agent to the latent fingerprints, iodine fuming is conceptually similar to dusting, with the iodine vapor acting as an extremely fine "dust."

Iodine fuming is still used today because it is nondestructive. Latent fingerprints revealed by iodine vapor gradually disappear as the iodine sublimates, leaving the sample in its original state. The revealed prints can be photographed before they disappear or they can be developed by treating them with a 1% aqueous solution of starch. The iodine-starch reaction turns the prints a darker blue color, which is persistent for weeks or months, depending on how the sample is stored. If preserving the original state of the sample is unimportant, the revealed prints can be developed with a dilute solution of benzoflavone, which reacts with the iodine to make the prints permanent.

Development with silver nitrate was another early method used to reveal latent fingerprints on paper and similar materials. A 3% solution of silver nitrate is sprayed or gently brushed onto the sample. The silver nitrate reacts with the salt present in the latent fingerprints to produce silver chloride.

$$NaCl(s) + AgNO_3(aq) \rightarrow AgCl(s) + NaNO_3(s)$$

The sample is sometimes then washed gently with distilled water to remove excess silver nitrate, after which it is exposed to sunlight or an ultraviolet lamp. The UV reduces the silver ions present in the silver chloride to metallic silver, which is visible as orange or black marks. The silver nitrate method is used less often than the iodine fuming method, because treating the sample with silver nitrate causes irreversible changes.

The final method of revealing latent fingerprints that we examine in this lab depends on the fact that fingerprints contain trace amounts of amino acids. Ninhydrin (triketohydrindene hydrate, $C_9H_6O_4$) reacts with the breakdown products of these amino acids to produce a deep

REQUIRED EQUIPMENT AND SUPPLIES

☐ goggles, gloves, and protective clothing

☐ gas burner or alcohol lamp

☐ ring stand

☐ support ring

☐ wire gauze square

☐ large beaker with watch glass or other cover

☐ stiff wire and paperclip (to suspend sample in beaker)

☐ magnifying glass or loupe to examine samples (optional)

☐ small sprayer bottles (3)

☐ samples of paper with latent fingerprints (3)

☐ iodine crystals (~1 g)

☐ starch solution (~1 g of cornstarch in 25 mL water; filter or decant off clear liquid)

☐ silver nitrate solution (0.3 g silver nitrate in 10 mL distilled water)

☐ ninhydrin solution (0.1 g ninhydrin in 10 mL of 95% ethanol)

SUBSTITUTIONS AND MODIFICATIONS

- You may substitute discarded decongestant nasal sprayers or similar for laboratory spray bottles.

blue or blue-purple dye called Ruhemann's Purple. A 0.5% to 1.0% solution of ninhydrin in ethanol or acetone is sprayed on the sample. The latent prints develop slowly at room temperature, requiring an hour or two to be fully revealed. This process can be hastened by heating the sample in an oven at 80°C to 100°C.

In practice, the order in which these methods are applied is critical. Iodine fuming is always used first, because it is completely reversible. If iodine fuming doesn't provide usable results, ninhydrin is used next. If ninhydrin fails, the traces of salt are still present in the latent fingerprints, so the forensic chemist can attempt to develop the latent prints with silver nitrate. The silver nitrate solution removes the fatty oils and amino acids from the sample, so if silver nitrate fails, no other method can be used subsequently to reveal the prints.

In this lab, we'll use all three methods to reveal latent fingerprints. We're not really forensic chemists, so we'll make it easier on ourselves by using three separate samples—one for each of the methods.

CAUTIONS

This experiment uses flame. Be careful with the flame, and have a fire extinguisher readily available. Iodine crystals stain skin and clothing. (Stains can be removed with a dilute solution of sodium thiosulfate.) Iodine vapors are toxic and irritating. Perform this experiment outdoors or under a fume hood or exhaust fan. Silver nitrate stains skin and clothing. Ninhydrin is toxic and irritating. Wear splash goggles, gloves, and protective clothing.

PROCEDURE

This laboratory is divided into three parts—one for each of the methods we cover. Before you dive into the procedures, you need to create some samples to be tested. We used ordinary copy paper for the samples. We found that the latent fingerprints were difficult or impossible to raise using any of the three methods if we washed our hands soon before touching the sample paper. Apparently, soap removes all or most of the skin oils, salt, and other residues that produce latent prints. Our best results were with samples that we'd touched when our hands were sweaty and hadn't been washed recently. There's a moral here. If you're going to forge a document without wearing gloves, wash your hands first. And afterwards.

PART I: IODINE FUMING

1. If you have not already done so, put on your splash goggles, gloves, and protective clothing.
2. Set up your tripod stand and gauze square, with the burner beneath it and the beaker resting on the stand.
3. Sprinkle a few crystals of iodine into the bottom of the beaker. You don't need much; a fraction of a gram is sufficient.
4. Use the stiff wire and paperclip to suspend a sample in the beaker, not touching the sides or bottom of the beaker. Alternatively, for a larger sample, simply place the paper against the side of the beaker, as shown in Figure 22-3. Place the watch glass or other cover on top of the beaker to contain the iodine fumes.
5. Observe the sample for 30 seconds or so. At room temperature, iodine vaporizes quickly enough to begin fuming the prints within a few seconds. (I didn't apply heat to the beaker for Figure 22-3, which shows the paper after about 30 seconds of fuming at room temperature.
6. If no prints have begun to develop after 30 seconds or a minute, heat the bottom of the beaker slightly to hasten vaporization of the iodine. Before you apply heat, if you're not working outdoors, turn on the fume hood or exhaust fan.

ANTIQUE FINGERPRINTS

I always smile when I read a mystery novel in which a detective finds "fresh fingerprints." As any forensic scientist will admit, there's no way to judge the age of a fingerprint.

In theory, the more volatile components of a latent fingerprint disappear at a rate that could be calculated by taking into account the temperature and other environmental factors, leaving only the nonvolatile sodium chloride. On that basis, you might expect that fingerprints that can be developed with iodine fuming or ninhydrin are "new" and those that can be developed only with silver nitrate are "old."

In practice, there's no feasible way to apply this reckoning because the type and amounts of residues laid down for a particular latent fingerprint vary so much from one to another. Fingerprints that are only hours old may respond only to silver nitrate, and fingerprints that are months or years old may respond to iodine fuming or ninhydrin. Fingerprints have been found in Egyptian pyramids. Although those latent fingerprints were probably thousands of years old (based on other evidence), no forensic scientist could swear under oath that they hadn't been left earlier that morning.

7. Light the burner and use it to gently heat the bottom of the beaker. Very soon, the iodine will start to sublimate, filling the beaker with violet iodine vapor. Continue heating the beaker gently. If the iodine vapor condenses on the sides of the beaker, heat the sides gently to force it back to vapor form.

8. After 30 seconds to a minute, the fuming phase should be complete. Allow the iodine vapor to condense on the sides of the beaker. Examine the sample, which should now have visible fingerprints. If not, continue fuming, perhaps after adding a bit more iodine to the beaker.

9. Remove the watch glass (being careful not to burn yourself) and lift the sample out of the beaker. Lay it on a flat surface and examine it carefully with the magnifying glass. You should see fingerprints revealed in considerable detail (Figure 22-4). Shoot an image of them for the record, if you wish.

10. Spray the sample with the cornstarch solution to develop the prints. You needn't drench the sample; a gentle misting is sufficient. As the iodine reacts with the starch, you should see the color of the prints change to a deep blue-black.

11. Record your observations and, if you wish, tape or staple the developed sample to your record.

PART II: NINHYDRIN SPRAY

1. If you have not already done so, put on your splash goggles, gloves, and protective clothing.

2. Place one of your samples on a flat surface and spray it with the ninhydrin solution. (Use several thicknesses of newspaper to protect the surface against the ninhydrin.) Again, a gentle misting is sufficient, just enough to dampen the sample without drenching it. Record the time when you sprayed the sample, and check the sample every few minutes to determine how far development has proceeded. At room temperature, ninhydrin normally takes an hour or two to develop most prints, although some prints may require 24 hours or more. Figure 22-5 shows a fully developed ninhydrin test.

3. Heat and humidity greatly increase the rate of print development with ninhydrin. If you want to examine the effect of heat on the development of the latent fingerprints, place the original or a second sample that has been sprayed with ninhydrin into an oven set to 200°F. Check the sample every minute or two to determine the reaction rate at the higher temperature. Alternatively, you can place the sample between sheets of paper towel or blotter paper and use a clothes iron on medium heat to develop the prints using steam, which requires only a few seconds.

FIGURE 22-3:
Iodine fumes in a beaker developing latent fingerprints

FIGURE 22-5:
Latent fingerprints after treating with ninhydrin

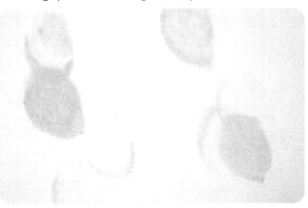

FIGURE 22-4:
Fingerprints developed by iodine fuming

PART III: SILVER NITRATE SPRAY

1. If you have not already done so, put on your splash goggles, gloves, and protective clothing.
2. Place one of your samples on a flat surface and spray it with the silver nitrate solution until it is just slightly damp.
3. Expose the treated sample to direct sunlight or a UV lamp, and observe it every few minutes to determine progress. The latent fingerprints should become visible, first as a pale yellow/purple color. With continued exposure, the prints gradually darken, eventually becoming black. The time needed to develop the prints completely may vary from a couple of minutes to an hour or more, depending on the particular fingerprint, the type of surface, and the intensity of the light. Exposing the sample too long will eventually cause the excess silver nitrate to be reduced to metallic silver, staining the entire sample.
4. Once the prints are fully developed, bathe the sample in water to remove the excess silver nitrate and "fix" the print.

DISPOSAL: Dilute all solutions with at least three or four times their volume of tap water and then flush them down the drain with plenty of water.

OPTIONAL ACTIVITIES

If you have time and the required materials, consider performing these optional activities:

- Use iodine fuming to treat a sample, and allow the revealed fingerprints to fade. (You can hasten this process by putting the sample in direct sunlight or by heating it very gently in the oven.) Retreat that sample using ninhydrin and/or silver nitrate to verify that the iodine fuming was fully reversible.

- Test one or more of the methods using samples on porous materials other than paper. For example, you might try raising fingerprints on samples of fabric, leather, or unpolished wood.

- Treat samples on paper or other porous materials after allowing different periods to elapse after creating the samples. For example, produce three similar samples on paper. Treat one of the samples immediately, treat one after a day has elapsed, and treat the third after one week. Do the results differ? If so, is one of the methods better than another at achieving results for older samples?

- Produce samples after liberally applying petroleum jelly or cream to your hands and then wiping them off. Are the latent prints in such samples easier or more difficult to raise? Is one or another of the methods better or worse at revealing such latent prints? If so, propose an explanation.

REVIEW QUESTIONS

Q1: What common household substance is also used to reveal latent fingerprints? (Hint: search the Internet for cyanoacrylate.)

Q2: Other than primates, very few mammals have fingerprints. There is one mammal, though, that is not a primate but produces fingerprints that are almost impossible to discriminate from human fingerprints, even using current instrumental methods. What is this mammal? (Hint: search the Internet for "mammal" and "fingerprints.")

LABORATORY 22.4:
PERFORM THE MARSH TEST

The Marsh Test provides an incredibly sensitive test for the presence of arsenic and the similar element antimony. Because arsenic and antimony are lethal in small doses and the only samples available may be body fluids in which the arsenic or antimony is extremely dilute, a forensic test for these elements must provide positive results when arsenic or antimony is present in the sample at microgram levels. The Marsh Test has that level of sensitivity.

The fundamental problem in devising a reliable test for these substances was that the tiny amounts likely to be present in a sample made it difficult or impossible to use standard chemical tests that depend on a precipitate forming or a color change occurring. At the time Marsh devised his test, it was well known that arsenic and arsenic compounds react with hydrochloric acid to form arsine gas and that antimony reacts similarly to form stibine gas. The problem was that the tiny amounts of arsenic or antimony likely to be present in a sample wouldn't form much gas.

Marsh's stroke of brilliance was to realize that he could "amplify" the effect of the hydrochloric acid on arsenic or antimony salts by adding very pure zinc metal to the reaction vessel. Zinc reacts with hydrochloric acid to form hydrogen gas, which would serve as a "carrier" for the tiny amounts of arsine or stibine that were produced by the reaction. Marsh directed the gases from the reaction to a nozzle, ignited them, and allowed the flame to impinge on an unglazed porcelain plate. The burning hydrogen left no mark on the plate, but the burning arsine or stibine gas left a metallic mirror of arsenic or antimony on the plate. Although the gas produced by Marsh's apparatus was typically 99.99% or more hydrogen, the tiny amount of arsine or stibine present in the gas was sufficient to mark the plate.

In this lab, we'll reproduce Marsh's original test, and discriminate between arsenic and antimony samples. Although the marks produced by arsenic and antimony have slightly different appearances, the sure way to discriminate is to treat the mark with sodium hypochlorite (chlorine bleach). A mark produced by arsenic disappears in the bleach solution, but a mark produced by antimony is unaffected.

REQUIRED EQUIPMENT AND SUPPLIES

- ☐ goggles, gloves, and protective clothing
- ☐ gas-generating bottle with drawn glass tube nozzle
- ☐ small unglazed porcelain plates (one per test)
- ☐ pliers, crucible tongs, or other gripping instrument
- ☐ grease pencil or other porcelain marker
- ☐ matches, lighter, or other source of flame
- ☐ arsenic and/or antimony samples (see text)
- ☐ hydrochloric acid, concentrated (a few mL per test)
- ☐ zinc metal, mossy or granular (a few g per test)
- ☐ laundry bleach (a few mL)

SUBSTITUTIONS AND MODIFICATIONS

- You may substitute a large test tube or flask with a rubber stopper and a drawn glass tube nozzle for the gas-generating bottle.

- You may substitute the unglazed bottoms of porcelain crucibles for the unglazed porcelain plates.

- Antimony trisulfide is used in the heads of ordinary safety (book) matches. Antimony sulfide is soluble in hydrochloric acid. You can obtain an antimony sample by crushing the heads from a book of matches and soaking the powder in sufficient concentrated hydrochloric acid to cover it.

CAUTIONS

This experiment involves burning hydrogen gas. Keep the quantities of reactants small, be careful with the flame, and have a fire extinguisher readily available. Arsenic and antimony compounds are toxic, as are the arsine and stibine gas produced during the experiment. Run the experiment outdoors or in a well-ventilated area. The odor of arsine gas is detectable at less than 1 ppm, and has a garlic smell. stibine gas is similar, with an undefinable but unpleasant odor. Hydrochloric acid is corrosive and toxic, and emits noxious fumes. Laundry bleach is toxic and corrosive. Wear splash goggles, gloves, and protective clothing.

PROCEDURE

This laboratory is divided into three parts. In Part I, we establish the purity of our reagents. In Part II, we run the Marsh Test on our samples. In Part III, we treat the porcelain plates with laundry bleach to observe the effects of bleach on the arsenic and antimony mirrors that the Marsh Test produces on the porcelain plates.

PART I: ESTABLISH REAGENT PURITY

1. If you have not already done so, put on your splash goggles, gloves, and protective clothing.
2. Set up your gas-generating bottle. Pour about 10 mL of concentrated hydrochloric acid into the bottle, add about 3 g of mossy or granular zinc metal, and replace the stopper.
3. The zinc metal begins reacting with the hydrochloric acid, producing hydrogen gas. Allow a few seconds for the air to be displaced from the bottle, and then apply a lighted match or lighter to the glass nozzle to ignite the hydrogen.

CAUTIONS

The hydrogen flame is extremely hot and may be invisible. Verify that the hydrogen is burning by placing the match or a wood splint near the nozzle to reveal the flame. When you heat a porcelain plate in the hydrogen flame, it may become extremely hot. (A hot porcelain plate looks exactly like a cold porcelain plate, as I can tell you from experience.) If you put down the porcelain plate, make sure to place it on a heat-proof surface, and ensure that it has cooled completely before you touch it with your fingers.

REAGENT PURITY

When forensic chemists run a Marsh Test, they use extremely pure hydrochloric acid and zinc, because less pure grades of hydrochloric acid and zinc may contain enough arsenic as an impurity to yield a positive result even if the sample contains no arsenic. (Yes, the Marsh Test is really that sensitive.) I used ordinary muriatic acid from the hardware store and lab-grade mossy zinc, at first reacting them without an arsenic sample present to determine whether they were contaminated with arsenic. As it turned out, my samples were not contaminated (or at least not with arsenic), but that's no guarantee that yours won't be. If you want to perform a definitive Marsh Test, either test your reagents first without a sample, or use reagent-grade hydrochloric acid and zinc. Or both.

ARSENIC AND OLD WOOD

Arsenic salts were formerly commonplace, found in nearly every household. Nowadays, though, arsenic and arsenic salts are very difficult to come by, unavailable even from many specialty chemical suppliers. The best way to obtain an arsenic sample for this lab is probably to find a piece of pressure-treated wood that was treated with CCA (chromated copper arsenate), which can be recognized by its green tint. Although CCA is still widely used worldwide, including the United States, recent restrictions on its use have made it harder to find. Any landscape timber more than a few years old was probably treated with CCA, and a small chip suffices for the purposes of this lab. (If possible, use a freshly sawn sample from the interior of the timber.) Antimony and its salts are readily available from specialty chemical suppliers. You can also get a sample by crushing the heads of safety matches (from a paper matchbook). Match heads contain antimony(III) sulfide.

4. Using pliers or another gripper to hold a porcelain plate, position the porcelain plate so that the hydrogen flame impinges directly on it. Allow the flame to touch the plate for 30 seconds to a minute, and examine the plate for any evidence of a deposit.

If your hydrochloric acid or zinc is contaminated with arsenic or antimony, there will be a visible mark on the plate. If so, you will need purer reagents for the actual test.

PART II: TEST THE SAMPLES

1. If you have not already done so, put on your splash goggles, gloves, and protective clothing.
2. Prepare the sample to be tested. Ideally, you want the sample to be as finely divided as possible, in the form of dust or small granules. A sample weighing two or three grams is sufficient. An even smaller sample is likely to provide positive results, but there's no reason to make things harder than necessary.

CAUTIONS

Be careful not to inhale any of the dust or expose your bare skin to it. Although the amount of arsenic contained in one of the sources described in this lab will be extremely small, you must develop habits that minimize your exposure to such chemicals. So even when handling trace amounts, take the same precautions you would for larger samples. Wear an inhaler or face mask when you are processing the samples.

3. Put the sample in the gas-generating bottle and add about 10 mL of concentrated hydrochloric acid. Swirl the bottle to mix the contents thoroughly, and allow the acid to react with the sample for a few seconds. Once the sample is thoroughly mixed, add about 3 g of mossy or granular zinc, recap the bottle, and repeat steps 3 and 4 from Part I.
4. Repeat the procedure until you have tested all of your samples. After each test run, discard the spent solution (flush it down the drain with copious water) and rinse the bottle thoroughly. Use a grease pencil or similar marker to label each test plate after it has cooled.

PART III: DISCRIMINATE BETWEEN ARSENIC AND ANTIMONY

1. If you have not already done so, put on your splash goggles, gloves, and protective clothing.
2. Set aside one of the porcelain plates that shows a positive Marsh Test for arsenic and one that shows a positive test for antimony. Put one or two drops of undiluted chlorine laundry bleach on each sample. Observe and record the results.

FIGURE 22-6: *A positive Marsh Test for arsenic*

DISPOSAL:
So much of the arsenic or antimony present in any sample will have been converted to arsine or stibine gas that it leaves only a trace quantity orders of magnitude smaller than the sample you started with. Even so, it would be poor practice to flush samples that contain arsenic or antimony into the public sewer system, so follow whatever hazardous waste disposal guidelines are in effect in your community.

REVIEW QUESTIONS

Q1: The mean lethal dose of arsenic in humans is variously estimated at between 1 mg/kg (0.001 g per 1,000 g) and 3 mg/kg. You know that the Marsh Test can detect arsenic in amounts as small as 1 μg (0.000001 g). A murder has taken place, and arsenic poisoning is suspected. You are tasked with using the Marsh Test to determine whether arsenic was the murder weapon. The victim weighed 100 kg, and you have only a 10 g sample of the victim's stomach contents to work with. If arsenic was used to kill this victim, is the Marsh Test sensitive enough to detect the arsenic? If so, and assuming that the victim died from a dose of only 1 mg/kg, how much smaller could your sample be and still yield usable results? If not, how much larger a sample would you require?

_____

_____

_____

_____

_____

Q2: Regulations restricting the use of CCA are becoming more common because of concerns that the CCA may leach out of the treated wood and into the surrounding soil. Tests have detected arsenic levels as high as 14 ppm in soil that is in contact with CCA-treated landscape timbers. If the Marsh Test can detect arsenic in amounts as small as 1 μg, how large a soil sample would you need to use the Marsh Test successfully?

_____

_____

_____

Index

droppers, 27
drops, partial, 79
drugs, illicit, 389–394
dry ice, safety, 254, 255
drying, 87–88
Dugan, Michael, 23
Dumas, Jen Baptiste André, 264
dust safety, 21

E

eating
    contaminated food, 295
    in lab, 11
    oxalic acid, 311
ebullioscopic constant, 147
economy-grade glassware, 14–15
educational-grade glassware, 14–15
efflorescence, 117
electric
    potential, 286
    receptacles, 42
    work, 285–286
electrochemistry
    battery, 304–306
    electrode potentials, 294–297
    electrolysis of water, 287–290
    energy transformation, 298–300
    everyday, 286
    iron oxidation, 291–293
    overview, 285–286
    voltaic cell, 301–303
electrode potentials, 294–297
electrodes, 36
electrolysis
    defined, 285
    electrode potentials, 294–297
    water, 287–290
electrolytic cell, 287–290,
  294–297, 304–306
Elemental Scientific, 66
elevation, boiling point, 147, 149–152
emulsion, 317–318
end point, 78
endothermic energy, 271
energy
    activation, 309
    electric potential, 285–286
    endothermic, 271
    exothermic, 271

    thermal, 280
    transformation, chemical to
      electric, 298–300
enthalpy
    change, chemical reactions,
      280–282
    change, of solution, 271–273
    of fusion, 274–275
equilibrium, chemical. *See* chemical
  equilibrium
equipment
    balances, 30–32
    batteries, 36
    corks, 36–38
    electrodes, 36
    glassware, 22–30
    group buying, 13
    heat sources, 32–33
    junk collecting, 41
    pH measuring, 34–35
    pipetters, 34
    recommended, 38–41
    ring stands, 33–34
    rubber stoppers, 36–37
    safety, 10
    scoops, 35–36
    spatulas, 35–36
    spoons, 35–36
    thermometers, 35
equivalence point, 78
Erlenmeyer flasks, 17, 19
errata, xiii
error, absolute *versus* relative, 81
esterification, 374
esters, 374
estimating, 69
ethanol
    mass-to-volume percentage
      solution, 140–141
    purifying, 97–100
evaporating
    dishes, 21
    procedure, 87–88
everyday
    acid-base chemistry, 192
    calorimetry, 270
    chemical equilibrium, 230
    chemical kinetics, 212
    chemical reactions, 162

    colligative properties, 148
    colloids, 320
    electrochemistry, 286
    gas chemistry, 247
    Le Chatelier's Principle, 230
    photochemistry, 310
    qualitative analysis, 331
    quantitative analysis, 356
    redox reactions, 182
    solutions, 125
    suspensions, 320
    synthesizing compounds,
      373–374
    thermochemistry, 270
exam, AP Chemistry, 2–4, 5
exothermic
    activation energy, 309
    dangers, 25
    energy, 271
explosion shield, 11
explosives, 11
extraction, 84, 105–107
eye protection, 9, 10
eyedroppers, 27
eyewash station, 10

F

FCC (Food Chemical Codex) grade
chemicals, 47
ferric chloride reagent, 390
filter
    flasks, 17–18, 19, 83
    paper, 36, 83–84, 108
filtering, barium sulfate, 369
filtrand, 83
filtrate, 83
filtration
    gravity, 83–84
    vacuum, 83
fingerprints, latent, 395–398
fire diamond, NFPA 704, 49–50
fire extinguisher, 10
fire hazards, 11
firefighting foam, 324–325
first-aid kit, 10
flame test, metal ions, 332–335
flames, open, 33